INTERNATIONAL CENTRE FOR THEORETICAL PHYSICS

INTERNATIONAL ATOMIC ENERGY AGENCY

UNITED NATIONS EDUCATIONAL, SCIENTIFIC AND CULTURAL ORGANIZATION

Proceedings of the Adriatico Workshop on

QUANTUM INTERFEROMETRY

2 – 5 March 1993, Trieste, Italy

Editors

F De Martini

Dipartimento di Fisica, Universitá "La Sapienza", Roma, 00185, Italy

G Denardo

International Centre for Theoretical Physics, Trieste, Italy

A Zeilinger

Institut für Experimentalphysik, Universität Innsbruck, A-6020 Austria

World Scientific

Singapore • New Jersey • London • Hong Kong

Published by

World Scientific Publishing Co. Pte. Ltd.
5 Toh Tuck Link, Singapore 596224
USA office: 27 Warren Street, Suite 401-402, Hackensack, NJ 07601
UK office: 57 Shelton Street, Covent Garden, London WC2H 9HE

British Library Cataloguing-in-Publication Data
A catalogue record for this book is available from the British Library.

QUANTUM INTERFEROMETRY
Proceedings of the Adriatico Workshop

ISBN-13 978-981-02-1517-0
ISBN-10 981-02-1517-7

FOREWORD

Quantum interferometry has a long history: its origins may be traced back to the first, low intensity, interferometric experiments by Taylor in the first decade of this century. The subject has become a fundamental one, at least at the stage of a set of sophisticated "gedanken" experiments, in the crucial period when the foundations of Quantum Mechanics were established. At that time an increasingly large number of scientists-philosophers, Einstein, Bohr, De Broglie, Pauli, Heisenberg, Schroedinger etc. started a passionate debate employing often gedanken quantum interferometry experiments. This debate is still alive in present times.

In the last one or two decades many of these "gedanken" experiments have become *real* experiments leading to results that have made some of the issues clearer and that have opened at the same time a wealth of new ideas and a rich perspective for new endeavours. The development is mostly signified by the fact that one can produce and manipulate quantum states of elementary particles such as photons, electrons, neutrons as well as of more complicated objects like atoms. At the same time the experimental technology has opened up new possibilities and a highly sophisticated level of precision was achieved. This opened up possibilities for new kinds of measurements and it offers new perspectives in the different areas of quantum interferometry. On the other hand, most of the basic ideas are common to the different areas and it is precisely the main purpose of this collection of papers to provide a forum for the exchange of these ideas and achievements.

The present book, which collects the text of the talks given at the Adriatico Workshop "Quantum Interferometry" held in Trieste in March 1993, comes right in the middle of this period of rapid development.

Francesco De Martini, Gallieno Denardo and Anton Zeilinger
Editors

CONTENTS

SQUEEZING

BASIC QUANTUM MECHANICS

QUANTUM INTERFEROMETRY

NEUTRONS

POSTSELECTION OF QUANTUM STATES IN NEUTRON INTERFEROMETRY AND EPR-EXPERIMENTS

H. RAUCH
Atominstitut der Österreichischen Universitäten, A-1020 Wien, Austria

ABSTRACT

Recent neutron interference experiments based on the post-selection of narrow momentum states behind the interferometer have shown that coherent wave packets remain totally coupled in configuration space, even when they become separated in ordinary space. Such non-overlapping wave packets in real space (Schrödinger-cat-like states) show a characteristical spectral modulation indicating its coherence structure. This shows analogies to EPR- experiments, with consequences for the debate about the non-locality phenomena of quantum mechanics.

1. INTRODUCTION

Widely separated coherent beams exist inside a perfect crystal neutron interferometer which overlap at the third crystal plate where the resulting interference pattern exhibits the coherence properties of the beams (e.g., [1-3]). In stationary situations, the related coherence function is given by the auto-correlation function of the overlapping wave functions which have be to be taken as three-dimensional wave packets:

$$\Gamma(\vec{\Delta}) = \langle\psi^*(0)\psi(\vec{\Delta})\rangle,$$
$$\psi(\vec{r}) \propto \int a(\vec{k})e^{i\vec{k}.\vec{r}}d\vec{k}. \qquad (1)$$

This shows that the coherence function is related to the Fourier-transformation of the momentum distribution of the beams $g(\vec{k}) = |a(\vec{k})|^2$, by

$$\langle|\Gamma(\vec{\Delta})|\rangle \propto \left|\int g(\vec{k})e^{i\vec{k}.\vec{\Delta}}d\vec{k}\right| \qquad (2)$$

with the normalization $\langle|\Gamma(0)|\rangle = 1$. In practice $\langle|\Gamma(0)|\rangle$ can be different from unity, due to an average over different beam paths with slightly different phase shifts, due to unavoidable imperfections of the crystal and the phase shifter, and of their dimensions and residual vibrations. In the following, $\Gamma(\vec{\Delta})$ denotes $\Gamma(\vec{\Delta}) = \langle|\Gamma(\vec{\Delta})|\rangle/\langle|\Gamma(0)|\rangle$. $\vec{\Delta}$ represents the

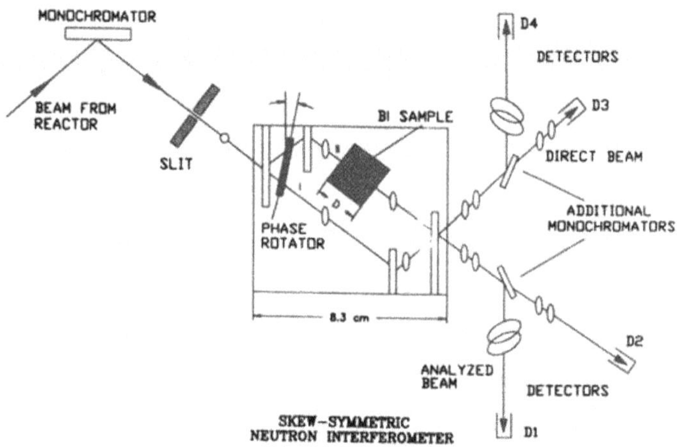

Figure 1: Scheme of a perfect crystal neutron interferometer with indications of the shapes of the wave packets in the case of large phase shifts.

spatial shift of the overlapping wave trains which can be calculated for a phase shifter with an index of refraction n, a thickness D_0 and oriented with a surface normal $\hat{s}$ as

$$\vec{\Delta} = \frac{(\vec{K}-\vec{k})}{k} D_0,$$

$$\vec{K}-\vec{k} = \frac{(1-n)\vec{k}\hat{s}}{(\hat{k}.\hat{s})} \tag{3}$$

(Fig.1), where $\vec{K}$ and $\vec{k}$ represent the wave vectors inside and outside the phase shifter, respectively. For Gaussian wave packets which describe Gaussian momentum-distributed beams with characteristic widths δk_i (i = x, y, z), one obtains

$$|\Gamma(\vec{\Delta})| = \prod_{i=x,y,z} \exp\left[-(\Delta_i \delta k_i)^2/2\right], \tag{4}$$

which shows that the characteristic length - the coherence length Δ_c^i - is determined by the momentum distribution of the beam in each direction. For Gaussian distributions, the relation $\Delta_c^i = 1/(2\delta k_i)$ holds, which is the minimum uncertainty relation.

The interference pattern behind the interferometer is given by the wave functions originating from beam paths I and II. For the forward direction (0) and the empty interferometer they are equal in amplitude and phase because the related beams are transmitted-reflected-reflected (TRR) and reflected-reflected-transmitted (RRT), respectively. A phase shifter with an effective thickness $D_{eff} = D_0/(\hat{k}.\hat{s})$ changes the related wave function by a phase factor $\exp\left[i(n-1)kD_{eff}\right] = \exp(i\vec{\Delta}.\vec{K})$.

$$I(\Delta) \propto |\psi_0^I + \psi_0^{II}|^2 \propto 1 + |\Gamma(\vec{\Delta})|\cos(\vec{\Delta}.\vec{K}) \tag{5}$$

shows the direct connection between interference pattern and the coherence function (eq.(1)). For the longitudinal direction (x) the spatial shift is given as $\Delta = 2\pi N b_c D_0/k^2$. This formula indicates that the interference pattern disappears when the spatial phase shift approaches the coherence length of the beam in a certain direction $\Delta_i \gg \Delta_i^c = 1/(2\delta k_i)$, a phenomenon which has been verified experimentally by several investigations[4-6]. Thus, higher order interference fringes can be observed for a higher monochromaticity of the incident beam, i.e., a better preselection of states, and/or phase shift in the direction of a high internal momentum selection (y-direction). This happens in the perfect crystal interferometer due to dynamic diffraction providing the possibility of a nondispersive phase shifter position[6]. The loss of interference fringes at high order can be understood as a loss of wave packet overlap at the exit crystal plate of the interferometer, but it should be mentioned that this loss of interference fringes does not indicate a loss of coherence, because the interference fringes can be restored - at least to a large extent - by proper phase-echo methods[5,8]. The question is whether the interference fringes can also be restored by experimental means behind the interferometer, i.e., by a postselection procedure.

2. POSTSELECTION MEASUREMENTS

In optical interferometry, some parts of the interference pattern have been restored behind the interferometer when an additional filtering process has been applied[9]. In more recent publications, some additional effects causing spectral changes in low resolution double-slit experiments have been discussed[10-12]. Related neutron interference experiments showed that interference phenomena can be restored if an additional monochromatization is applied behind the interferometer (Fig.2)[9,13,14]. The related set-up corresponds to a double crystal arrangement between the interferometer crystal (mosaic spread $\eta_i \to 0$, Bragg angle Θ_i) and the analyser crystal (mosaic spread η, Bragg angle Θ_A) whose momentum spread is given as in[13]:

$$\left[\frac{\delta k'}{\delta k}\right]^2 = \frac{(\delta k_A)^2}{(\delta k)^2 + (\delta k_A)^2},$$

$$\left[\frac{\delta k_A}{k_0}\right]^2 = \eta^2 \left|\frac{\cos\Theta_i \cos\Theta_A}{\sin(\Theta_i \pm \Theta_A)}\right|^2, \tag{6}$$

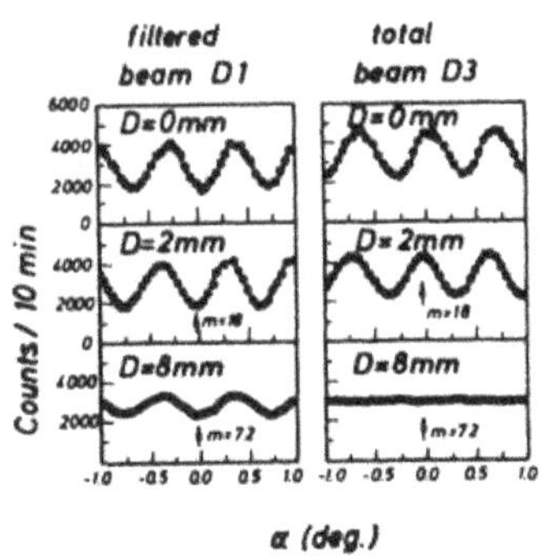

Figure 2: Results of a recent neutron interference post-selection experiment where the enhanced visibility of the filtered (monochromatized) beam has been observed[13].

where the positive sign corresponds to the parallel setting and the minus sign for the anti-parallel configuration. Figure 2 shows the results obtained with a pressed silicon analyser crystal (η_A = 0.0035 rad) in the anti-parallel position ($\delta k'$ = 1.87x10^{-2} Å^{-1}, where δk = 3.2x10^{-2} Å^{-1}). Even less attenuation of the interference pattern at high order occurred with a nearly perfect crystal post-analyser, also in the anti-parallel position (η_A = 0.00015 rad, $\delta k'$ = 8.0x10^{-4} Å^{-1}). The related amplitude of the interference pattern is shown for both cases in Fig.3[13] as a function of the phase shift which represents the coherence function defined in eqs. (2) and (4). In a first approximation, one can assume that the wave packet does not change in the x-direction when going through the interferometer, because the tangential components of the momentum do not change at all during reflection from the lattice planes. If one further assumes that the analyser crystals cut out a narrower band width causing a more extended wave packet, one ends up with a mutual overlap of wave functions from both beam paths, as indicated in Fig.2. This describes only partly the physical situation. At high order, the momentum dependence of the interference pattern becomes distinctly influential[15,16]. In this case, we have to rewrite eq.(5) in a purely k-dependent form:

$$I(k) \propto |a(k)|^2\left[1+\cos(\Delta.k)\right]$$

$$\propto \exp\left[-(k-k_0)^2/2\delta k^2\right]\left\{1+\cos\left[\chi_0 \frac{k_0}{k}\right]\right\} \qquad (7)$$

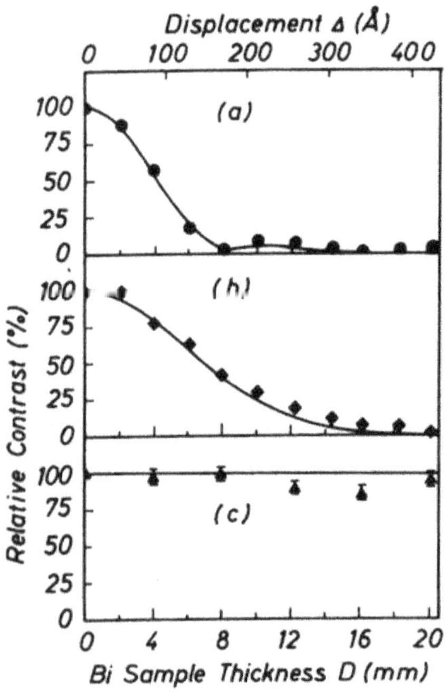

Figure 3: Reduction of the interference contrast (= coherence function) at high order for different degrees of post-selection of momentum states[12]. (a) direct beam C3, (b) pressed silicon (111) - monochromator in antiparallel position, (c) nearly perfect silicon (111) - monochromator in antiparallel position.

where the mean phase shift is introduced ($\chi_0 = \Delta_0.k_0 = Nb_c\lambda_0 D_{eff}$). Formula

(7) can also be obtained from the Fourier-transformation of the two spatially (by $|\Delta_x|$) separated (Gaussian) wave functions. This function exhibits, at high order, a marked oscillating structure in momentum space sweeping intensity between the forward and deviated beam (Fig.4). Thus, the intensity behind the interferometer is by no means the intensity sum arising from both beam paths. This indicates that each momentum component has a very distinct behavior just determined at the exit of the interferometer where the wave functions do not overlap. This underlines that the plane-wave components create a real action even outside the packet. Thus, complete interference properties can be restored in both out-going beams when proper measurement methods are applied. The vanishing interference pattern at high interference order is accomplishied with the appearance of spectral modelling in momentum space, indicating the persistent coupling in configuration space where the whole information about the history of both coherent beams remains stored, even in cases where the narrow oscillation cannot be resolved experimentally anymore. This spectral modelling effect might also have practical use for light and particle optics in achieving a spectral narrowing by interferometric means only.

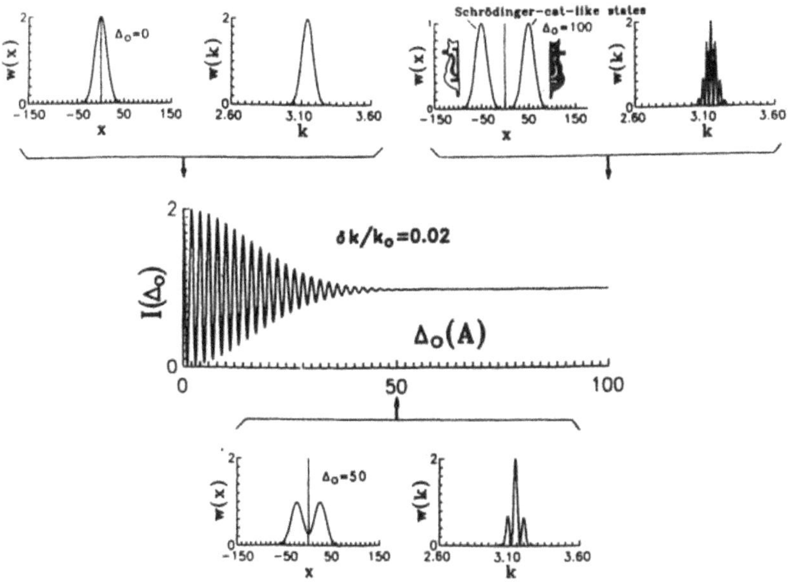

Figure 4: Appearance of spectrum modulation when the interference pattern disappears at high order. The spectral modulation occurs in an alternative sense in both emerging beams and is sensitive to small variations of the phase shift.

The spatial distribution of the packets arising from beam paths I and II can also be obtained from eq.(5) or, equivalently, from the Fourier- transform of the momentum distribution (eq.(7)).

$$I_0(x) \propto \exp\left[-x^2/2\delta x^2\right] + \exp\left[-(x+\Delta)^2/2\delta x^2\right] \qquad (8)$$

which exhibits two separate peaks for large phase shifts (Fig.4) and which represents the amplitude function of the related wave function (e.g.[17]). For Gaussian packets, δx corresponds to the coherence length Δ_c and fulfills the minimum uncertainty relation $\delta x \delta k = 1/2$. For an appropriate large dis- placement ($\Delta >> \Delta_c$), the related state can be interpreted as a superposition state of two macroscopically distinguishable states, that is, a stationary Schrödinger-cat-like state[18,19], but here the first time for massive particles. These states - separated in ordinary space and oscillating in momentum space - seem to be notoriously fragile and sensitive to dephasing effects[20-23]. Spatially separated packets remain entangled in phase space and nonlocality appears as a result of this entanglement. The analogy with optical experiments performed in the time-frequency domain is striking[12].

The new quantum states created behind the interferometer can be analyzed with regard to their uncertainty properties. Analogies between a coherent state behavior and a free but coherently coupled particle motion inside the interferometer have been addressed previously[24]. In such cases, the dynamical conjugate variables x and p minimize the uncertainty product with identical uncertainties $(\Delta x)^2 = (\Delta k)^2 = 1/2$ (in dimensionless units). Using $I_0(k)$ and $I_0(x)$ (eqs. (7) and (8)) as distribution function we get in our case

$$<(\Delta x)^2> = <x^2> - <x>^2 = (\delta x)^2 + \left[\frac{\Delta}{2}\right]^2 \qquad (9)$$

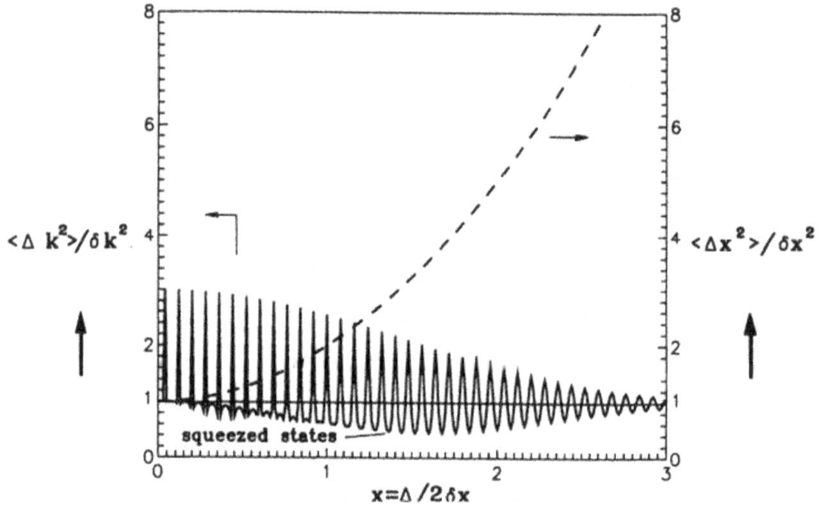

Figure 5: Spatial and momentum uncertainties of the outgoing beams with the indication of squeezing.

and (for $\delta k/k_0 << 1$)

$$<(\Delta k)^2> = <k^2> - <k>^2$$

$$= (\delta k)^2 \left[1 + \left(\frac{\Delta_0}{2\delta x} \right)^2 \frac{e^{-(\Delta_0 2\delta x)^2} \cos(\Delta_0 k_0) + e^{-(\Delta_0/2\delta x)^2}}{1 + e^{-(\Delta\ /2\delta x)^2/2} \cos(\Delta_0 k_0)} \right] \tag{10}$$

These relations are shown in Fig.5 indicating that for $(\Delta k)^2$ a value below the coherent state value can be achieved, which in quantum optic terminology means squeezing[25-27]. One emphasizes that a single coherent state does not exhibit squeezing, but a state created by superposition of two coherent states can exhibit a considerable amount of squeezing. Thus highly nonclassical states can be made by the power of the quantum mechanical superposition principle.

3. EPR-CONFIGURATION SPACE COUPLING

In the EPR-situation[28-30,16], pairs of correlated photons (or particles) are produced which form an entangled quantum state (Fig.6):

$$\psi \propto |-k\rangle_1 |k\rangle_2 + |-k\rangle_2 |k\rangle_1. \tag{11}$$

When the photon pairs are produced by a down conversion cascade, as shown in Fig.6, the momenta of the photons are correlated by the relation

$$k_1 + k_2 = k_{01} + k_{02} = \text{const.} \tag{12}$$

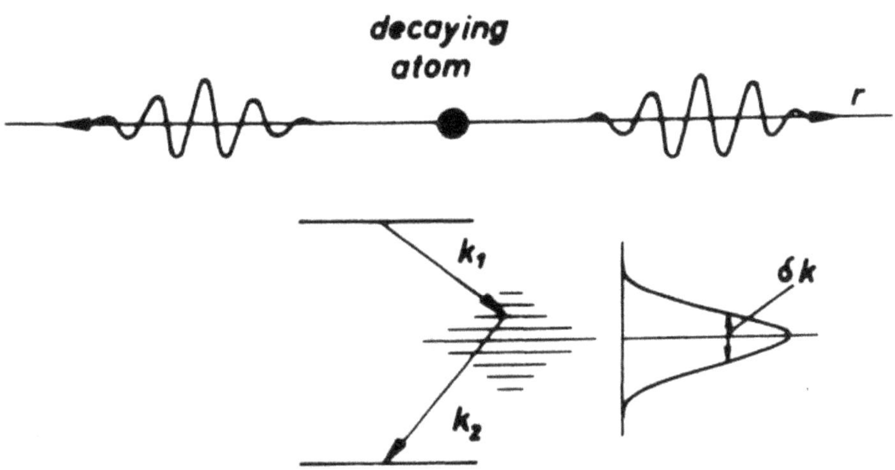

Figure 6: Scheme of an EPR-arrangement and a simplified double photon decay scheme

For a Gaussian distribution the wave function reads as

$$|k\rangle_1 = \exp\left[-(k_1-k_{01})^2/4\delta k^2\right] \exp\left[i(k_1 r - \omega_1 t)\right]$$

$$|k\rangle_2 = \exp\left[-(k_2-k_{02})^2/4\delta k^2\right] \exp\left[i\left(k_2 r - \omega_2 (t+\Delta t)\right)\right] \tag{13}$$

The momentum and space dependent intensity can be written in the form[16]

$$I(k_1,k_2,r) = |{}_{\mathsf{I}}{-k}\rangle_1{}_{\mathsf{I}}k\rangle_2 + {}_{\mathsf{I}}{-k}\rangle_2{}_{\mathsf{I}}k\rangle_1|^2$$

$$= 2|a(k_1)|^2\ |a(k_2)|^2 \left[1 + \cos 2(k_2-k_1)r\right] \qquad (14')$$

which shows for each pair a characteristic intensity modulation (Fig.7). This formula can again be seen as Fourier-transform of spatially separated Gaussian packets, separated by 2r. It should be mentioned that the intensities $I(k_1,r)$ and $I(k_2,r)$ are smooth functions on both sides of the source determined by their distribution functions but the correlated intensity $I(k_1,k_2,r)$ exhibits marked, and for each pair characteristical, modulations. This indicates that individual $|k\rangle$-states remain interacting even at arbitrarily large spatial separations of the wave packets. The coupling in configuration space persists unless a random interaction is applied to erase the fine structure of the wave function mentioned above. All EPR-experiments performed up till now were performed without momentum resolution and/or within the near field zone $r < 1/2\ \delta k$, thus averaging over the characteristical modulation shown in Fig.7. Therefore, it is

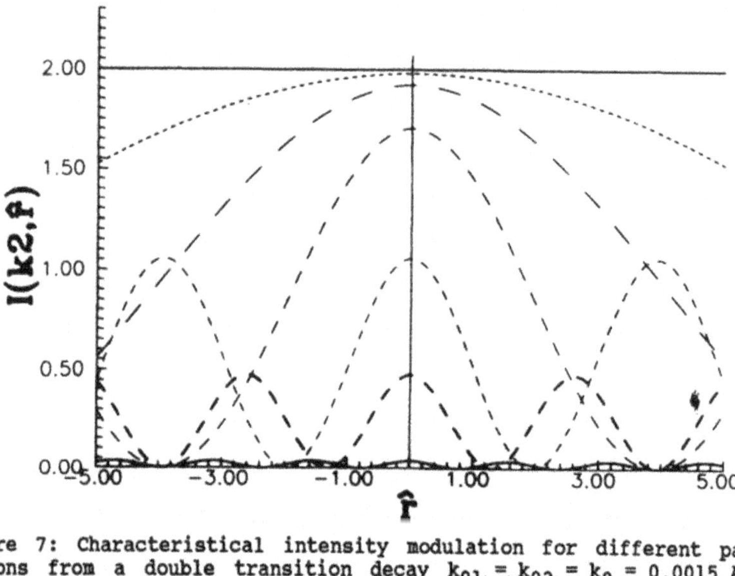

Figure 7: Characteristical intensity modulation for different pairs of photons from a double transition decay $k_{01} = k_{02} = k_0 = 0.0015\ \text{Å}^{-1}$ and $\delta k_0 = 0.05\ k_0$ for $|k_1-k_2| = 0.2\ n\delta k_0$ ($n = 0$, 0.5, 1, 2, 4, 6 and 10). $\hat{r} = r/r_c$, where $r_c = 1/2\ \delta k_0$ denotes the distance where the packets become separated.

recommended to repeat these experiments with a proper momentum resolution and to include the phase space coupling in the general discussion about EPR-phenomena. General conclusions about wave function properties can only be drawn if all accessible information about it is included. Thus the completeness of a quantum experiment has to be seen in a new light. The nonlocality phenomenon of quantum mechanics could be understood as the far reaching action of the plane wave components of the wave function as well. The band width of these plane wave components is determined by the momentum resolution of the measurement with an upper limit defined by the inverse of the source-detector distance.

4. DISCUSSION

An analogy has been shown between neutron interference (single particle) and EPR-experiments (two or more particle systems). It has been demonstrated that coherently separated wave packets remain coupled in momentum space, which can be attributed to the far reaching action of the plane-wave components constituting the wave packets. Therefore, many previously performed experiments devoted to this topic appear as incomplete experiments because not all information inherently contained in the related distribution function has been measured and used for interpretative questions. This puts a new light on the problem of non-locality measurements in quantum mechanics.

Work supported partly by Fonds zur Förderung der Wissenschaftlichen Forschung (project no. P8456).

5. REFERENCES

1. U. Bonse and H. Rauch (Eds.), *Neutron Interferometry* (Clarendon Press, Oxford, 1979).
2. A. G. Klein and S. A. Werner, *Rep. Prog. Phys.* **46** (1983) 259.
3. G. Badurek, H. Rauch and A. Zeilinger (Eds.), *Matter Wave Interferometry* (North-Holland, Amsterdam, 1988).
4. H. Rauch, in 1., p. 161.
5. H. Kaiser, S. A. Werner and E. A. George, *Phys. Rev. Lett.* **50** (1983) 563.
6. S. A. Werner, R. Clothier, H. Kaiser, H. Rauch and H. Wölwitsch, *Phys. Rev. Lett.* **67** (1991) 683.
7. H. Rauch, E. Seidl, D. Tuppinger, D. Petrascheck and R. Scherm, *Z. Physik* **B69** (1987) 313.
8. R. Clothier, H. Kaiser, S. A. Werner, H. Rauch and H. Wölwitsch, *Phys. Rev.* **A44** (1991) 5357.
9. L. Mandel, *J. Opt. Soc. Am.* **51** (1961) 1342.
10. D. F. V. James and E. Wolf, *Opt. Comm.* **81** (1991) 150.
11. D. F. V. James and E. Wolf, *Phys. Lett.* **A157** (1991) 6.
12. X. Y. Zou, T. P. Grayson, L. Mandel, *Phys. Rev. Lett.* **69** (1992) 3041.
13. H. Kaiser, R. Clothier, S. A. Werner, H. Rauch and H. Wölwitsch, *Phys. Rev.* **A45** (1992) 31.
14. D. Jacobson, S. A. Werner and H. Rauch, *Phys. Rev. A*, in print.
15. H. Rauch, Proc. *"Quantum Physics and the Universe"*, ed. M. Namiki, Waseda University, Tokyo, August 1992 (in print).
16. H. Rauch, *Phys. Lett.* **A173** (1993) 240.
17. J.-H. Levy-Leblond, f. Balibar, *Quantics* (North-Holland, 1990).
18. A. Legett, Proc. *"Found. Quantum Mechanics"*, ed. S. Kamefuchi, *Phys. Soc. Japan*, 1984, p.74.
19. B. Yurke, W. Schleich, D. F. Walls, *Phys. Rev.* **A42** (1990) 1703.
20. D. F. Walls, G. J. Milburn, *Phys. Rev.* **A31** (1985) 2403.
21. R. J. Glauber, *"New Techniques and Ideas in Quantum Measurement Theory"*, ed. D. M. Greenberger, (N.Y. Acad. Sci. 1986) p.336.
22. M. Namiki, S. Pascazio, *Phys. Rev.* **A44** (1991) 39.
23. H. Zurek, *Physics Today*, Oct. 1991, p.36.
24. H. Rauch, J. Summhammer, M. Zawisky, E. Jericha, *Phys. Rev.* **A42** (1990) 3726.
25. W. Schleich, M. Pernigo, Fam Le Kien, *Phys. Rev.* **A44** (1991) 2172.
26. D. F. Walls, *Nature* **306** (1983) 141.
27. S. L. Braunstein, R. I. McLachlan, *Phys. Rev.* **A35** (1987) 1659.

NEUTRON MANY-PHOTON EXCHANGE AND ION INTERFEROMETRY

Johann Summhammer
Atominstitut der Österreichischen Universitäten
Schüttelstraße 115, A-1020 Vienna, Austria

ABSTRACT

We present progress reports on two interferometry experiments we are currently carrying out. The first involves single crystal neutron interferometry in a measurement of the amplitudes of various many-photon exchange processes between a neutron and particular configurations of oscillating magnetic fields. The other is an attempt at a first realization of double slit interference with protons and light ions. Finally we give a general limit of how many particles must pass one arm of an interferometer in order to transfer one bit of information by phase shifting.

1. Neutron Many-Photon Exchange

The neutron is a chargeless massive fermion of spin 1/2. It interacts with the electromagnetic field via its magnetic moment μ and in the case of very strong electric fields, as for instance close to a nucleus[1], also via an induced electric dipole moment. The latter can generally be neglected in fields obtainable in a laboratory such that only the interaction with the magnetic field remains. The potential is

$$H = \mu\sigma B(r,t), \tag{1}$$

where $\mathbf{B}(\mathbf{r},t)$ is the space and time dependent magnetic field vector and σ is the vector of the Pauli spin matrices. This potential is analogous to that of a two-level atom in a radiation field for which many treatments exist[2-4]. We are particularly interested in the case of a monoenergetic free neutron which originally propagates in free space and then traverses a sharply bounded region with a spatially homogeneous field modulated sinusoidally in time. How are the energy and the spin state of the neutron affected? The neutron is assumed non-relativistic and the potential energy (eq.1) is much smaller than the neutron's kinetic energy. These are the conditions met in single crystal neutron interferometry, where typical energies of the neutron are around 25 meV, and where magnetic field strengths of some 10^{-3} to 10^{-1} T at frequencies between 10 kHz and several MHz can be achieved with ac-driven electric coils. What we want to find out is the state of the neutron after the field region. When H is periodic with ω it follows from the continuity of the Schrödinger equation that the energy of the neutron can only change by multiples of $\hbar\omega$. This can be interpreted as absorption of photons from or emission of photons into the field. In addition the spin of the neutron may change. Let the incident neutron have energy E_0 and momentum $\mathbf{p}_0$ and let it be in the spin up state with respect to the z-axis. Then its wave function is given by

$$\psi_{in}(r,t) = \begin{pmatrix} 1 \\ 0 \end{pmatrix} e^{\frac{i}{\hbar}(p_0 r - E_0 t)}. \tag{2}$$

Without having to know any details of the oscillating field we can write the neutron's state after the interaction with the field as

$$\psi_{out}(r,t) = \sum_{n=-\infty}^{\infty} \begin{pmatrix} T_n^+ \\ T_n^- \end{pmatrix} e^{\frac{i}{\hbar}(p_n r - E_n t)} \tag{3}$$

with

$$E_n = E_0 + n\hbar\omega \tag{4}$$

and

$$p_n = \sqrt{p_0^2 + \frac{n\hbar\omega}{2m}}. \tag{5}$$

Here, n is a positive or negative integer indicating how many field quanta the neutron has absorbed or emitted, respectively. The possible energies and the momenta after the exchange process are E_n and p_n, respectively. The T_n^+ and T_n^- are the corresponding probability amplitudes. The superscripts "+" and "-" refer to non-spin flip and spin flip, respectively. For instance, T_5^- means the amplitude for absorbing five photons from the field and simultaneously undergoing a spin flip. These amplitudes depend on the strength and frequency of the oscillating field, the field configuration, and the passage time of the neutron through the field region.

In our ongoing project two field configurations deserve special attention, because they require different mechanisms for the conservation of angular momentum during the photon exchange process. The first one is familiar from nuclear magnetic resonance and has also been used in earlier neutron interferometry experiments.[5,6] This is a static magnetic field and an oscillating one orthogonal to it such that one gets

$$\boldsymbol{B}(t) = (B_1 \cos\omega t,\ 0,\ B_0). \tag{6}$$

For this case the transition amplitudes cannot be calculated analytically, but there exist very good approximate methods.[4,7] With these one can clearly establish that only the even non-spin flip amplitudes and the odd spin flip amplitudes do not vanish. This implies that whenever the neutron exchanges an even number of photons with the field it leaves the oscillating field region in the same spin state it entered it, provided its initial spin state was an eigenstate with respect to the direction of the static component of the field (B_0). And conversely the neutron flips its spin only when it exchanges an odd number of photons with the field. This permits a simple picture for understanding angular momentum conservation: The photon has spin 1. When emitting or absorbing such a particle the neutron, having spin 1/2 must flip its spin, otherwise total angular momentum before and after the process would not be the same. In this picture one need therefore not consider the field as a whole. This is no longer possible in the second field configuration:

$$\boldsymbol{B}(t) = (0,\ 0,\ B_0 + B_1 \cos\omega t). \tag{7}$$

Here, only the strength of the field is modulated in time, but its direction is not altered. When inserting eq.7 into eq.1 the resulting interaction matrix has no off-diagonal elements. Therefore a neutron originally in an eigenstate with respect to the direction of the field, as e.g. the state given by eq.2, will never change its spin state. Nevertheless a detailed calculation shows that the neutron can exchange an even as well as an odd number of photons with the field. (An excact solution can be obtained in this case and the transition amplitudes as a function of the oscillating component of the field are Bessel functions.[7]) To accomodate angular momentum conservation in this situation it is necessary to impart some angular momentum to the field as a whole, or to the device producing it, e.g. a particular configuration of electric coils, as in our experiments. In a recent neutron interferometry experiment to test the scalar Aharanov-Bohm effect a similar field was used, but it had rectangular instead of sinusoidal time dependence.[8]

Figure 1 shows transition amplitudes for the field configuration with the oscillating component orthogonal to the static one (eq.6). Parameters have been chosen to be typical for what can easily be implemented with coils mounted in one path of a single crystal neutron interferometer. The dashed line indicates the situation of an experiment where such a field was used as a resonance spin flip device just as in nuclear magnetic resonance.[6] One notes that for certain values of the oscillating component of the field the output state is the same as the input state, except for a possible phase factor. This means that all photon exchange is suppressed. However, this is not a peculiarity of our choice of passage time, but can occur for any passage time. Furthermore it should be noted that appreciable amplitudes exist even for the exchange of five or more photons. The corresponding probabilities would be small, however. This is where the importance of interferometry comes in, because the modulation of the interfering intensity is directly proportional to the amplitudes rather than the probabilities.

The lay out of our experiment is shown in Fig.2. A stationary monochromatic

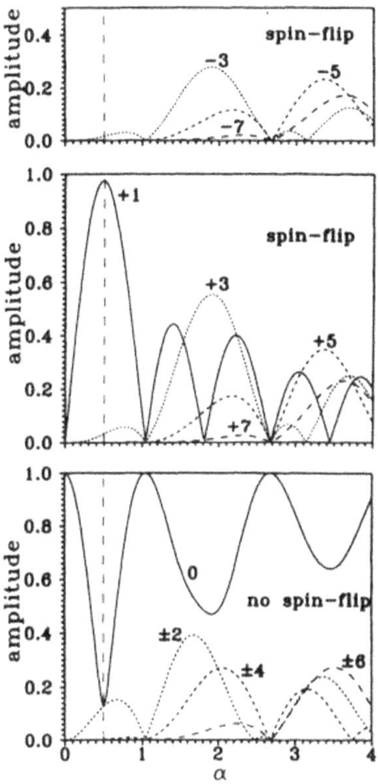

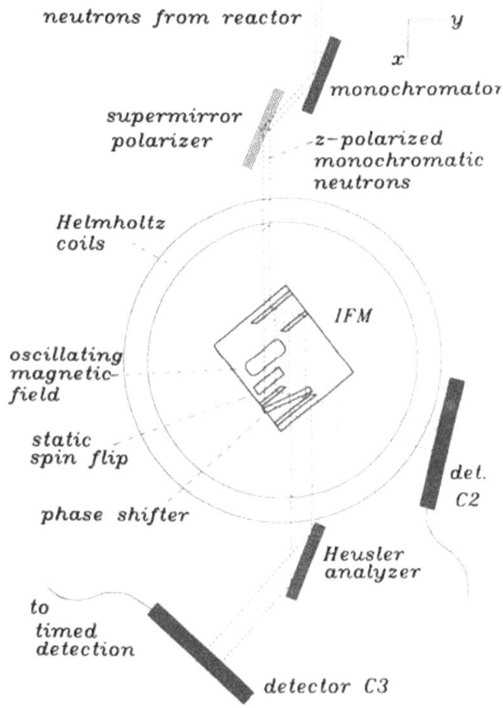

Fig.1: Amplitudes for photon exchange with and without spin flip when the neutron passes through a magnetic field region consisting of a static field and an oscillating one orthogonal to it. As a fucntion of normalized strength of oscillating field ($\alpha := \mu B_1/\hbar\omega$). Time of flight through field region is one period of oscillating field. Static field component is set to resonance for one-photon exchange ($2\mu B_0 = \hbar\omega$). Numbers indicate the number of photons exchanged. Amplitudes for emission or absorption of an even number of photons are always equal in this configuration (bottom), but this is not so for an odd number of photons (middle and top). Taken from ref.7.

Fig.2: Lay out for interferometric measurement of many-photon exchange. The single crystal silicon interferometer is of the skew symmetric type. The phase shifter is rotatable around an axis normal to the plane such that different optical path lengths for the left and the right beam can be adjusted. The interferometer is typically 10 cm long and 6 cm wide, the four crystal slabs for beam separation, reflection and superposition jutting up from a common base. The Heusler analyzer reflects only the same spin state as that of the incident beam into the detector.

beam of polarized neutrons is split into two in the interferometer. In one arm it first traverses a region with the kind of oscillating field given by eq.6 or by eq.7 and then goes through a static spin flipper that is either turned on or turned off. Finally it passes an adjustable phase plate. The beam in the other arm is not affected, aside from also passing through the phase plate. After recombination one of the resulting interference beams is Bragg-reflected by a polarizing crystal (a so called Heusler crystal), such that only neutrons with the same spin state as the incident beam reach the detector. The latter registers the intensity as a function of phase of the oscillating field, i.e. as a function of time. This enables the necessary frequency analysis of the intensity. When the static spin flipper is turned off the signal at the detector is essentially given by (setting r=0)

$$I_{det}(t) = \left| \begin{pmatrix} 1 & 0 \\ 0 & 0 \end{pmatrix} (\Psi_{in} + \Psi_{out}) \right|^2 = 1 + 2\sum_{j=-\infty}^{\infty} \sum_{n=-\infty}^{\infty} |T_j^+ T_n^+| \cos[(n-j)\omega t] + 2\sum_{n=-\infty}^{\infty} |T_n^+| \cos(n\omega t + \chi) \tag{8}$$

The interference signal thus only contains non-spin flip amplitudes T_n^+. (We have neglected their phases here.) The magnitude of each amplitude can be obtained by determining the coefficients of the last sum with the help of the adjustable phase shift χ. One measures for a series of different phase shifts evenly distributed between 0 and 2π and each time obtains a frequency spectrum by Fourier analysis. From the difference between minimum and maximum for a given frequency $n\omega$ one gets directly the corresponding amplitude T_n^+. In order to obtain the amplitudes for exchange of photons plus undergoing a spin flip one must turn on the static spin flipper. Then a formula similar to eq.8 results, except that it contains the T_n^-.

The experiment was originally planned for the high flux reactor at the Institut Laue Langevin in Grenoble but is now carried out at the Missouri University Research Reactor in collaboration with the group of S. A. Werner. The incident neutron beam has a wavelength around 2.3 Å and is polarized by a magnetic supermirror. The incident beam cross section is 2 mm wide and 8 mm high. The static component of the magnetic field is normal to the plane of the interfering beams and is produced by a pair of Helmholtz coils. The magnitude will be between 20 and 80 Gauss. The oscillating component of the magnetic field will be produced by little coils driven at frequencies between 10 kHz and 50 kHz. The magnitude of the oscillating field will be up to 100 Gauss. Analysis of the time dependent intensity will only be done at the forward beam of the two exit beams because it has the better signal to noise ratio. The intensity is below 1 count per second. Currently (June 1993) no data are available yet, as only the polarization of the incident beam has been started.

2. Ion Interferometry

According to quantum theory two path interference should occur for all objects suffciently separated from the environment. Nevertheless it has been observed in a controlled way only for electrons and neutrons, and with atoms for He, Na and Ca. No molecules nor any ions have been investigated in this direction. With ions, one of the reasons may have been the lack of a proper source. But with the recent development of very sharp tips the situation has changed. Tips made from single crystal tungsten wire grown in the <111>-direction can terminate in a cone and ultimately end in only a few atoms or even a single atom. Such tips have been reported by Fink[9] and by Binh[10], and have recently also been produced in our laboratory. Such a tip can be operated at a positive voltage in a thin atmosphere of a gas and will then create positive ions of that gas by field ionization. The advantage of the small radius of curvature of such a tip is that field ionization starts as low as a few hundred volts and is confined to the few atoms right at the top. Thus one has virtually a point source of ions of relatively large de Broglie wavelength. The brightness of this source is also very high, because the strong electric field around the tip captures many gas atoms or molecules by polarizing and subsequently attracting them. A field ion image of such a tip from our laboratory is shown in Fig. 3.

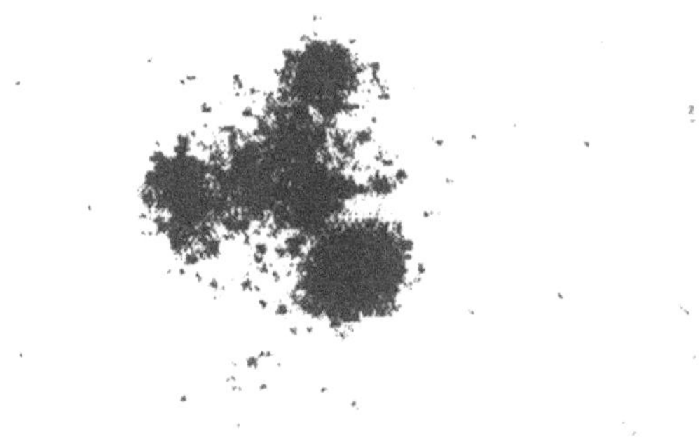

Fig.3: Field ion image of a single crystal tungsten tip of <111>-direction. One can discern three inner atoms and three outer atoms aligned along the three crystallographic axes. Imaged with Ne at some 10^{-4} mbar at a tip voltage of 1.8 keV. Tip is at room temperature. Therefore such a configuration is stable only for a few seconds. Tip to screen distance was 2.5 cm. Diameter of this image of 6 atoms is about 1 cm. Size of the images of the individual atoms is not determined by the wave packet of the Ne ions hitting the screen but by their transverse thermal velocity distribution.

In a current project we want to use such a source to realize the double slit experiment for protons (H^+), H_2^+ and possibly also for H_3^+. All three kinds of ions are generated by field ionization of hydrogen gas at small metallic tips.[11] The relative abundances depend on the field around the tip, the temperature of the tip and the pressure of the gas. The tip will be operated at around 1 kV at the temperature of liquid nitrogen. The pressure of hydrogen around the tip will be between 10^{-5} and 10^{-3} mbar. The ion beam emerging from the tip has an aperture of only a few degrees. Already after about two centimeters from the tip the projected image of an atom has a diameter between one and two millimeters. At that position a screen with a small hole will be placed to select a single atom of the tip as our actual ion source. After that a combination of crossed electric and magnetic fields will filter out the desired species of hydrogen ions. After some 20 cm of free flight the beam will be separated into two by an electrostatic biprism whose central wire has a diameter of about 2 μm. This will be our original double slit separation. In principle a much larger separation is possible because the width of the proton wavepacket at the biprism is much larger.[12] Assuming a source width of 4 Å, an energy of 1 keV and a distance from source to biprism of 25 cm the wavepacket is around 380 μm wide when it will be cut into two. However, after a further distance of some 20 cm the two beams must be recombined. And then it turns out that the fringe spacing of the interference pattern is only some 0.1 μm. This must be magnified by electrostatic quadrupol lenses before it can be imaged on a screen with the help of micro channel plate amplifiers. It is mainly the problem of magnifying the interference pattern by a three to four orders of magnitude which keeps us from starting with a larger beam separation.

Two path interference with a new kind of object always has a fascination in itself, and a number of fundamental questions can be raised once it has become possible. However, especially with protons we want to emphasize a possible practical application. This is holography with protons. It can be done in analogy to holography with electrons,[13] but one can expect very high resolution already at moderate energies, as protons of e.g. 100 eV have a wavelength of only 0.03 Å. This could enable imaging biological samples with almost no radiation damage.

3. Information transfer by phase shifting

The neutron experiment of sec. 1 and many other interferometry experiments and in particular holography share one common feature. Information imprinted on the wavefunction in one beam must be retrieved from the interference signal. Before this background we want to ask how weak we can make the signal carrying beam in one arm of the interferometer and still obtain information about the phase shift. The question has also practical importance as radiation damage to a sample is in general proportional to the flux through it. And it is related to the question of the information carrying capacity of the free electromagnetic field.[14] Consider the experiment shown in Fig. 4. An interferometer is used to send information in a bit by bit fashion. The sender transmits

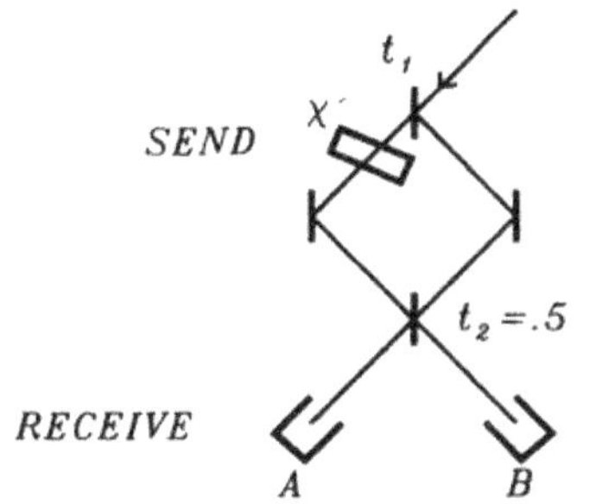

Fig.4: Transfer of one bit by setting χ to either 0 or π. Transmission of beam splitter, t_1, is made very small to reduce flux through phase shifter.

a bit by letting N particles pass through the interferometer. If the value of the bit is 0 the phase shifter is set to $\chi_0=\pi$, if it is 1, the phase shifter is set to $\chi_1=0$. The receiver registers the particles in the detectors A and B and knows that the information of one bit is available every time N particles have been detected. (For the sake of simplicity we assume no losses and also neglect noise.) The probability for a particle to arrive at detector A when the phase shift is χ_0 is given by

$$p_A(0) = \frac{1}{2} - \sqrt{t_1(1-t_1)}\,, \tag{9}$$

and when the phase shift is χ_1 it is

$$p_A(1) = \frac{1}{2} + \sqrt{t_1(1-t_1)}\,. \tag{10}$$

The corresponding probabilities at detector B are determined by $p_A+p_B=1$. The receiver has now but one rule for interpreting the data: The bit shall be understood as having value 0 when the counts in detector A are less than N/2, otherwise the bit is assigned the value 1. (In order to have the same number of possible outcomes when the bit shall mean "0" and when it shall mean 1 the sender and the receiver agree on an odd N. The possible counts at detector A are then 0, 1, 2, ... , N.) Of course, as this is a quantum mechanical way of transmitting information the counts at the detectors are subject to statistical fluctuations. Therefore the probability of interpreting the bit correctly is always less than 1. We have binomial statistics and hence we get for the probability of correctly interpreting the bit, which we call the probability of success,

$$P_{success} = \sum_{L=0}^{L<N/2} \frac{N!}{L!(N-L)!}\, p_A^L\,(1-p_A)^{N-L}. \tag{11}$$

It is straightforward to show that this probability of success applies to both values of the bit. Note that $P_{success}$ depends on two quantities controllable by the experimenters. One is the number of particles sent per bit, N. The other is the transmission of the beamsplitter, t_1. Clearly, when N gets large, we expect $P_{success}$ to approach 1. On the other hand, when t_1 is at its optimum value of 0.5 this will also give the best $P_{success}$. But note that we want both N and t_1 to be very small, because we want the flux through the phase shifter, which here is symbolic for a sample, to be small.[15] And this flux through the phase shifter per bit is given by

$$I_\chi = N t_1. \tag{12}$$

In Fig. 5 we plotted $P_{success}$ as a function of N for various I_χ. The surprising result is that

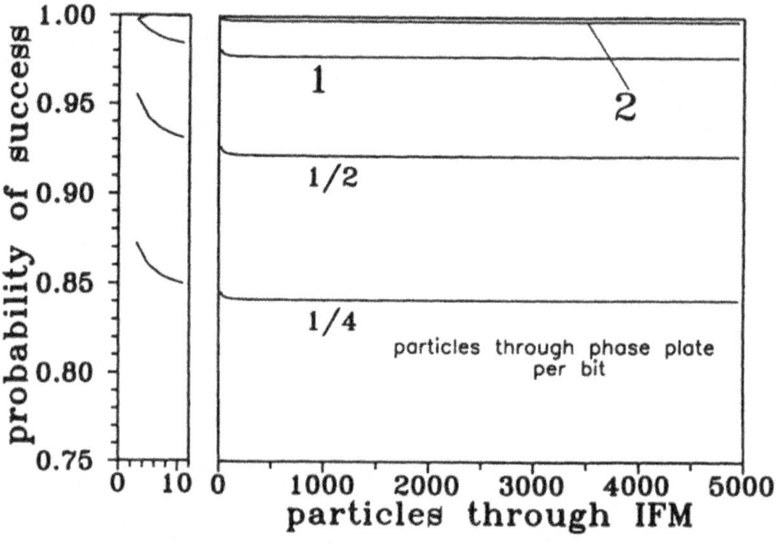

Fig.5: Probability of successful transmission of 1 bit as a function of total particles through interferometer (IFM) per bit. Numbers indicate flux of particles through phase plate per bit.

in order to have a certain probability of success there must have been a certain flux through the phase shifter which is almost independent of the choice of N and t_1 as long as their product is the same. A small deviation towards the better exists for N less than 10, where peculiarities of the binomial distribution come into play. In principle this can be exploited, however we did not take into account noise in our considerations, and when doing so this small advantage may be much reduced. But generally we can say that in order to transmit a bit with a given probability of success in an interferometric set up a minimum number of particles must have "sensed" the phase shifter. For holographic applications, e.g. in electron holgraphy, this means that a minimum radiation damage to the sample cannot be avoided when a certain information about the sample shall be obtained. However, it is not yet clear whether Fig. 5 already shows a general limit. Perhaps a lower limit is possible with more interference loops. This comes to mind because a beam splitter actually increases the sum of the amplitudes, which is the sum of the square roots of the probabilities (leaving aside the phases). Therefore when repeatedly splitting a weak beam into many subbeams the sum of the amplitudes can be very large. When bringing these many subbeams into superposition with a reference beam one might obtain a large modulation of the probabilities at the various detectors (which would then also be many). This is a question we are currently investigating.

4. Acknowledgment

The author would like to acknowledge financial support from the Austrian Fonds zur Förderung der Wissenschaftlichen Forschung which made possible both the neutron experiments (grant P9266-PHY) and the ion experiments (grant P8781-PHY). Thanks also go to G. Krenn for preparing the W-tips and recording the field ion microscopy images.

5. References

1. J. Schmiedmayer, P. Riehs, J. A. Harvey and N. W. Hill, *Phys. Rev. Lett.* **66** (1991) 1015.
2. S. Stenholm, *Phys. Rep.* *C6* (1973) 1.
3. H.-I. Yoo and J. H. Eberly, *Phys. Rep.* **118** (1985) 239.
4. J. H. Shirley, *Phys. Rev.* **138** (1965) B979.
5. G. Badurek, H. Rauch, J. Summhammer, *Phys. Rev. Lett.* **51** (1983) 1015.
6. G. Badurek, H. Rauch, D. Tuppinger, *Phys. Rev.* **A34** (1986) 2600.
7. J. Summhammer, *Phys. Rev.* **A47** (1993) 556.
8. B. E. Allman, A. Cimmino, A. G. Klein, G. I. Opat, H. Kaiser, and S. A. Werner, *Phys. Rev. Lett.* **68** (1992) 2409.
9. H. W. Fink, H. Schmid, H. J. Kreuzer, and A. Wierzbicki, *Phys. Rev. Lett.* **67** (1991) 1543.
10. V. T. Binh, S. T. Purcell, N. Garcia and J. Doglioni, *Phys. Rev. Lett.* **69** (1992) 2527.
11. E. W. Müller and T. T. Tsong, *Field Ion Microscopy* (American Elsevier Publishing Corporation, Inc., New York, 1969) p. 49.
12. J. Summhammer and J. Schmiedmayer, *Physica Scripta* **42** (1990) 124.
13. Current status of electron holography is reported in Q. Fu, H. Lichte and E. Völkl, *Phys. Rev. Lett.* **67** (1991) 2319.
14. H. P. Yuen and M. Ozawa, *Phys. Rev. Lett.* **70** (1993) 363.
15. Neutron interferometry experiments with effectively very small t_1 are reported in J. Summhammer, H. Rauch and D. Tuppinger, *Phys. Rev.* **A36** (1987) 4447.

ELECTRONS

Electron Holography and Its Application to Flux Line Obsevation

AKIRA TONOMURA
Advanced Research Laboratory, Hitachi, Ltd.
&
Tonomura Electron Wavefront Project, ERATO, JRDC
Hatoyama, Saitama 350-03, Japan

ABSTRACT

An electron wavefunction's phase distribution can be precisely measured to 1/100th of the electron wavelength using both electron holography and a "coherent" field-emission electron beam. This technique is used to directly observe flux lines penetrating superconducting films. These flux lines are detected as phase shifts due to the Aharonov-Bohm effect of an electron beam passing through the films.

1. INTRODUCTION

While electron microscopy uses the intensity of an electron beam to observe specimens, electron holography can be used to measure the relative phase of the electron wavefunction. This has become feasible with the development of "coherent" field emission electron beams.[1] The precision of this phase measurement is 1/100 of the wavelength[2]. This technique makes it possible to both observe and masure microscopic distributions of electromagnetic fields, especially flux lines in superconductors.

Although the behavior of flux lines plays a key role in superconductivity and much effort has been expended on developing methods to observe them, until recently it was not possible to observe flux lines because they are shaped like extremely thin threads and have a small flux value. In 1967, Essman and Träuble[3] developed the Bitter technique to directly observe the flux-line lattice predicted by Abrikosov.[4] This technique has recently been used to elucidate the microscopic characteristics of high-Tc superconductors.[5] However, this technique is essentially static and cannot determine the dynamic behavior of flux lines. New techniques for observing flux-lines have therefore been developed.[6,7] For example, Hess, *et al.*[6] used a scanning tunneling microscope to observe the flux-line lattice of $NbSe_2$. However, dynamic observation is still not feasible with these techniques.

Using transmission electron microscopy for direct observation has been theoretically investigated making use of the fact that an electron beam is deflected,[8-10] or phase-shifted by flux lines. Trials have determind, however, that the deflection angle is too small for flux lines to be observed as a Lorentz micrograph (a greatly defocused electron micrograph). We have developed of a "coherent" and "penetrating" electron beam that opens the way to direct observation of flux lines with the help of electron holography.[11-14] In this method, a single flux line leaking from a superconductor surface can be observed directly and even dynamically as a contour fringe in an interference micrograph. Furthermore, for the first time, flux lines can be observed in the transmission mode.

2. ELECTRON HOLOGRAPHY

In our experiment, electron holography, which is a two-step imaging method, plays an important role. An object wave and a reference wave are overlapped to form an interference pattern, which is recorded on film. This film, called a hologram, is subsequently illuminated by a collimated laser beam. The exact image is then reproduced three-dimensionally in a diffracted beam. Once electron wavefronts have been reproduced as light wavefronts, more versatile optical techniques can be used to supplement electron optics.

An interference micrograph, or contour map of the wavefront, can be obtained simply by overlapping an optical plane wave with this reconstructed wave (see Fig. 1(a)). If a conjugate wavefront instead of a plane wave overlaps this wavefront, the phase difference becomes twice as large, as if the phase distribution were amplified twice, as shown in Fig. 1(b). By repeating this process, even a phase shift as small as 1/100 of a wavelength can be detected.[2]

This phase-amplified interference electron microscopy provides information about the microscopic distribution of the electric [15] and magnetic[16] fields.

3. FLUX LINE OBSERVATION

3.1 Experimental Method

3.1.1 Interference Microscopy

The observation principle for directly observing flux lines in a twice phase-amplified interference micrograph is shown in Fig. 2. When an electron beam is incident to a uniform magnetic field, it is deflected to the left by the Lorentz force. To view electrons as waves, the introduction of a "wavefront" perpendicular to the electron trajectory will suffice. The incident electron beam is a plane wave, but the outgoing beam becomes a plane wave with its left side tilted up. In other words, the wavefront is viewed as rotating around a rotating axis: the magnetic line of force. From a contour map of this wavefront, it can be seen that the contour lines follow the magnetic lines of force. This is because the height of the wavefront is the same along

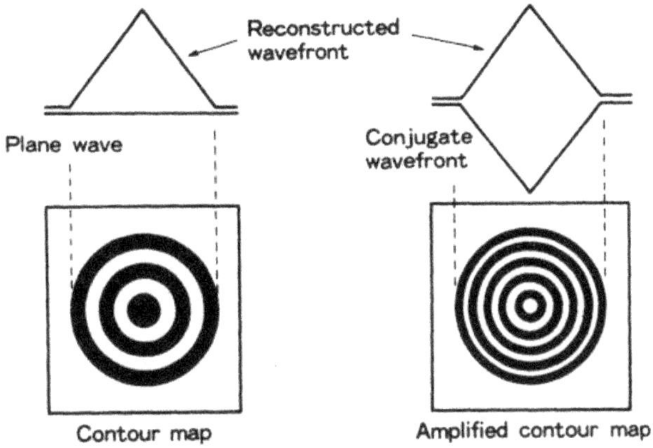

Fig. 1: Principle behind electron holography

a magnetic line. Thus, a very simple conclusion can be reached: when a magnetic field is observed in an interference electron micrograph, the contour fringes can be considered to represent the magnetic lines of force.

The interference fringes are also quantitative. A simple calculation convinces us that a constant minute amount of magnetic flux, h/e, is flowing between adjacent contour fringes. It can be concluded then that a contour fringe in a twice phase-amplified interference micrograph indicates a single flux line.

3.1.2 Lorentz microscopy

A Lorentz micrograph is a greatly defocused electron micrograph. The principle behind it is shown in Fig. 3. When an electron beam is incident to a ferromagnetic thin film, which has two magnetic domains, the beam is deflected by the magnetization, and the deflection directions are different for the two domains. Therefore, when the electron intensity distribution is observed in the lower plane, the domain wall can be observed as a line of weak intensity. Thus, Lorentz microscopy is effective when the magnetic field changes suddenly, such as at a domain boundary in a ferromagnetic film. However, it is not easy to observe flux lines in free space using Lorentz microscopy, since those magnetic fields are distributed smoothly in a harmonic form.

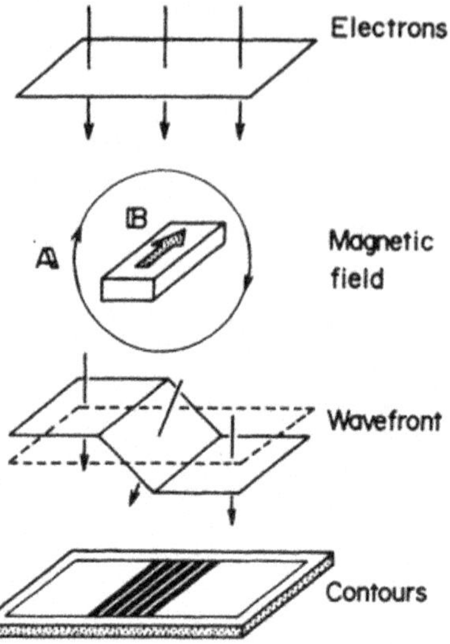

Fig. 2: Principle behind magnetic flux observation.

3.2 EXPERIMENTAL RESULTS

Individual flux lines were statically and dynamically observed using holography electron microscopes.

3.2.1 Observation in the Profile Mode

The flux lines leaking out from a superconductor surface can be directly observed as contour fringes in a twice phase-amplified interference micrograph through electron holography. A magnetic field of a few Gauss or less was applied perpendicularly to a lead film, which was then cooled to 4.5 K. The magnetic lines of force of the flux lines were observed through electron holography by applying an electron beam parallel to the superconductor surface. Figure 4(a) shows the result[17]. In this photograph, the phase difference is amplified by a factor of two. Therefore, one interference fringe corresponds to one flux line. A single flux line is captured on the right side. The magnetic line of force is produced from an extremely small area of the lead surface, and then spreads out into free space.

In addition to observing isolated flux lines, a pair of flux lines were observed that were oriented in opposite directions and connected by magnetic lines of force (Fig. 4 (a) left). This may be because that when the specimen is cooled below the critical temperature, the lead becomes superconductive. During the cooling, however,

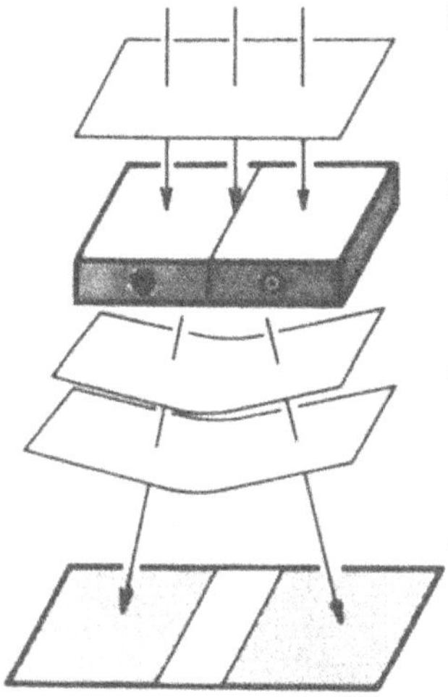

Fig. 3: Lorentz microscopy.

the specimen experiences a state where the flux-line pair appears and disappears repeatedly due to thermal excitation [18] and is pinned by some imperfection in the superconductor, eventually resulting in the flux being frozen.

What happens when the thickness of the superconducting thin film is increased? Figure 4(b) shows the state of the magnetic lines of force when the thickness is 1μm. It can be seen that the state changes completely. Magnetic flux penetrates the superconductor, not as individual flux lines, but in a bundle. The figure does not show any flux-line pairs.

Our explanation for this phenomenon is as follows. Since lead is a type-I superconductor, a strong magnetic field applied to it partially destroys the superconductive state in some parts of the specimen (intermediate state). Figure 4(b) shows that the magnetic lines of force penetrate the parts of the specimen where superconductivity has been destroyed. However, since the surrounding parts are still superconductive, the total amount of penetrating magnetic flux is an integral multiple of the flux quantum, $h/2e$. Thin superconducting films (Fig. 4(a)) are an exception. They behave like a type-II superconductor and the flux penetrats the superconductor in the form of individual flux lines.

Since the flux itself can be observed using electron holography, its dynamic behavior can also be observed.[19] In this case, after electron holograms were dynamically recorded on videotape, a twice phase-amplified contour map of each frame was numerically reconstructed, and again recorded on videotape. After this process, flux

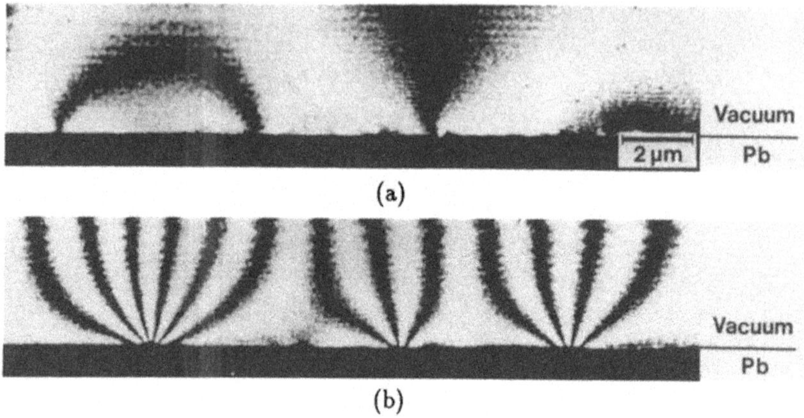

Fig. 4: Interference micrograph of flux lines leaking from Pb film (Phase amplification: ×2).
(a) Thickness = 0.2 μm. (b) Thickness = 1 μm.

dynamics could be observed with a time resolution of 1/30 of a second.

The experiment is carried out as follows. While trapped fluxes in a Pb thin film remain stationary at 5 K, when the sample temperature is raised, the flux line diameter gradually increases. Just below the critical temperature, the flux lines begin to move. Figure 5 shows a section from a videotape of this movement.

Three upward direction flux lines were trapped in the superconductor and their magnetic lines of force can be seen in Fig. 5(a). At 0.13 seconds, the flux lines moved suddenly to the left after the lapse of only a single frame. Two upward flux lines and two downward flux lines were connected by magnetic lines. At 1.33 seconds, the downward flux lines moved to the right and only a broad single magnetic line remained.

Although this flux movement due to thermal activation is random, a similar experiment was made where a current was applied to the superconductor. Flux lines received a Lorentz force determined by the current, but with opposite directions for upward and downward flux lines. The pinning force at each pinning site could thus be measured.

3.2.2 Observation in the Transmission Mode

Flux lines have recently been observed in the transmission mode.[20,21] A two-dimensional distribution of flux lines was seen dynamically using Lorentz microscopy with a 300-kV holography electron microscope.

The experimental arrangement is shown in Fig. 6. A Nb thin film, set on a low-

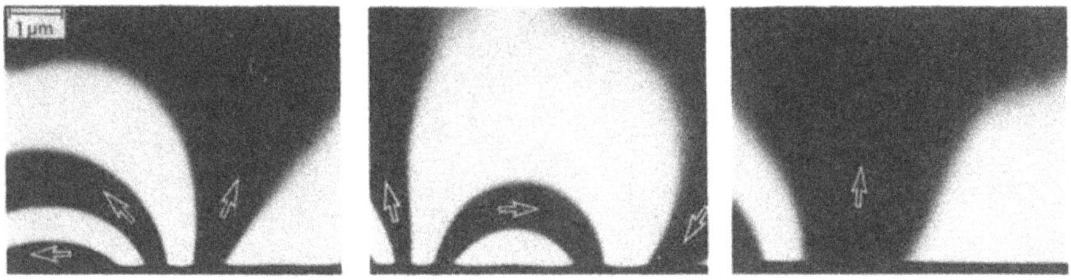

(a) (b) (c)

Fig. 5: Dynamic observation of trapped flux lines near Tc.
(a) 0 seconds (b) 0.13 seconds later (c) 1.33 seconds later

temperature stage, was tilted at 45° to a vertieel incident beam of 300-keV electrons, so that the electrons could receive the flux-line magnetic fields penetrating the sample perpendicularly to its surface. An external magnetic field of up to 150 Gauss was applied horizontally.

The information about the flux lines is contained in the phase distribution of the transmitted electron beam. This information cannot be read from a conventional electron micrograph, but reveals itself in a defocused image, *i.e.*, a Lorentz micrograph, in which a flux line can be seen as a tiny spot, one half bright and the other half dark.

The sample was first cooled down to 4.5 K and the applied magnetic field B was gradually increased. As B increased, flux lines suddenly began to penetrate the film at $B = 32$ Gauss, and their number increased with B. Their dynamic behavior was quite interesting: at first, only a few flux lines appeared here and there in the field of view, $15 \times 10\mu m^2$, oscillating around their own pinning centers and occasionally hopping from one center to another. These movements continued as long as the flux lines were not closely packed ($B \leq 100$ Gauss).

An equilibrium Lorentz micrograph at $B = 100$ Gauss is shown in Fig. 7. The film has a fairly uniform thickness in the region shown, but is bent along the black curves, called bend contours, which are due to Bragg reflections at the atomic plane brought to a favorable angle by bending. Each spot with a black and white contrast is the image of a single flux line. This contrast reversed, as expected, when the applied magnetic field was reversed. The tilt direction of the sample can be read from the line dividing the black and white part of the spots. Since the black part is on the same side for all the spots, the polarities of all the flux lines as seen in the region are the same.

At low B, *i.e.*, up to 30 - 50 Gauss, the flux lines are too scarce to form a lattice, even in equilibrium. At $B = 100$ Gauss, where the flux-line density is so high that it

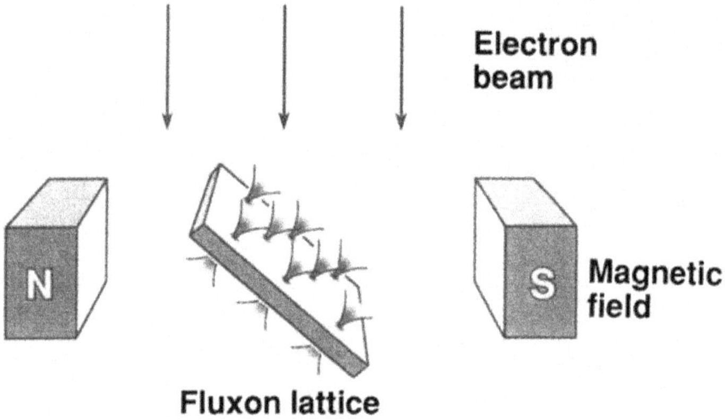

Fig. 6: Schematic for flux-line lattice observation.

cannot be anything but a hexagonal lattice, the flux-line configuration and movement are influenced by structure defects.

4. CONCLUSION

Electron holography has opened up a new window to direct, real-time observation of the microscopic dynamics of individual superconducting flux lines, such as in flux creep and pinning, which up to now have only been observed in macroscopic experiments. This technique can be effectively employed to elucidate the fundamentals and practical applications of superconductivity, especially in the field of high-Tc superconductors.

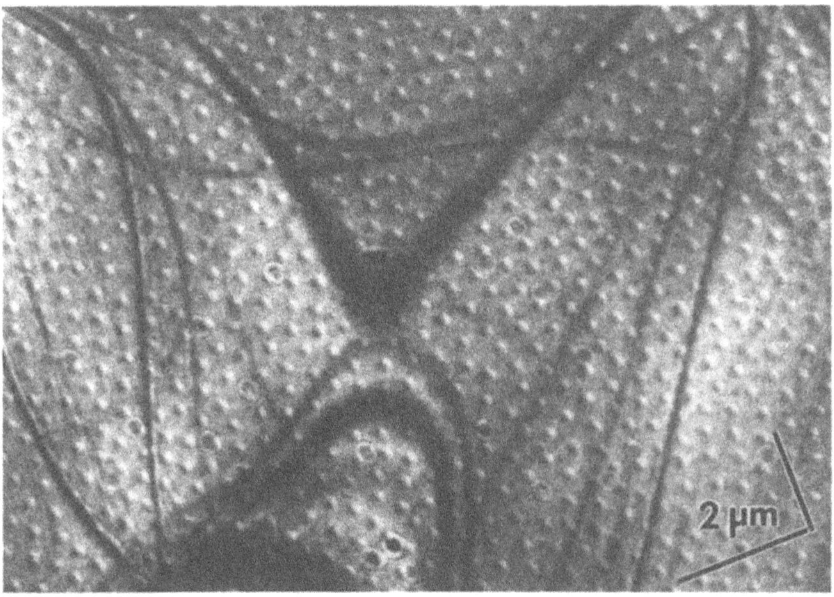

Fig. 7: Lorentz micrograph of a two-dimensional array of flux lines in superconducting Nb film.

6. REFERENCES

1) A. Tonomura et al., *J. Electron Microsc.* **28** (1979)1.
2) A. Tonomura, et al.: *Phys. Rev. Lett.* **54** (1985) 60.
3) V. Essman and H. Träuble: *Phys. Lett.* **A24** (1967) 526.
4) A. A. Abrikosov, *Zh. Eksp. Teor. Fiz.* **32** (1957) 1442 [Sov. Phys. JETP 5 (1957) 1174].
5) For example, D. J. Bishop, et al.: *Science* **255** (10 January 1992) 165.
6) H. F. Hess, et al.: *Phys. Rev. Lett.* **62** (1989) 214.
7) J. Mannhart, et al.: *Phys. Rev.* **B35** (1987) 5267.
8) H. Yoshioka: *J. Phys. Soc. Jpn.* **21** (1966) 948.
9) C. Colliex: *Acta Crystallogr. Sect.* **A24** (1968) 692.
10) C. Capiluppi, G. Pozzi, and V. Valdre: *Phil. Mag.* **26** (1972)865.
11) D. Gabor: Proc. *R. Soc. London,* Ser. **A197** (1949) 454.
12) A. Tonomura, *Rev. Mod. Phys.* **59** (1987) 637.
13) A. Tonomura: *Physics Today* **22** (April 1990) 22.
14) A. Tonomura: *Adv. Phys.* **41** (1992) 59.
15) S. Frabboni, G. Matteucci & G. Pozzi, *Phys. Rev. Lett.* **55** (1985) 2196.
16) A. Tonomura, et al., *Phys. Rev. Lett.* **44** (1980) 1430.
17) T. Matsuda et al., *Phys. Rev. Lett.* **62** (1989) 2519.
18) J. M. Kosterlitz and D. Thouless, *J. Phys.* **C6** (1973) 1181.
19) T. Matsuda, et al., *Phys. Rev. Lett.* **66** (1991) 457.
20) K. Harada, et al.: *Nature* **360** (5 November 1992) 51.
21) J. Bonevich et al., *Phys. Rev. Lett.* **70** (1993) 2952.

INTERFEROMETRY AND HOLOGRAPHY WITH ELECTRONS

HANNES LICHTE

Institut für Angewandte Physik, Universität Tübingen, Auf der Morgenstelle 12
D-7400 Tübingen, Germany

ABSTRACT

Matter Wave Interferometry started in 1954 when Möllenstedt and Düker developed the electron biprism to produce two-beam interference pattern with electrons. Thus they gave birth to the new field of interferometry and holography in electron optics which substantially contributes to the investigation of the basic physics of electron waves and their applications to problems e.g. in materials science.

1. The electron biprism - the basic tool for electron interferometry

Möllenstedt and Düker [1] invented the electron biprism as an electron optical analogue of the Fresnel-biprism in light optics:

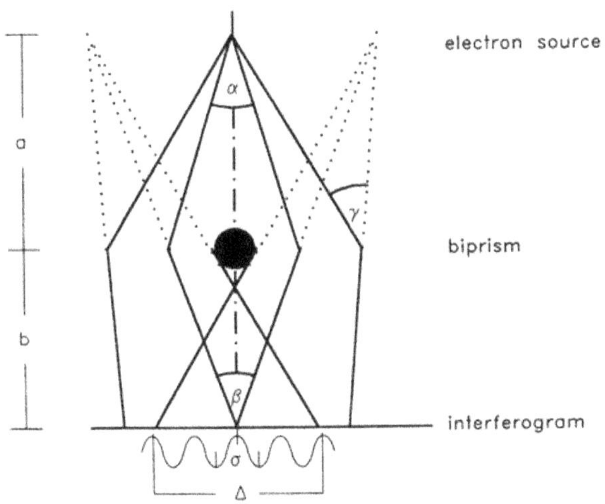

Fig. 1 Scheme of the electron biprism interferometer

In analogy to the Fresnel biprism in light optics, the two partial electron waves passing the positively charged biprism filament on the left and on the right side seem to come from two virtual sources. α determines the degree of lateral coherence of the two waves, i.e. the contrast of the interference fringes. β gives the spacing σ of the fringes, Δ is the width of the interferogram.

A thin conductive filament splits the wave front of an incoming electron wave in two coherent partial waves passing on either side of the filament. If a positive voltage is applied to the filament, the two waves are deflected towards each other such that they overlap at some distance behind the filament. There they built up a pattern of equidistant and parallel interference fringes given by

$$I(x) = A_1^2 + A_2^2 + 2A_1A_2 \cdot C \cdot \cos(2\pi x / \sigma) \qquad (1)$$

where the filament is assumed in y-direction. A_1 and A_2 are the amplitudes of the partial waves. The fringe spacing σ is given by

$$\sigma = \lambda_0 / \beta \qquad (2)$$

with $\lambda_0 = h / \sqrt{2em_0U^*}$ the de Broglie wavelength of the electrons of energy eU; the asterisk denotes relativistically corrected voltages, h is Planck's constant and e and m_0 are electron charge and mass. β means the angle of superposition of the two partial waves. C is the contrast of the fringes mainly determined by the degree of coherence between the partial electron waves.

2. Coherence of electron waves

Since a LASER-like electron source is not available, the coherence conditions usual with incoherent sources have to be met. The coherence properties of electron sources have been discussed in more detail by Lenz [2]. The longitudinal coherence generally does not pose a severe problem since the energy width of about 1eV at an energy of e.g. 100keV results in a coherence length of about 10^5 wave lengths.

Most critical, however, is the fulfillment of the lateral (sometimes called the angular) coherence. Following the Zernike-van Cittert theorem, the degree of lateral coherence of the superimposed waves leaving the source with an angular difference α is given by

$$\mu(\alpha) = \int i(q) \exp(i2\pi k \alpha q) dq \qquad (3)$$

with $k = 1 / \lambda_0$ and $i(q)$ the intensity distribution in the electron source.

The contrast in the interference pattern follows as $C = |\mu(\alpha)|$. An estimation shows that the mean current density in the interferogram at an axial distance l=a+b from the electron source is given by [3]

$$j = -2B \ln(C) / (k^2 \pi \alpha^2 l^2), \qquad (4)$$

where B means the brightness of the electron gun. In particular in the early days of electron interferometry, long exposure times up to minutes were needed to collect a sufficient number of electrons in the interferogram, since the brightness of the electron guns was comparably poor. Today with modern field emission technology, high-brightness electron guns allow to take electron interferograms at a good contrast with exposure times of only a few seconds.

3. Phase shift of electron waves

A wave traversing through an area with an index of refraction $n = \lambda_0 / \lambda$ experiences a phase shift

$$\varphi = \frac{2\pi}{\lambda_0} \int n ds \qquad (5)$$

where the integral is taken over a trajectory orthogonal to the wave front. This relation holds for electrons with the electron optical index of refraction

$$n(r, \vec{s}) = \sqrt{\frac{(U + V(r))^*}{U^*}} - \sqrt{\frac{e}{2m_0 U^*}} (\vec{A}(r), \vec{s}) \qquad (6)$$

given as early as 1933 by Glaser [4]; $V(r)$ and $\vec{A}(r)$ mean the distribution of the electric potential and of the magnetic vector potential, respectively. Evidently, n depends not only on the coordinate r but also on the direction of propagation $\vec{s}$ of the wave. For the usual case of $V \ll U$, the index of refraction can be written as

$$n(r, \vec{s}) = 1 + \frac{V(r)}{2U^*} \frac{m}{m_0} - \sqrt{\frac{e}{2m_0 U^*}} (\vec{A}(r), \vec{s}). \qquad (7)$$

In an biprism interferometer, each of the two coherent waves sums up the optical path $\int n(r, \vec{s}) ds$ on its respective way from the source to the observation plane. If one of them travels through vacuum with $V(r) = 0$, in the interference plane a phase difference

$$\delta\varphi = \frac{2\pi}{\lambda_0} \frac{m}{m_0} \frac{1}{2U^*} \int V(r) ds - \frac{e}{\hbar} \Phi \qquad (8)$$

arises with $\Phi = \oint \vec{A}d\vec{s}$ the magnetic flux enclosed between the two partial waves. Consequently, in the interference pattern

$$I(x) = A_1^2 + A_2^2 + 2A_1A_2 \cdot C \cdot \cos(2\pi x / \sigma + \delta\varphi) \qquad (9)$$

the interference fringes are shifted aside by $\Delta x = \delta\varphi \cdot \frac{\sigma}{2\pi}$ allowing to measure $\delta\varphi$ easily.

4. Applications of electron biprism interferometry

Since the development of the electron biprism, electron interferometry was steadily improved and applied for the analysis of properties of electron waves and for measuring properties of matter, such as

- Determination of the electron wavelength in agreement with de Broglie from the macroscopic data of the interferometer geometry [5]
- Investigation of angular coherence [6,7] and longitudinal coherence [8] of electron waves
- Measurement of the phase shift due to macroscopic electric potential field V(r) [8], and due to the Mean Inner Potential of Thin Films [9]
- Experimental proof of the phase shift of electron waves due to an enclosed magnetic flux Φ [10], as theoretically predicted by Ehrenberg and Siday [11] and by Aharonov and Bohm [12].
- Determination of the Flux Quantization in superconducting hollow cylinders [13,14].
- Measurement of atomic scale height differences on surfaces by electron mirror interference microscopy [15]
- Measurement of the Doppler shift of electron waves reflected at a moving electron mirror [16]
- Proof of the Sagnac effect with electrons [17]
- Development of Electron Holography as described in more detail in the following.

5. Electron Holography

Basically, holography is an interferometric method combined with a highly sophisticated technique of evaluation of the interferogram: Again by means of an electron biprism, a wave modulated by an object $A(r) \cdot \exp(i\phi(r))$ is superimposed on a plane reference wave amplitude 1, and this yields an interference pattern

$$I_{hol}(x,y) = 1 + A^2(r) + 2C \cdot A(r) \cdot \cos(2\pi R_c x + \phi(r)) \qquad (10)$$

which, in the cosinoidal term with carrier frequency Rc, contains both amplitude A and phase ϕ. Therefore, such an interferogram can be considered as a hologram of the wave in the sense of Gabor [18]. Using the electron hologram as a diffraction grating for light, one of the diffracted light waves turns out to be a light optical copy of the electron wave recorded in the hologram; this is due to the reciprocity between interference and diffraction in wave optics. Therefore, the reconstructed wave can be processed by wave optical means on the light optical bench.

Gabor´s idea was to use this lensless imaging technique in electron microscopy to avoid the use of an electron objective lens which, in principle, cannot be free from spherical aberration hence restricts resolution in a conventional electron image. Furthermore, the wave optical facilities, not fully developed in electron microscopy, would widely be extended by application of the light optical tools on the reconstructed light wave. In fact, Möllenstedt and Wahl recorded Fresnel-off-axis electron holograms and reconstructed the corresponding light wave [19]. However, it turned out that, due to the restricted coherence of electron sources, the resolution in the reconstructed wave is restricted, if the hologram is taken at a distance from the object.

Therefore, Wahl developed "Image-Plane-Off-Axis-Electron-Holography" [20]: In an electron microscope equipped with an electron biprism, a reference wave is superimposed on the image wave b(r) produced by the aberrated objective lens.

The image wave is blurred with respect to the object wave $o(r)$ according to

$$b(r) = o(r) \otimes PSF(r) \qquad (11)$$

by convolution with the "point spread function" PSF(r) given by

$$PSF(r) = FT^{-1}[\exp(-i\chi(R))]. \qquad (12)$$

The function $\chi(R)$ represents a phase plate in the back focal plane of the objective lens strongly depending on the spatial frequency R as

$$\chi(R) = 2\pi k\left\{0.25Cs\left(\frac{R}{k}\right)^4 - 0.5Dz\left(\frac{R}{k}\right)^2\right\} \qquad (13)$$

which describes the effect of spherical aberration by the coefficient Cs, and defocus Dz.

The idea was to correct for the aberration during the subsequent reconstruction step. In the beginning, however, Wahl applied holography to medium resolution problems, i.e. small spatial frequencies R, where aberrations can be neglected. There he showed the advantage of holography by utilizing the light optical facilities for imaging the object structure at phase contrast, e.g. by means of the Zernike phase plate, which cannot be applied in the electron microscope, or by means of holographic interferometry. The technique of image plane holography was taken up by the groups of Tonomura [21], Pozzi [22] and Hanßen [23] and further developed for revealing the phase modulation of different objects in electron microscopy.

6. Electron Holography at atomic dimensions

In Tübingen, since 1980 we tried to enter the high resolution domain of electron microscopy with electron holography. As Wahl pointed out, the carrier spatial frequency Rc of the hologram fringes has to be larger by a factor 2÷3 than the maximum spatial frequency one wants to reconstruct from the hologram. Consequently, to reach about 0.1nm resolution, the fringe spacing has to be smaller than 0.05nm. This establishes a main barrier for high resolution holography since, for that purpose, the stability of electron microscopes is usually not sufficient, and disturbances arising from mechanical vibrations or AC stray fields in usual laboratories are too high. However, after some years of experience, we learned how to record holograms with fringes smaller than 0.05nm [24].

The reconstruction of the electron wave from high resolution holograms is preferably performed by means of numerical image processing because it offers a high degree of flexibility, accuracy and reproducibility [25]: The hologram is fed into a computer where, by means of the laws of wave optics, the electron object wave is determined numerically. The computer, wave-optically coupled to the electron microscope via the hologram, is the most flexible and accurate wave-optical tool one can think about.

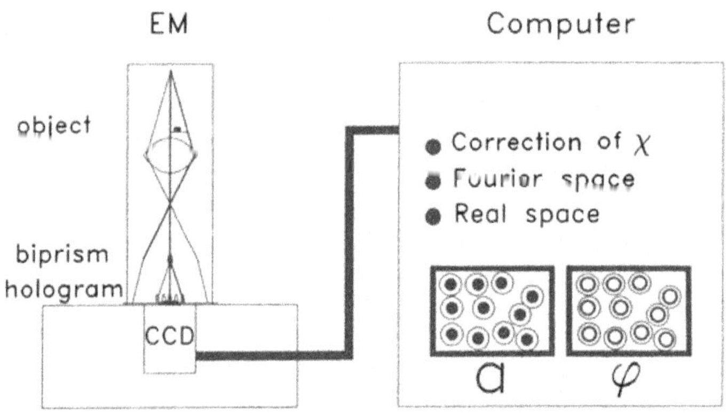

Fig.2 Schematic arrangement for electron holographic analysis in electron microscopy.

The computer, wave optically coupled to the electron microscope via the hologram, allows to perform the correction of aberrations of the microscope and the wave optical analysis of amplitude a and phase φ of the electron wave for a thorough investigation of the object structure

Meanwhile, the following reconstruction procedures are established [26]:

• *Correction of aberrations*

During reconstruction the Fourier spectrum of the recorded image wave

$$FT(b) = FT(o) \cdot \exp(-i\chi(R)) \qquad (14)$$

is determined, which contains the wave aberration $\chi(R)$ of the electron objective lens. For correction of aberrations, the spectrum is divided by a numerically generated phase plate $\exp(-i\chi(R))$. Then the Fourier spectrum of the object wave is at hand. This procedure is easy to perform, if one knows the wave aberration with sufficient accuracy.

• *Analysis of Fourier space*

The Fourier spectrum of the reconstructed object wave can be used for analysis of the object structure, since it agrees with the electron diffraction pattern found in the electron microscope, i.e. it reflects e.g. crystal symmetry relations, crystal orientations with respect to the electron beam, and so on. Holography offers additional benefits from evaluation of the phases of the spectrum: In holography there is no phase problem restricting the analysis of conventional diffraction patterns.

• *Analysis of amplitude and phase of the object wave in real space*

Amplitude and phase of the reconstructed object wave can uniquely be determined and displayed on the monitor; an example is shown in fig.3. Therefore the interpretation in terms of the underlying object structure is easier than with a conventional micrograph, where, due to the convolution with the point spread function PSF(r), amplitude contrast and phase contrast mix up and blur each other.

• *Nanodiffraction*

For a highly localized analysis of the object structure, a numerical aperture can be applied to isolate a small area of the object - typically few nm wide - and to determine the correponding diffraction pattern by Fourier transformation. Shifting the aperture over the field of view allows one to determine variations in thickness or orientation of the object.

Limits of electron holography

In principle, the limits of electron holography in terms of lateral resolution are given by the incoherent aberrations, e.g. the chromatic aberration, of the microscope used for recording the hologram, and by the pixel capacity of the computer used for reconstruction [27]. With the installation and the test of our new Philips CM30 300kV-electron microscope we are optimistic about approaching the resolution limit of 0.1nm in near future with the holographic method.

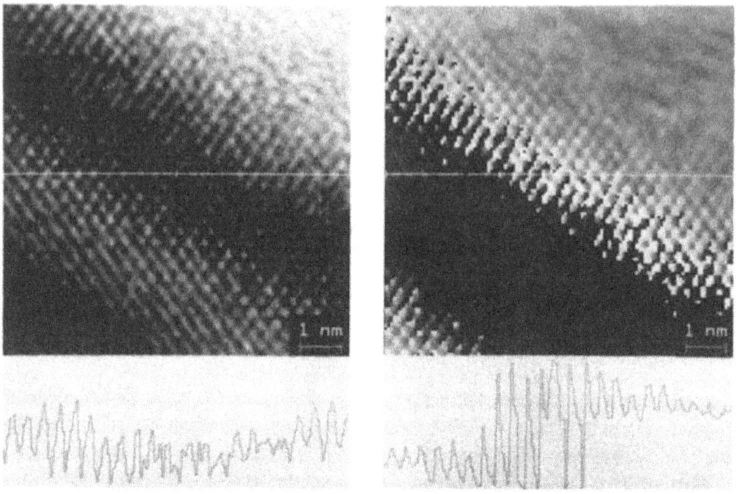

Fig.3 Object wave of a wedge-shaped Silicon crystal in [110] orientation reconstructed from a hologram taken with a 100kV-electron microscope (from 28).

The crystal thickness increases from top right to bottom left. Amplitude (left) and phase (right) are modulated on an atomic scale in a different way. Additionally, the dark band in the amplitude and the sudden jump in the phase running diagonally across the field of view show up; they mark the extinction thickness of the crystal according to the dynamical interaction of the electrons with the crystal.

7. References

1. G. Möllenstedt and H. Düker, Naturwiss. 42(1955),41

2. F. Lenz, 2nd Int. Symp. Foundations of Quantum Mechanics, Physical Society of Japan, Tokyo 1986, 112

3. H. Lichte, P. Kessler, F. Lenz and W.D. Rau, *0.1nm Information Limit with the CM30FEG-Special Tübingen*, submitted to Ultramicroscopy

4. W. Glaser, Z. Physik 80(1933), 451

5. G. Möllenstedt and H. Düker, Z. Physik 145(1956), 377

6. T. Hibi and S. Takahashi, Z. Angew.Phys. 27(1969),132

7. R. Speidel and D. Kurz, Optik 49(1977), 173

8. H. Schmid, *Ein Elektronen-Interferometer mit 300μm weit getrennten kohärenten Teilbündeln zur Erzeugung hoher Gangunterschiede und Messung der Phasenschiebung durch das magnetische Vektorpotential bei metallisch abgeschirmtem Magnetfluß*. Thesis Tübingen 1985

9. G. Möllenstedt and M. Keller, Z. Physik 148(1957),34

10. W. Bayh, Z. Physik 169(1962), 492

11. E. Ehrenberg and R.E. Siday, Proc. Phys. Soc. London B62(1949), 8

12. Y. Aharonov and D. Bohm, Phys.Rev. 115(1959), 485

13. H. Wahl, Optik 30(1970),577

14. B. Lischke, Z. Physik 239(1970), 360

15. G. Möllenstedt and H. Lichte, J.Phys.E.:Sci.Instr. 12(1979),1979

16. G. Möllenstedt and H. Lichte, 9th Int. Congr. on Electron Microscopy, Toronto 1978, vol.I, p.178

17. M. Nicklaus, *Ein Sagnac-Experiment mit Elektronenwellen*, Thesis Tübingen 1989

18. D. Gabor, Nature 161(1948), 777

19. G. Möllenstedt and H. Wahl, Naturwiss. 55(1968),340

20. H. Wahl, *Bildebenenholographie mit Elektronen*, Thesis Tübingen 1975

21. A. Tonomura, e.g. this conference

22. G. Pozzi, e.g. this conference

23. K.-J. Hanßen, J. Phys.D (Appl.Phys.) 19(1986),373

24. H. Lichte, Ultramicroscopy 20(1986),293

25. F.J. Franke, K.-H. Herrmann and H. Lichte, Scanning Microscopy Suppl.2(1988), 59

26. H. Lichte, E. Völkl and K. Scheerschmidt, Ultramicroscopy 47(1992),231

27. H. Lichte, Ultramicroscopy 47(1992),223

28. H. Lichte, A. Orchowski and K. Scheerschmidt, 10th Europ. Congr. on Electron Microscopy, Granada 1992, Last Minute Brochure, p31

INVESTIGATIONS OF ELECTROSTATIC FIELDS BY ELECTRON INTERFEROMETRY AND HOLOGRAPHY

GIORGIO MATTEUCCI
Department of Physics, University of Bologna,
via Irnerio 46, 40126 Bologna, Italy.

and

GIULIO POZZI
Department of Materials Science, University of Lecce,
via Arnesano, 73100 Lecce, Italy.

ABSTRACT

The most recent results obtained by our group in the application of electron interferometry and holography to the investigation of electric fields are reviewed. They range from the quantum phase shift due to a line of dipoles, which represents the electrostatic version of the Aharonov-Bohm effect, to the holographic mapping of the fields associated to reverse-biased p-n junctions, charged dielectric particles and microtips.

1. Introduction

The birth of electron interferometry dates back to 1953, when the first interference pattern was obtained by Marton and coworkers, who conceived and realized the three-crystal interferometer[1]. This device, which has found a wide application in X-ray and neutron interferometry[2], has been superseded in electron optics by the electron biprism, realized two years later by Möllenstedt and Düker[3], which is up to now the most reliable, versatile and widely used interferometry device. Thanks to the introduction of high brightness electron sources, such as field emission guns[4] (FEG) electron interferometry and holography experiments can be carried out today by commercial transmission electron microscopes (TEM) and do no longer require dedicated instrumentation[5–8].

The aim of this work is to review the most recent results obtained by our group in the application of these techniques to the investigation of electrostatic fields[9–12]. First some general considerations are given about the interaction of the electron beam with the electrostatic field present in the specimen and its detection by means of electron interferometry and holography experiments. Then, the various cases investigated so far will be considered, starting from the rather basic experiment in which a constant quantum phase difference is introduced between two interfering electron beams by their interaction with the macroscopic electrostatic field associated with a bimetallic wire[11]. This set-up may be regarded as the electrostatic version of the time-dependent scalar potential Aharonov-Bohm effect[13].

The problems associated with the mapping of electrostatic fields by means of electron holography methods are then illustrated with several examples, reverse-biased p-n junctions [14], electrostatic charges trapped in dielectric particles [9] and the electric field around microtips [12].

2. General considerations

The interaction of the electron beam with the electrostatic potential $V(x,y,z)$ associated to the specimen can be described through a complex transmission function (object wavefunction) $T(\mathbf{r})$ which represents the ratio between the amplitudes of the out-going and in-going electron wavefunctions; $\mathbf{r} = (x,y)$ is a bidimensional vector perpendicular to the optic axis z which is parallel to and in the same direction as the electron beam.

In the standard phase-object approximation[5−8] $T(\mathbf{r}) = \exp(i\phi(\mathbf{r}))$, where the phase term $\phi(\mathbf{r})$ is given by

$$\phi(\mathbf{r}) = \frac{\pi}{\lambda E}\int_l V(\mathbf{r},z)dz \qquad (1)$$

The integral is taken along a trajectory l parallel to the optical axis z inside and outside the specimen to include stray fields, E is the accelerating voltage in the non-relativistic approximation and λ is the electron wavelength. According to equation (1), the phase is proportional to the projected potential.

Information about the field distribution, via the phase difference which is the observable quantity, can be obtained by interference electron microscopy experiments, according to the set-up shown in Figure 1. The interferometry device is an electron biprism, realized by placing a thin charged wire W between two earthed plates P and is located under the back focal plane of the objective lens Ob. The wire splits the wavefront of the incoming electron beam and its electrostatic field produces a deflection and a subsequent overlapping in the plane OP (observation plane) below the wire. As the width of the overlapping region (hologram) H and the interference fringe spacing can be controlled by varying the charge on the wire by means of an external voltage supply, it is possible to obtain electron interferograms and holograms. In particular, if the interference region contains a large number of fringes (above $100 \div 200$), then the interferogram can be considered an off-axis electron hologram. The preferred arrangement is to use only one half of the specimen plane for the specimen S and the other for the reference beam RB, as shown in Figure 1. In fact, if the specimen occupies the whole plane, the interpretation of the pattern is more difficult because the information available concerns the phase difference between points in the specimen.

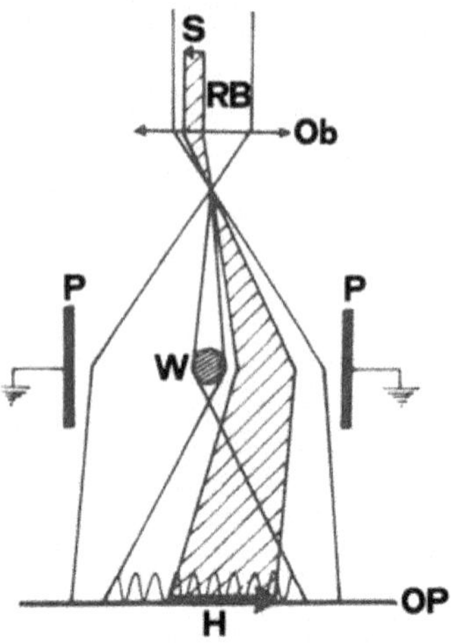

Fig.1. Schematic arrangement for interferogram and hologram formation with an electron biprism. S: specimen placed off-axis; RB: reference beam; Ob: objective lens; W and P: wire and earthed plates, respectively, of the electron biprism; OP: observation plane; H: hologram.

While the information stored in electron interferograms can be directly interpreted, electron holograms require an optical or digital process. Although the method seems at first sight rather cumbersome, it presents invaluable advantages due to the fact that in the optical reconstruction step (usually done in a coherent Mach-Zehnder optical bench) or in the digital one we can apply techniques which have no counterpart on the electron microscope. The capabilities include bright field, dark field, phase contrast imaging, optical interferometry and contour maps, diffraction image filtering, low-angle diffraction, phase detection and amplification. However, the most useful methods for investigating holograms of electric fields are optical interferometry and contour maps. The contour lines that appear in the image are lines of equal phase difference, which, according to (1), give a map of the projected potential.

3. Quantum interference by a macroscopic line of dipoles

The contact potential difference due to a cylindrical bimetallic wire with no net charge introduces a constant quantum phase difference which is equivalent to that produced by a macroscopic line of dipoles[13] and which can be detected by means of diffraction and interference experiments in an electron microscope. In the actual experiment a thin platinum wire, less than 1 μm in diameter, was coated on one side with a thin layer of a different metal (gold or aluminum). The bimetallic wire was mounted on an aperture inserted at the standard specimen level and the existence of the phase difference was detected by the technique of image interference electron microscopy or by observing the Fresnel and Fraunhofer patterns and their variations when the bimetal was rotated around its axis.

However, it has been shown by another group that the non-uniformity of the radius of curvature of the wire of the electron biprism is able to introduce an asymmetry in its Fresnel pattern in a way quite similar to our effect[15].

Therefore, previous model calculations were extended to include boundary conditions, non-cylindrical wire shape and possible charging up of the bimetallic wire, in order to ascertain whether and how much they can mask the basic effect. Moreover further experiments were carried out by both techniques in order to ascertain if they are able to discriminate between the two different contributions to the phase difference[11].

The main results are shown in the following. In Figure 2 the shadow Fresnel patterns are shown for four different rotation angles of the goniometer. The intensity distribution within the shadow region is due to the two beam interference involving only the Young edge diffraction waves. It may be noticed that in Figure 2 (a) the low-intensity fringes within the shadow are symmetrical with a dark central minimum, corresponding to a phase difference of $(2n+1)\pi$. Figure 2 (b) shows an intermediate case with a somewhat asymmetrical intensity distribution, corresponding to a non-integral multiple of π for the phase difference, whereas in Figure 2 (c) the pattern displays again a symmetrical trend with a central maximum, corresponding to a phase difference of $2n\ \pi$. Between Figures 2 (a) and 2 (c), the phase difference has varied by π , whereas between Figures 2 (a) and 2 (d), this latter having nearly the same intensity distribution as (a), the phase difference varied by $2\ \pi$.

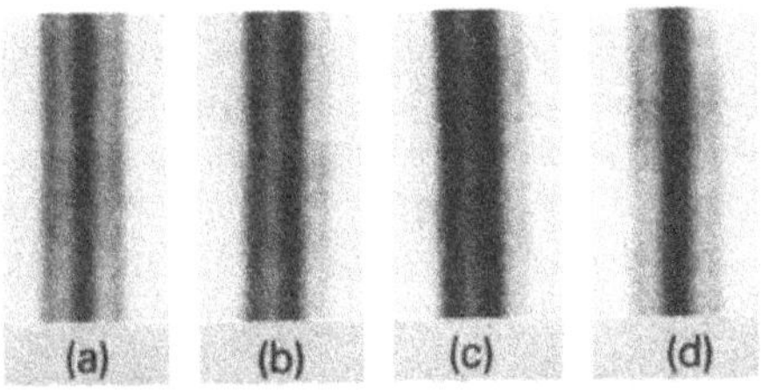

Fig.2. Images of the shadow region of Fresnel patterns of a bimetallic wire obtained at the rotation angles: (a) -40°; (b) -26°; (c) -12°; (d) +12°.

An alternative way for displaying phase differences is to use image interference electron microscopy, a technique which allows, in our case, the recording of an interference pattern by coherently superimposing the electron wavefronts passing on both sides of the bimetallic wire and, at the same time, the observation of the in-focus image of the wire itself. The interference pattern is produced with the experimental set-up of Figure 1. Figure 3 reports two interference images taken with the same bimetallic wire used for the Fresnel diffraction experiments reported in Figure 2. The bimetallic wire was rotated by the angles -40°, Figure 3 (a), and -12°, Figure 3 (b), which correspond to those of Figure 2 (a) and 2 (c), respectively.

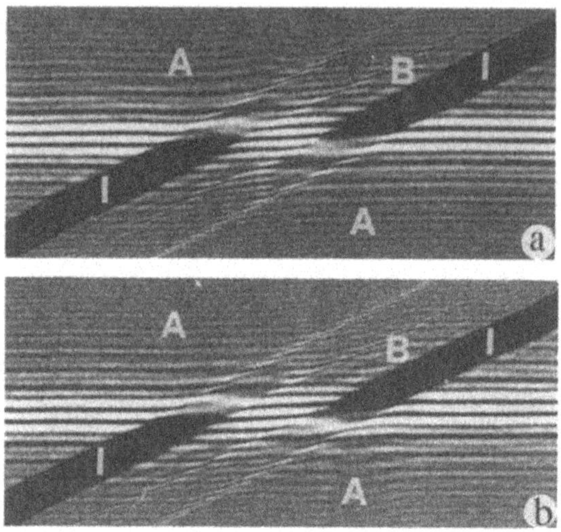

Fig.3. Electron interference micrographs of the same bimetallic wire of Fig.2. Rotation angles: (a) -40°; (b) -12°.

The overall interference region is divided into three parts by the two images I of the bimetallic wire. In the region of interest B, where the waves coming from the opposite sides of the bimetal overlap, the phase difference can be detected as a lateral displacement of the interference fringes with respect to those in the outer regions A which remain stationary. The different symmetry of the intereference fringe contrast in B is in agreement with the Fresnel diffraction results.

From the analysis of the whole set of experimental data[11] it can be seen that the wire is not cylindrical (by comparing the widths of the shadow at different angles) and is charged (as revealed by the non-alignement of the interference fringes in the outer regions A). Nonetheless, the interpretation of the interferometric experiments allows the fixing of an upper limit for the effect due to the non-cylindrical shape and charging up of the wire which amounts to $\pi/4$ and therefore cannot account for the observed variation of the phase difference, which is larger than 2π.

This discrimination is not possible in the Fresnel diffraction experiments.

However, a more striking and direct proof of the existence of the electrostatic Aharonov-Bohm effect is represented by the experiments carried out with a true macroscopic line of dipoles, realized by two parallel conducting wires carrying opposite charges, i.e. two opposite biased biprisms.

Figure 4 reports a magnified view of an interference pattern obtained by rotating the two wires so that the interference pattern and the phase shift which takes place between them is displayed. In the region R between the two biprism images I the linear increase of the phase difference can be directly observed, and the total phase difference is estimated to be larger than 40 π.

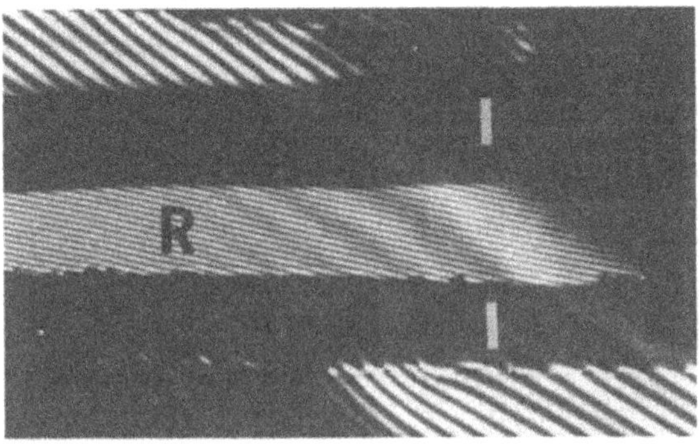

Fig.4. Electron interference image showing the region R between the two oppositely charged biprisms I.

4. Holographic mapping of electrostatic fields

The potentialities of electron holography, with respect to conventional methods, in the investigation of micro electric and magnetic fields has been demonstrated in a number of applications[6,7]. In the following, we will briefly report about the problems which arise in the observation of electrostatic potential distributions.

These problems were first brought to our attention during the observation of reverse-biased p-n junctions[14]. Figure 5 shows an interferogram obtained by the simultaneous reconstruction of one hologram at 2V reverse bias and another without specimen. The optical fringes propagate in the whole interference field and fan out through the hole connecting two neighbouring junctions, whereas within the specimen they are influenced also by the thickness variations and topographical features.

However, it turned out that in this case several equivalent interferograms could be obtained. Another particularly interesting aspect is represented by the fact that the projected potential lines between the junctions are closed. Thanks to the finding of a model for the field associated to an array of p-n junctions [16], it has been possible to calculate the projected potential and to establish that this effect is not due to the field itself, but is an artefact due to the perturbation of the reference wave by the long-range tail of the electrostatic field.

Fig.5. Optical interferogram of the specimen at 2V reverse bias. The depletion layer regions are indicated by the solid lines.

This violates the basic assumption of electron holography that the reference wave is an unperturbed plane wave. The consequences of this fact are analysed in paper [10], which presents a detailed and comprehensive discussion of the whole issue, by comparing experimental and theoretical results for specimens which are easily prepared and which can be described by analytical models.

As regards the electrostatic field case, dielectric latex microspheres 0.31μm in diameter deposited on a carbon film were taken as specimen. These become charged under the action of the electron beam and give rise to an external electric field which also extends in the spatial region crossed by the reference beam. The field can be modelled by that of an equally charge-valued point sphere located in front of an infinite conducting plane at a distance equal to the radius of the sphere itself[9].

Figure 6 shows a four-time phase difference amplification in a contour map of an assembly of six latex spheres. The dark curves represent the projected potential distribution around the dark images of the spheres, one of which, rather far from the others and from the edge, has been labelled S. From the simulation of the intensity variations of the equipotential lines for an isolated sphere and their comparison with the experimental pattern at S, it was possible to evaluate the total charge on the sphere itself, which amounted to 400e, with an accuracy of 20%, being the sphere diameter known.

By investigating spheres of smaller radii, owing to the nature of the Coulomb potential, it should be possible by electron holography to detect even the presence of single electrons. This is a future challenge which may show the ability of electron holography for quantum measurements.

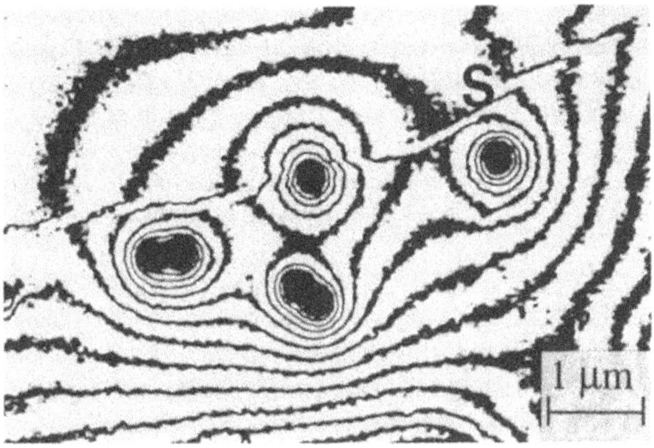

Fig.6. Four-time phase difference amplification in the holographic contour map of charged dielectric spheres.

The usefulness of electron holography has been further tested to map the electrostatic field around a charged microtip. The following results constitute an intermediate but important step towards the investigation of the newly developed monoatomic point sources[17]. The phase term, equation 1, can be found from a previous calculation of an analytical expression of the electric potential in the outer space of a charged tip. To this end, a simple model is first considered, made by two linear segments with the same length and whose centres are at a given distance, placed along a common axis, in a symmetric position with respect to a plane[12]. The holographic method reveals the loci of points with constant phase shift as a set of curves with a phase difference of 2π between two successive dark and white ones.

Fig.7. (a) Computer simulation of the equiphase lines around the tip T. (b) Collage of three (1,2,3) double-exposure holograms taken from adjacent regions showing the trend of the equiphase lines. (c) Computer simulation of the equiphase lines obtained with a perturbed reference wave.

Figure 7 (a) shows the computer simulation of the equiphase lines obtained by the coherent superposition of the object wave and a plane reference wave. While the equipotential surfaces around the tip apex can be envisioned approximately as a family of hyperboloids of rotation, the interpretation of Figure 7 (a) , where the equiphase lines seem to enter the tip shadow T, is less intuitive, because the phase shift is related to the potential distribution around the tip integrated along the electron trajectories. However, when experimental observations are made of the field close to the tip apex, we must consider also that the reference beam is modulated by the field of the tip, which extends a few microns away from the tip itself.

Therefore, the final contour map will show the loci of constant phase difference between the perturbed reference wave and the object wave and does not exactly represent the object phase variations. However from the knowledge of the phase expression and by taking into account the distance between the interfering points (in the electron microscope), the perturbed reference wave can be calculated and the actual expected map displayed directly in the computer.

Simulated maps were then compared with experimental ones. In order to experimentally display equiphase lines in a large enough area, three double exposure holograms were taken from parallel and adjacent regions and then mounted together. Figure 7 (b) shows a montage of these three regions (labelled 1,2 and 3) in which the useful interference field extending along the tip axis is about 15 μm. The three strips are of different width since the overlapping regions were removed. It can be noted that in this hologram, recorded with a perturbed reference wave, the equiphase lines circle around the vertex of the tip and then join the tip itself, a behaviour that could not have been inferred previously.

Figure 7 (c) reports the computer simulation obtained by the coherent superposition of the object wave and a perturbed reference wave and by adjusting the parameters of the tip model and the charge density in order to fit the experimental data. The satisfactory agreement between experimental and theoretical results is evident. The comparison between Figures 7 (a) and (c) clearly shows the difference between the trend of the distribution displayed by a hologram recorded with an unperturbed reference wave and a perturbed one.

5. Conclusions

The results reported in this paper show that a quantum phase shift can be introduced by the electric potential difference originated by a bimetallic wire or by a macroscopic line of dipoles and can be detected by Fresnel diffraction and interference electron microscopy. Although this effect has not the same paradoxical content as the magnetic effect, and can also be accounted for by means of a classical lag [13], it represents a clear demonstration of the fact that a change of phase for the quantum state of the electron can be induced by its interaction with a particular class of fields, which have the peculiarity of changing the semiclassical action into a way independent of the trajectory [18]. The possibility of obtaining qualitative and quantitative information about the distribution of electric fields on a microscopic scale by electron holography has also been demonstrated. However, we have shown that in order to extract reliable field configurations from the experiments, some *a priori* information about the field sources is necessary, in order to estimate the influence of the field on the reference wave with a theoretical model.

6. Acknowledgements

Useful discussions with Professor G. F. Missiroli and the skillful technical assistance of S. Patuelli are gratefully acknowledged. This work has been supported by funds from MURST coordinated by Consorzio INFM and CNR-GNSM.

7. References

1. L. Marton, J.A. Simpson and J. A. Suddeth, Phys. Rev. 90 (1953) 490.
2. U. Bonse and W. Graeff, in X-ray Optics, H. J. Queisser, ed. (Springer, Berlin, 1977), p. 93; W. Rauch, these Proceedings.
3. G. Möllenstedt and H. Düker, Z. Phys. 145 (1956) 377.
4. J.F. Hainfield, in Scanning Electron Microscopy 1977, ed. O. Johary (IITRI, Chicago, 1977) Vol. 1, p. 591.
5. G. F. Missiroli, G. Pozzi and U. Valdrè, J. Phys. E 14 (1981) 649.
6. K.J. Hanszen, Adv. Electron. Electron Phys. 59 (1982) 1; J. Phys. D 19 (1986) 373.
7. A. Tonomura, Prog. Opt. 23 (1986) 185; Rev. Mod. Phys. 59 (1987) 639, Adv. Phys. 41 (1992) 59.
8. H. Lichte, Adv. Optical Electron Micr. 12 (1991) 25.
9. J. W. Chen, G. Matteucci, A. Migliori, G. F. Missiroli, E. Nichelatti G. Pozzi and M. Vanzi, Phys. Rev. A 40 (1989) 3136.
10. G. Matteucci, G. F. Missiroli, E. Nichelatti, A. Migliori, M. Vanzi and G. Pozzi, J. Appl. Phys. 69 (1991) 1835.
11. G. Matteucci, F. F. Medina and G. Pozzi, Ultramicroscopy 41 (1992) 255.
12. G. Matteucci, G. F. Missiroli, M. Muccini and G. Pozzi, Ultramicroscopy 45 (1992) 77.
13. T. H. Boyer, Phys. rev D 8 (1973) 1679; Il Nuovo Cimento 100 B (1987) 685.
14. S. Frabboni, G. Matteucci and G. Pozzi, Ultramicroscopy 23 (1987) 29.
15. A. Oshita, Y. Numata, Y. Mametani and H. Tomita, J. Electron Microscopy 37 (1988) 47.
16. G. Matteucci, A. Migliori, G. Pozzi and M. Vanzi, EUREM 1988, Inst. Phys. Conf. Ser. No.93 (IOP, York, 1988) Vol. 1, p.195.
17. H. W. Fink, Phys. Scripta 38 (1988) 260.
18. Y. Aharonov, in: Proc. Int. Symp. on Foundations of Quantum Mechanics, Eds. S. Kamefuchi et al. (Physical Society of Japan, Tokyo 1984) p. 10.

ATOMS

Diffraction of atoms from optical potentials

T. Pfau, A. Schnetz, C. S. Adams, C. Kurtsiefer, M. Sigel and J. Mlynek
Fakultät für Physik, Universität Konstanz, D-78434 Konstanz, Deutschland

Introduction

Atom optics, in analogy to electron or neutron optics, is concerned with the manipulation of atomic matter waves [1-3]. Experiments on atomic beams have been performed since the nineteen-twenties, but recently the improvement of tools, like free-standing microstructures and tunable laser light, has lead to a rapid development of the field. Beams of atoms have been split coherently, focussed and reflected from mirrors. "Optical instruments" for atoms have a number of potential applications: Atom interferometers could be well suited as gravitational and inertial sensors and may allow precision measurements of atomic properties. A number of such measurements have already been carried out. An "atom-microscope" - investigating the interaction of atoms with surfaces with high spatial resolution - may provide an interesting new tool for surface physicists. "Atom lithography" may allow the deposition of atoms on surfaces with high precision. Cavities for atoms could be interesting to store cold atoms and for the study of quantum-statistical effects.

In the first part of this article we attempt to provide a very brief introduction to atom optics. In the second part we report on a recent experiment in our group on an atom-optical element that may be useful in atom interferometers and atom cavities: the demonstration of a new beam splitter based on the diffraction of atomic matter waves from a "magneto-optical grating" .

1 Atom optics

Particle beams exhibit many phenomena well known from classical optics like diffraction and refraction. Particle optics based on electrons and neutrons [4] are well established fields. Experiments with these particles have played an important role in exploring the puzzling wave-particle duality. Both neutron and electron interferometers have been used in many beautiful experiments, which have improved our understanding of quantum mechanics. The interactions of neutron and electron beams with matter have become indispensable tools in the analysis of structures and surfaces, consider e.g. neutron scattering and the electron microscope.

As atoms are different from electrons and neutrons, particle optics with atoms promises a wealth of new effects and applications. Like neutrons, atoms are neutral and therefore cannot be manipulated as easily as electrons or ions using static electromagnetic fields. In return beams of atoms and neutrons are less susceptible to electromagnetic stray fields. Atoms are heavier, which makes them more suitable as gravitational and inertial sensors and means that very short de Broglie wavelengths can be obtained for lower particle energies. Atoms come in many species, which can be either bosons or fermions, and may have a total angular momentum or magnetic moment much larger than neutrons or electrons. Beams of atoms are easier and cheaper to produce than neutron beams as there is no need for a nuclear reactor. And maybe most importantly, atoms have a complex internal structure that can be probed and manipulated using resonant laser light or static electromagnetic fields.

As in classical optics, the building blocks of any experiment are optical elements e.g. beamsplitters, mirrors and lenses. Light can be manipulated by transmission through interfaces between materials with different refractive indices. Electrons, due to their charge, couple strongly to static electric or magnetic fields. Neutrons penetrate through solids and can be diffracted from crystal lattices.

As atoms experience much smaller forces in static fields and do not penetrate through matter, none of these approaches are suitable for atom optics.

So far optical elements for atoms have been demonstrated using diffraction, refraction or the recoil due to the stimulated emission or absorption of a single photon. "Diffractive optical elements" achieve the desired distortion of an incoming wave front by interaction with small structures, thus exploiting the wave nature of the atomic centre of mass motion. "Refractive optics" is based on introducing phase shifts by a spatial modulation of the index of refraction experienced by the atoms. In analogy to standard optics the index of refraction is defined as the ratio of a particle's group velocity and the vacuum group velocity. The index of refraction can be adjusted by shifting the potential energy of the atom. Intuitively this means: If the internal energy of an atom is e.g. decreased, energy conservation demands an acceleration of the centre of mass motion (i.e. an increase in the group velocity of the atomic matter wave). The phase velocity on the other hand is decreased in this case. The phase fronts are consequently retarded with respect to the unperturbed propagation. The principle of refractive optical elements is to choose the spatial dependence of these phase shifts to obtain the desired distortion of the incoming wave fronts. In static electric or magnetic fields the atomic internal energy can be changed by the Zeeman or Stark shift, in near-resonant light fields by the light shift.

The diffraction of atoms from the surface of single crystals was studied by Stern et al. as early as 1929 [5]. In 1969 Leavitt and Bills [6] observed a single-slit diffraction pattern using a thermal potassium beam. The progress in micro-fabrication technology now permits the production of structures sufficiently fine to diffract thermal atoms, whose de Broglie wavelengths are typically below 1 Å. The diffraction of thermal atoms from free-standing micro-fabricated gratings has been studied by several groups [7,8]. The first deflection of an atomic beam by the light pressure force was already demonstrated by Frisch in 1933 [9]. In recent years the availability of intense tunable laser radiation has again focussed interest on the mechanical effects of atom-light interactions. The feasibility of such experiments has stimulated considerable theoretical interest. In addition to the conservative interaction using the dipole force, the spontaneous decay of excited electronic states can be utilized for the preparation of atomic beams by laser cooling techniques to obtain well-collimated, dense atomic beams with a well defined velocity.

2 The magneto-optical beamsplitter

2.1 Introduction

A coherent beam splitter couples an incoming momentum state to a coherent superposition of outgoing momentum states. Beam splitters for atoms are not only important for atom interferometry, they also could be used for correlation measurements or as in/outcouplers for atomic cavities. Beam splitters have been produced using the magnetic [10] or optical Stern-Gerlach effect [11], reflection from crystalline surfaces [12], the recoil due to a single absorption or stimulated emission [13,14], diffraction from micro-fabricated transmission gratings (amplitude gratings) [7,8] and the diffraction from standing wave light fields (phase gratings) [15,16].

We present a new type of beam splitter which is a modification of the diffraction of atoms from an optical standing wave. This diffraction can be looked upon as a beam splitter both in the regime of long and short interaction times. For short interaction times the change in "transverse" kinetic energy can be neglected. In the following we are concerned with this regime, which is called the Raman-Nath regime.

As for the diffraction of two-level atoms from a standing wave the magneto-optical beamsplitter is based on introducing position dependent phase shifts. The advantages of the diffraction from a phase grating over diffraction from an amplitude grating are that there is no absorption and that by changing the phase modulation amplitude it is possible to change the envelope of the diffraction pattern. For a two-level atom the eigenenergies of the two energy eigenstates in a near-resonant light field with Rabi frequency $\omega_R = \langle e|\mathbf{d}\cdot\mathbf{E}|g\rangle E_0/\hbar$ and resonance detuning Δ are (using the rotating wave approximation)

$$E_{\pm}(x) = -\frac{\hbar\Delta}{2} \pm \frac{\hbar}{2}\sqrt{\omega_R(x)^2 + \Delta^2} \ .$$

In a standing wave light field with large detuning these eigenergies display a roughly sinusoidal modulation. In the Raman-Nath regime and if neither spontaneous emission nor non-adiabatic following induce transitions between the eigenstates the phase modulation $\Delta\Phi(x) = \int E(x,t)/\hbar \, dt$ displays the same sinusoidal modulation. The momentum distribution and therefore the far-field spatial distribution (for a monochromatic beam) is given by the square of the Fourier transform of $e^{i\Delta\Phi(x)}$ with respect to x. For a sinusoidal phase modulation the population of the nth diffraction peak is given by the square of the nth order Bessel function $J_n(\Delta\Phi_{max})$ wherein the argument is the amplitude of the phase modulation [17]. Diffraction from a standing wave can therefore be looked upon as an effective beam splitter only for momentum transfers of a few $\hbar k$. In order to obtain a phase grating which diffracts all intensity into a few high order momentum states a triangular phase modulation is necessary. Such a triangular phase modulation can be obtained if a combination of light shifts and Zeeman shifts are used to control the spatial modulation of the eigenenergies [18,19]. We consider the experimental situation shown in figure 1: a V-type three-level system interacts with two counterpropagating crossed linearly polarized light beams and a static magnetic field applied parallel to the light beams. The level scheme contains the Zeeman shifts but not the light shifts.

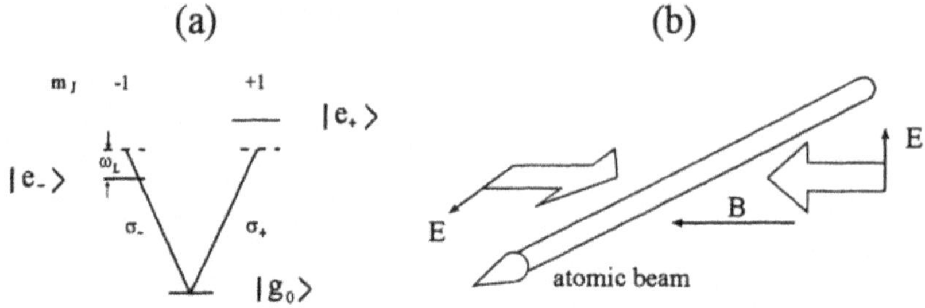

Figure 1: (a) Level scheme for a J=0 to J'=1 transition with the quantization axis chosen parallel to the magnetic field. (b) Configuration of the counterpropagating light beams which orthogonally intersect the atomic beam. The light fields are linearly polarized and the polarization vectors enclose an angle of 90°. A magnetic field is aligned parallel to the light beams.

In figure 2 the energy eigenvalues of this three-level atom are displayed as a function of the transverse position in the light field for three ratios of the Larmor frequency $\omega_L = g_J \mu_B B / \hbar$ and the Rabi frequency $\omega_R = \langle e_\pm | \mathbf{d} \cdot \mathbf{E} | g_0 \rangle E_0 / \hbar$. For small Rabi frequencies the levels are split by the Zeeman effect, for large Rabi frequencies by the light shift – for $\omega_R = 2\omega_L$ the Zeeman shift and the light shift at positions of circularly polarized light are equal and the eigenstate $|2\rangle$ displays the desired triangular modulation. For the V-system the ground state $|g\rangle$ evolves adiabatically into this eigenstate $|2\rangle$. If nonadiabatic processes and spontaneous emission do not populate the two other eigenstates the diffraction pattern obtained from this phase grating should display two distinct peaks.

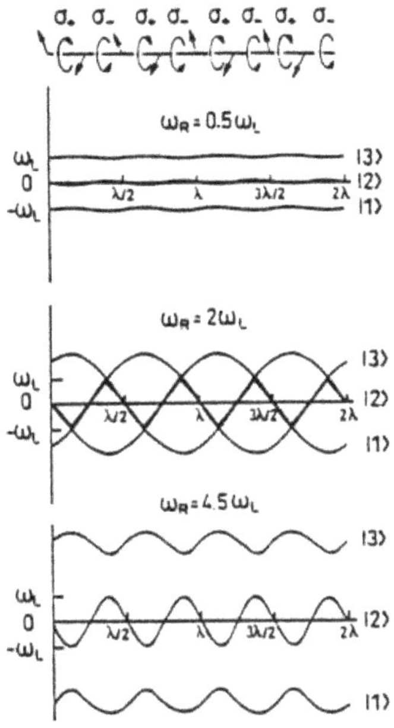

Figure 2: Plots of the spatial dependence of the eigenenergies of a three-level atom in a light field created by two counterpropagating orthogonally linear polarized light beams and a static magnetic field for three different ratios of Rabi frequency ω_R and Larmor frequency ω_L. a) $\omega_R = 0.5\ \omega_L$, b) $\omega_R = 2\ \omega_L$ and c) $\omega_R = 4.5\omega_L$

2.2 Experiment

We investigated the magneto-optical beamsplitter in an experiment using helium atoms in the triplet state. The light field was resonant with the transition at 1083 nm from the metastable 2^3S_1 state to the 2^3P_1 state. If the incoming metastables are optically pumped into the $m_J=0$ substate a V-type three-level system as depicted in figure 1a is obtained. A schematic of our experimental setup is displayed in figure 3. The laser beam, which was derived from a Ti:sapphire ring laser, was focussed to a waist of 21 μm in the beam direction. The polarization of the back-reflected beam was rotated by double pass through a λ/4-plate. As a consequence of the large mean velocity in the supersonic beam of 2000 m/s and the long lifetime of the excited state of 100 ns it was possible to largely avoid spontaneous emission. The superimposed magnetic field was produced by a pair of Helmhotz coils. The far-field diffraction pattern of the supersonic helium beam was observed with a scanning slit detector.

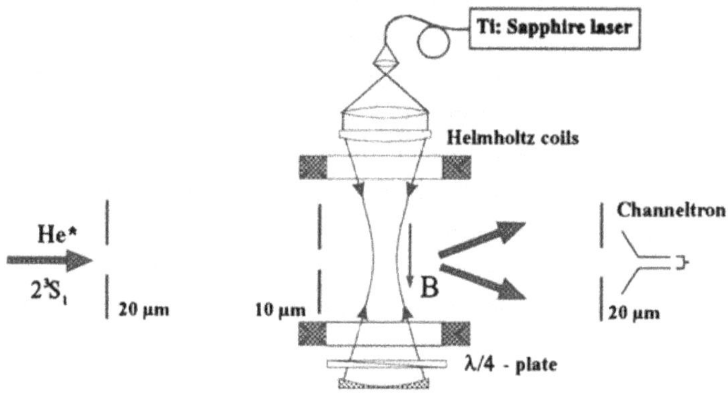

Figure 3: Experimental setup of the magneto-optical beamsplitter. A beam of triplet helium atoms interacted with two counterpropagating orthogonally linear polarized laser beams and a static magnetic field. The far-field diffraction pattern was detected with a scanning slit and a channeltron detector.

In Figure 4 the atomic intensity is plotted as a function of transverse position in the detection plane 0.85 m downstream from the interaction zone. The choice of parameters was such that the Rabi frequency ω_R in the centre of the Gaussian beam was twice the Larmor frequency ω_L. It can be seen that the incoming atomic matter wave is largely diffracted into two groups of diffraction orders as expected from theory. The finite velocity distribution leads to a broadening of the split beams in the real space distribution. The splitting of 1.5 mm between the two peaks observed in the detection plane corresponds to a momentum splitting of 42 $\hbar k$. This is a large splitting compared to the splitting typically achieved by diffraction from an amplitude grating or compared to the 1 ħk achieved with a single π/2-pulse.

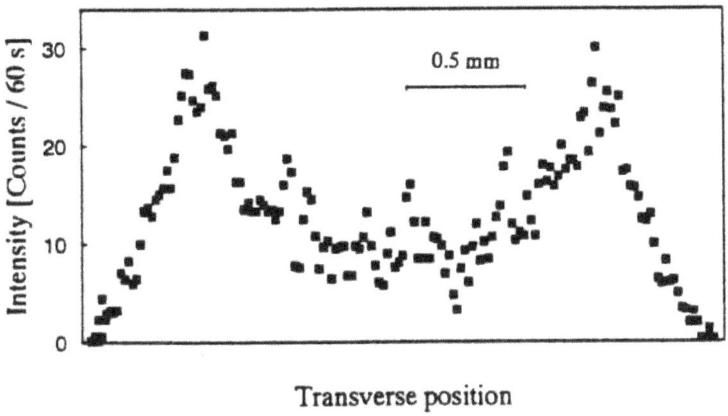

Figure 4: Measured far-field intensity distribution for the diffraction of triplet helium atoms from a magneto-optical grating for $\omega_R = 2\,\omega_L$.

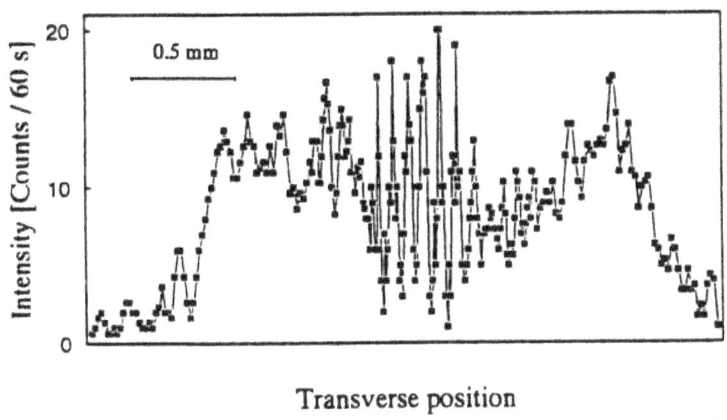

Figure 5: Measured far-field intensity distribution for the diffraction of triplet helium atoms from an optical standing wave.

For comparison we also examined the diffraction of atoms from an "ordinary" standing wave. Figure 5 shows a far-field diffraction pattern we obtained for a laser intensity comparable to the one used in figure 4. It displays the expected broad, Bessel function shaped envelope. In the real space distribution the $2\hbar k$ structure is only resolved for low diffraction order. For larger diffraction order the finite width of the velocity distribution smears out distinct diffraction peaks.

3 Conclusion

After a short introduction in the field of atom optics we presented experimental results for a new type of beamsplitter. The beamsplitter relies on the diffraction of atomic matter waves from a phase grating with triangular phase modulation produced by a combination of a light field and a magnetic field. We observed a momentum splitting of $42\hbar k$. To our knowledge this is the largest splitting in momentum space observed so far. A next step could be to investigate the coherence of this beamsplitter by recombining the two output ports in a Mach-Zehnder type interferometer. Another interference effect should occur in the near field of the magneto-optical grating, where the triangular shape of the phasemodulation acts as an array of Fresnel biprisms.

Acknowledgement

C.S.A. acknowledges support of the Royal Society's European Science Exchange Programme. This work was supported by the Deutsche Forschungsgemeinschaft.

References

[1] Special Issue of Appl. Phys. B (April 1992), *Optics and Interferometry with Atoms*, J. Mlynek, V. Balykin and P. Meystre (eds.).
[2] M. Sigel, C. S. Adams and J. Mlynek, to be published in *Frontiers in LaserSpectroscopy*, T. W. Hänsch, M. Inguscio (eds.), Proceedings of the International School of Physics "Enrico Fermi", Course CXX, Varenna, 1992.
[3] M. Sigel and J. Mlynek, *Atom Optics*, Physics World, February 1993.
[4] V. Sears, *Neutron Optics* (Oxford University Press, New York, Oxford, 1989).
[5] O. Stern, Naturwissensch. **17**, 391 (1929).
[6] J. A. Leavitt and F. A. Bills, Am. J. Phys. **37**, 905 (1969).
[7] D. W. Keith, M. L. Schattenburg, H. I. Smith and D. E. Pritchard, Phys. Rev. Lett. **61**, 1580 (1988).
[8] O. Carnal, A. Faulstich and J. Mlynek, Appl. Phys. B **53**, 88 (1991).
[9] R. Frisch, Zeits. f. Phys. **86**, 42 (1933).
[10] W. Gerlach und O. Stern, Zeits. f. Phys. **8**, 110 (1922).
[11] T. Sleator, T. Pfau, V. Balykin, O. Carnal and J. Mlynek, Phys. Rev. Lett. **68**, 1996 (1992).
[12] I. Estermann und O. Stern, Zeits. f. Phys. **61**, 95 (1930).
[13] F. Riehle, Th. Kister, A. Witte, J. Helmcke and Ch. J. Bordé, Phys. Rev. Lett. **67**, 177 (1991).
[14] M. Kasevich and S. Chu, Phys. Rev. Lett. **67**, 181 (1991).
[15] P. E. Moskowitz, P. L. Gould, S. R. Atlas and D. E. Pritchard, Phys. Rev. Lett. **51**, 370 (1983).
[16] P. L. Gould, G. A. Ruff and D. E. Pritchard, Phys. Rev. Lett. **56**, 827 (1986).
[17] R. J. Cook and A. F. Bernhardt, Phys. Rev. A **18**, 2533 (1978).
[18] T. Pfau, C. S. Adams and J. Mlynek, Europhys. Lett. **21**, 439 (1993).
[19] C. S. Adams, T. Pfau, Ch. Kurtsiefer and J. Mlynek, "Interaction of atoms with a magneto-optical potential" to be published in Phys. Rev. A.

ATOM INTERFEROMETRY

David E. Pritchard, Troy D. Hammond, Jörg Schmiedmayer,
Christopher R. Ekstrom and Michael S. Chapman
Department of Physics and Research Laboratory of Electronics
Massachusetts Institute of Technology, Cambridge, Massachusetts 02139, USA.

ABSTRACT

We have constructed an atom interferometer using three 200 nm period transmission gratings. The two interfering portions of each atom's wave pass on opposite sides of a stretched metal foil. A fringe amplitude of up to 900 cps is observed, which allows us to determine the phase to 15 milliradians in 1 minute. By applying different interactions to the two sides of the interferometer, we have determined the electric polarizability of the ground state of sodium, the coherence length of our beam, and the differential phase shifts and coherences of the different Zeeman ground states in a magnetic field. Some future prospects of separated beam experiments are discussed.

1. Introduction

Interference has a long tradition that takes us from Young's demonstration of two slit interference of light[1] in the 19th century to the application of optical interferometers to a number of important scientific problems by Fizeau[2], Michelson[3] and Rayleigh[4] among others. The rise of quantum mechanics, with its central tenet of wave-matter duality, led directly to the idea of matter wave interference devices such as the electron interferometer[5,6] and the perfect crystal neutron interferometer[7] both well known for their contribution to precision experiments and fundamental physics.[8]

Given the many demonstrations of matter wave interference devices and interferometers with electrons and neutrons, the concept of atom interferometers requires only a small conceptual step. Indeed, an atom interferometer was patented in 1973.[9] The billiard ball model[10] of coherent transients emphasized the spatial separation of the two coherent, internal states, and several papers discussed the connection between multiple pulse laser spectroscopy and atom interferometers.[11,12] About half a dozen experiments that demonstrate atom interference and atom interferometers have been performed during the last five years[13-19] (cf. overviews[20-22]). In this paper about the new subfield of atom interferometers we stress their areas of application, rather than review the various recent demonstrations. We will, as an example of new measurements, focus on recent experiments performed at MIT, beginning with a brief discussion of our atom interferometer.

2. The MIT Atom Interferometer

Our interferometer is a Mach-Zender type built with three nanofabricated transmission gratings[23] mounted on separate translation stages inside the vacuum envelope (Fig.1). This configuration produces a robust white fringe.[24] Our setup has been considerably improved since the first demonstration experiment.[13] We now use 200 nm period gratings which completely separate the centers of the interfering beams (trapezoids with 20 μm flat top, 37 μm FWHM and 54 μm width on the bottom) by 55 μm at the position of the second grating. These new finer and taller gratings do, however, require much better alignment and vibration control. We have a supersonic sodium beam with argon as a carrier gas ($v = 1000$ m/s, $\lambda_{dB} \sim 0.17$ Å). By varying the gas pressure in our new source we can now change the FWHM of our velocity distribution from $\Delta v/v = 70\%$ (no seed gas) to $\Delta v/v < 8\%$ at 3 bar Ar pressure. The mean velocity (deBroglie wavelength) of the beam can be changed by using different seed gases. A new Rhenium hot wire detector has higher detection efficiency, faster time response (~1 msec) and background as low as 10 cps.

Our interaction region consists of a stretched metal foil positioned symmetrically between two side electrodes, each spaced 2 mm from this septum. It was inserted behind the second grating so that the portions of the interfering atom wave in the two arms of the interferometer pass on opposite sides of the foil. The septum was up to 10 cm long and 10 μm thick, but the shadow it cast on the detector was typically 20 - 30 μm wide due to slight deviations of the stretched foil from perfect flatness and imperfections at the ends.

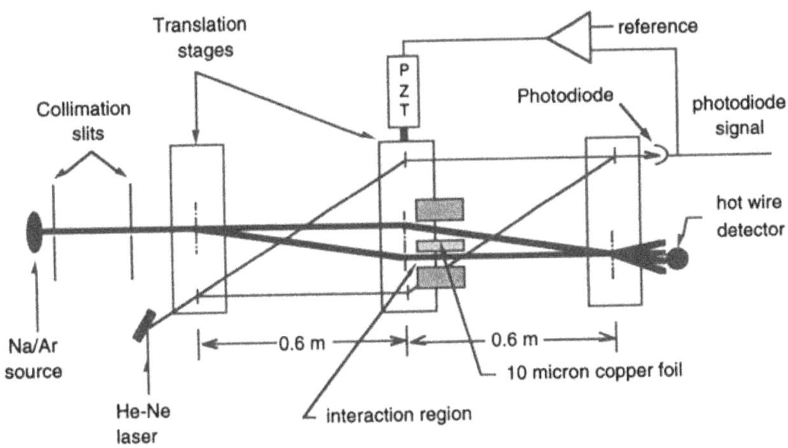

Figure 1: A schematic, not to scale, of our atom interferometer (thick lines are atom beams). The 0th and 1st order beams from the first grating strike the middle grating where they are diffracted in the 1st and -1st orders so that they form an interference pattern in the plane of the third grating, which acts as a mask to sample this pattern. The 10 cm long interaction region with the 10 μm thick copper foil between the two arms of the interferometer is positioned behind the 2nd grating. An optical interferometer (thin lines are laser beams) measures the relative position of the 200 nm period atom gratings (which are indicated by vertical dashed lines).

Since we have a conducting, physical barrier between the separated beams, we can apply different, nominally uniform electric and magnetic fields to the portions of the atom wave on each side of the interferometer. The fringe amplitude with the septum between the two arms of the interferometer is typically 700 cps, which allows us to determine the phase to 15 milliradians in 1 minute (Fig.2). It is interesting to note that the deBroglie wavelength $\lambda_{dB} \sim 0.16$ Å and the coherence length $l_{coh} \sim 0.65$ Å (1.6 Å FWHM) were both smaller than the size of the atom (3 Å) whose motion they characterize. The use of a compound particle with such large dimensions in an interferometer was unprecedented before the advent of atom interferometers.

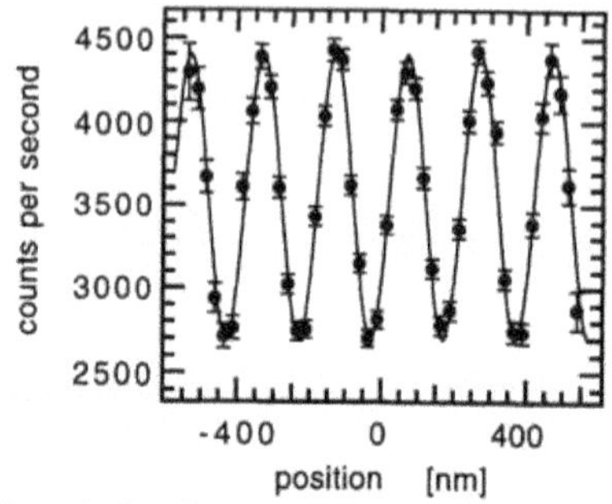

Figure 2: Interference pattern from 40 seconds of data (1 second per point). The contrast is 25% and the phase uncertainty is 17 milliradians.

With the two beams now completely separated, let us consider the effect of an interaction in only one arm of the interferometer. Applying a potential *V(x)* in one arm changes the *k*-vector, $k(x) = \frac{1}{\hbar}\sqrt{2m(E - V(x))}$. The phase difference, $\Delta\varphi$, between the two arms will be:

$$\Delta\varphi(k_0) = \int k(x)dx - \int k_0 dx = \int \Delta k(x)dx$$

where $k_0 = \frac{1}{\hbar}\sqrt{2mE}$ is the unperturbed *k*-vector. If the potential *V* is much smaller than the energy of the atom *E* (as in all the cases discussed here), the phase shift can be evaluated by expanding the square root, yielding

$$\Delta\varphi(k_0) = \frac{m}{\hbar^2 k_0}\int V(x)dx = \frac{1}{\hbar v}\int V(x)dx.$$

The effect of the potential *V(x)* can also be viewed as an refractive index $n = k/k_0$. Using the approximation $V << E$ one finds $n = k/k_0 = 1 - (V/2E)$ and the phase shift $\Delta\varphi(k) = k\int(1 - n(k))dx$. In our interferometer a phase shift of 1 radian corresponds to $|n - 1| = 2.7 \times 10^{-11}$.

The discussion above shows that the phase shift for a potential that is independent of velocity and time depends on k_0. Since our beam is not monochromatic, the observed phase shift is an average over the initial *k*-vector distribution *f(k)*:

$$\tan(\Delta\varphi) = \frac{\int f(k)\sin(\Delta\varphi(k))dk}{\int f(k)\cos(\Delta\varphi(k))dk}.$$

This average over the final velocity width of the beam leads to a reduction of the interference contrast for dispersive phase shifts like the Stark phase shift or the phase shift from the magnetic interaction. In the most general case, the interference contrast $C = \frac{I_{max} - I_{min}}{I_{max} + I_{min}}$ is given by:

$$C = C_0 \sqrt{\left[\int f(k) \sin(\Delta\varphi(k)) dk\right]^2 + \left[\int f(k) \cos(\Delta\varphi(k)) dk\right]^2}$$

where C_0 is the contrast at zero phase shift. This reduction of the contrast can be parameterized by the coherence length l_{coh} of the beam (Fig. 3a). The above equation reduces to $C = C_0 e^{-(\Delta\varphi)^2 \sigma_k^2 / 2k_0^2}$ in the case of linear dispersion if $f(k)$ is a gaussian distribution with rms width σ_k centered at k_0. This form is reasonable because seeded nozzle beams give a flux density which varies as $v^3 e^{-(v-v_0)^2/2\Delta v}$, which is nearly Gaussian when $\Delta v << v_0$. Our single grating diffraction data are fit within estimated total errors by this expression (Fig. 3b). The coherence length is given by $l_{coh} = \frac{1}{\sigma_k}$.

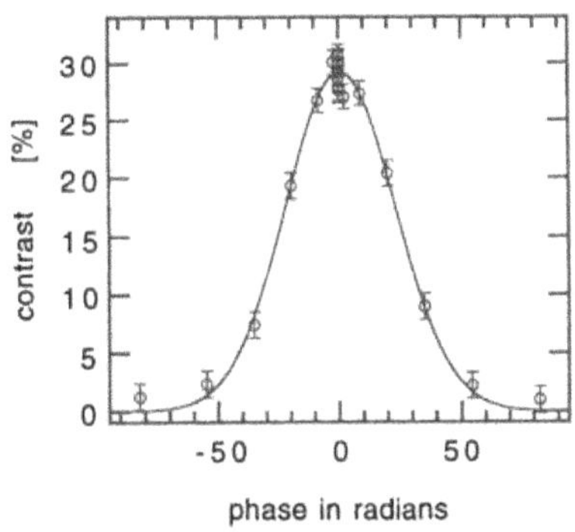

Figure 3a: Reduction of the interference contrast with applied Stark phase. From the with of the contrast curve we evaluated a coherence length of 0.65(3) Å in good accordance with an evaluation of the velocity from a measurement of the diffraction pattern.

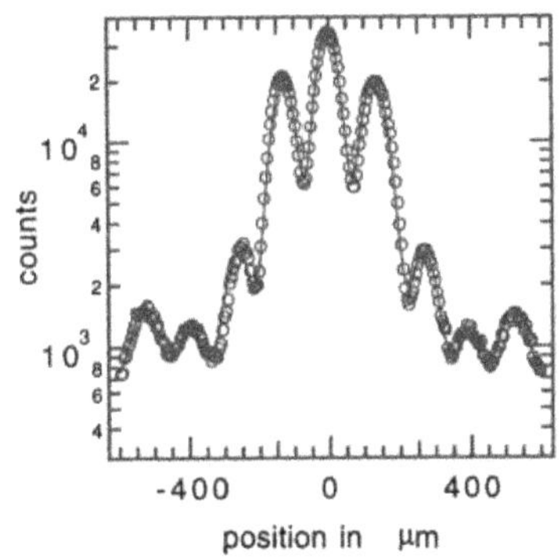

Figure 3b: Diffraction pattern of sodium atoms from a 200 nm period grating. The solid line is calculated for a grating with average open fraction of 40 % with 9% rms relative variation. The sodium atoms have a mean $\lambda_{dB} = 0.165$ Å and a rms velocity distribution of 3.7(3)%. This implies a coherence length of 0.71(6) Å.

It is important to note that, in contrast to photons in a coherent state in vacuum, the coherence length and wave packet size for matter waves are never the same, except perhaps at specific points in time. This is because matter waves, even in a vacuum, are dispersive. In our beam the coherence length is about 0.65 Å (1.6 Å FWHM), but the length of our wave packet at the third grating (where the interference "occurs") is on the order of 10 cm.

To conclude this cursory description of our interferometer, let us compare its sensitivity with spectroscopic methods. A good comparison is with separated oscillatory field resonance experiments, where the resonance pattern may be viewed as an interference between two different internal states of the atom. In both these instruments the fringe separation is set by the inverse of the interaction time. Our 0.1 m long interaction region and the 1000 ms^{-1} velocity results in an intrinsic line width of 10 kHz. Equivalently, a potential of 4×10^{-11} eV gives a phase shift of 2π radian. In both instruments, neglecting systematic errors, the fringe can be split by a factor proportional to $\sqrt{NC}$ where C is the contrast and N is the total number of atoms counted in the measurement. For our interferometer a typical one minute measurement gives a sensitivity better than 15 mrad (2 millifringes) corresponding to an energy sensitivity of roughly 10^{-13} eV or about 20 Hz.

In contrast to an oscillatory field measurement, the separated beam interferometer allows one to do experiments using only *one* state instead of relying on difference measurements between *two different* states. This will be very important in applications of the atom interferometer to measurements of ground state atomic and molecular properties. In addition the interferometer gives us the opportunity to subject part of the atom wave to an interaction which causes a phase shift without a classical force and then to measure this phase shift by interference with the unshifted part of the atom wave in the same state. The latter is important for clear demonstrations of topological quantum phases.

3 Atomic Properties

Separated beam atom interferometers are ideally suited for absolute measurements of atomic and molecular properties for *single* states. The key point is that the separated beams in the interferometer present the opportunity to subject part of the atom wave to an interaction which causes a phase shift, and then to measure this phase shift by interference with the unshifted freely evolving part of the atom wave in the same state.

3.1 The Electric Polarizability of the Sodium Ground State

An obvious application of our interferometer is a precision measurement of the polarizability of the ground state of the sodium atom, by subjecting one part of the atom wave to a uniform electric field. This is intrinsically a higher precision approach than measuring the deflection in a field gradient, whether this is measured by conventional[25] or interference techniques.[26] We also note that the polarizability *differences* between two internal states of an atom have long been measured using optical resonance techniques, and can also be measured in Chebotayev type interferometers, even though the two legs of the interferometer are not spatially resolved.[19,27]

A homogeneous electric field on one side of the interaction region induces a potential due to the DC Stark effect $V_{Stark} = -\frac{1}{2}\alpha\mathbf{E}^2$, where α is the electric polarizability, resulting in a differential phase accumulation relative to the wave that passes on the other side of the septum where there is no field. Since V_{Stark} is about eight orders of magnitude smaller than the kinetic energy E the phase shift is $\Delta\varphi_{Stark} = -\frac{l}{\hbar v}\frac{1}{2}\alpha\mathbf{E}^2 = \frac{t}{\hbar}V_{Stark}$ where l is the length of the interaction region and v is the velocity of the atoms, giving a transit time t. We verified that the phase shift is a quadratic function of the applied electric field (Fig.4) and determined the appropriate coefficient with a statistical error of less than 0.1%. After investigating several systematic errors, we have determined the absolute value of the polarizability to 0.4%, a factor of five better than previous results.

3.2 Magnetic Rephasing

We have also investigated the magnetic interaction of the sodium ground state with a magnetic field. By running a current down the metal septum, perpendicular to the plane of the interferometer, we generate a magnetic field in the interaction region along the beam direction with the same magnitude, but opposite sign for the two arms of the interferometer. When an additional uniform magnetic field is applied along the beam axis, it adds to the field from the current in the septum giving different field magnitudes on both sides of the septum. In this case one gets a differential Zeeman energy, $\vec{\mu}\cdot\vec{\mathbf{B}}$, and therefore different accumulated phases, for the two paths around the septum. The phase difference is proportional to the current passing through the septum and the projection of the magnetic moment $g_F\mu_o$ along the quantization axis, $\Delta\varphi = \frac{1}{\hbar}\frac{l}{v}g_F\mu_o m_F \mathrm{B}$.

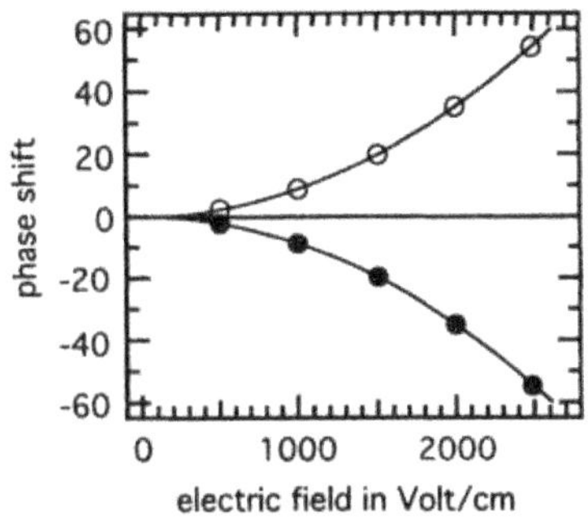

Figure 4: Stark phase shifts for voltages applied to the left (o) and the right (•) side of the interaction region. The phase shift per applied electric field squared in $(\mathrm{volt/cm})^2$ is $1.220(7)\times10^{-5}$ for the left and $1.224(7)\times10^{-5}$ for the right side.

Because our beam is unpolarized, the observed interference pattern is a sum of interference patterns for each of the eight sodium ground states. Since the g-factors of the F = 1 and F = 2 hyperfine levels have equal magnitude (but opposite sign), there are only five different projected magnetic moments. At low fields, these are proportional to 0, $\pm\frac{1}{2}$, and ±1 times a Bohr magneton. Consequently, as the current is increased, the independent interference patterns periodically rephase constructively to produce a high contrast interference pattern with the same phase as the pattern at zero field. The first revival of contrast is the point where the phase shifts are 4π for the $|m_F| = 2$ states, 2π for the $|m_F| = 1$ states, and 0 for the $m_F = 0$ states. Therefore in this experiment the important observable is the contrast of the interference pattern vs. magnetic field.

We have observed this periodic rephasing of the 8 independent magnetic substates in an unpolarized sodium beam (see Fig.5) as an increasing differential magnetic field is applied to opposite sides of the septum. The fit to the data is from models which account for the 8 magnetic states, the effects of our finite velocity distribution and the misalignment of the uniform magnetic field that determines the quantization axis. The relative positions and widths of the contrast maxima and the contrast degradation of subsequent rephasings due to the finite coherence length are well accounted for by the model. An application of this rephasing experiment is that (since the value of the Bohr magneton is accurately known) one of the fit parameters is the average velocity of the atoms that contribute to the interference pattern. This can be exploited to eliminate systematic effects arising from processes, like the clipping of part of the beam by the

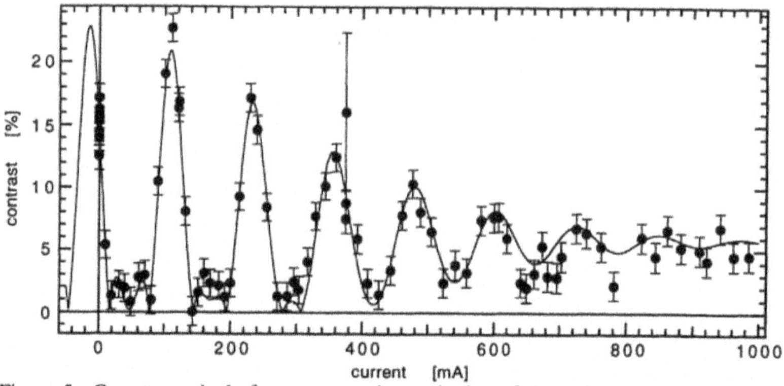

Figure 5: Contrast revivals from constructive rephasing of the independent interference patterns of the 8 different magnetic sub-states of sodium. These patterns are dephased by a current flowing down the septum which alters the magnitude of the uniform magnetic field on the two sides.

septum, that cause this velocity to differ from the average velocity of the atoms in the beam upstream of the interferometer as determined by a measurement of the diffraction pattern (Fig.3a).

For large currents through the foil, the average over the velocity distribution of the atom beam reduces the contrast in the interference pattern of all atoms except those in the two $m_F = 0$ states, which experience no Zeeman phase shift. This results in a contrast one-fourth of that observed for no current. At this point, any small phase shifts observed from additional interactions are those of only the $m_F = 0$ states. By applying a large Stark phase shift to all of the substates, the contrast of these $m_F = 0$ states could be reduced to nearly zero while another polarization state was shifted back into coherence with itself (Fig.6). This allows experiments to be performed with polarized atoms in a unpolarized beam without the difficulty of optical pumping (but without the gain in intensity which such optical pumping could bring).

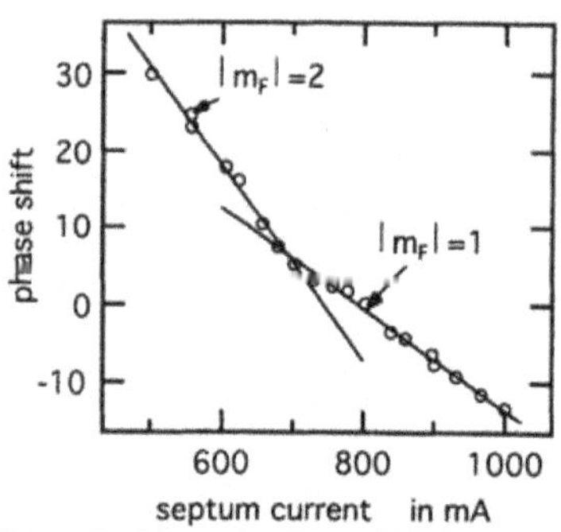

Figure 6: Magnetic phase shift for $|m_F|=2$ and $|m_F|=1$ states as observed with an unpolarized beam in a *"magnetic rephasing"* experiment. An additional DC stark phase shift of 65 rad is applied in one arm of the interferometer. For 400 - 700 mA septum current one of the $|m_F|=2$ states, and between 700 - 1050 mA two of the $|m_F|=1$ states are shifted back in coherence with itself.

3.3 Refractive Index of an Atom in a Gas

Another experiment currently pursued with our interferometer is the effect of a gas medium on the atomic wave function. From the perspective of wave optics, the passage of a wave through a medium is described in terms of an index of refraction $n = 1 + \frac{2\pi}{k^2} N f(k,0)$ where

N is the density of the medium and $f(k,\theta=0)$ is the forward scattering amplitude.[28] The real part of n is proportional to the phase shift of the wave and the imaginary part is proportional to the attenuation of the wave by scattering into undetected directions. The atom wave function evolving in a gas with density N is given by (x is the distance traveled in the gas)

$$\Psi_{Atom}(x)=\Psi_{Atom}(0)e^{ik_0x}e^{i\frac{2\pi}{k}Nx\,\mathrm{Re}(f(k,0))}e^{-\frac{2\pi}{k}Nx\,\mathrm{Im}(f(k,0))}.$$

We see that the amplitude of Ψ_{Atom} is attenuated as one would expect from the total scattering cross section $\sigma_{tot}=\frac{4\pi}{k}\mathrm{Im}(f(k,0))$. In addition, there is a phase shift $\Delta\varphi(k,x)=\frac{2\pi}{k}Nx\,\mathrm{Re}(f(k,0))$ proportional to the real part of the forward scattering amplitude. This is analogous to optics with photons or neutrons. In contrast to neutron optics,[29] where scattering is dominantly s-wave, up to a few hundred partial waves participate in atom-atom scattering at thermal energies.

For our experiment we add gas (He, Ne, Ar, Kr, Xe) to one side of our interaction region. We have three observables: the transmission through our sample, the interference amplitude, and the phase shift of the interference pattern. From the measured attenuation and the phase shift, a velocity average of the phase and amplitude of $f(k,0)$ can be determined. Furthermore, the scattering cross section is not a constant with respect to the center of mass energy and, therefore, different velocities have different attenuation. Therefore the average velocity of the transmitted wave differs for different gas densities. This may enable us to gain additional information on the energy dependence of $f(k,0)$.

Interferometry is the only direct way to measure the real part of the forward scattering amplitude. Therefore the separated beam atom interferometer offers the possibility of obtaining additional information about the atom-atom scattering process. This is especially interesting for low energy collisions, because knowing $f(k,0)$ means that both the magnitude and sign of the scattering length can be determined. A knowledge of the sign of the scattering length, previously not measurable, is critical to predicting low temperature collective behavior[30] of dense atomic gases.

4 Fundamental Measurements

The inherent precision available with phase sensitive measurements makes atom interferometers ideal instruments for a variety of fundamental measurements, like inertial[15,31] and gravitational effects[31,32] and the photon momentum,[33] or for "null" tests, like a search for any residual charge of neutral atoms. The sensitivity to phase will allow atom interferometers to probe physical processes that generate phase shifts such as Berry's (and other) topological phases (cf. a recent related proposal[34]), the passage of atoms through a wave guide, and the phase shift which accompanies surface bounces. In general, it has not previously been possible to observe these phase-generating effects.

A recent proposal by Anandan[35] and Aharonov and Casher,[36] an analog to the Aharonov-Bohm effect but for the magnetic moment, is a good example of a fundamental test using an atom interferometer: it is a topological phase which tests a fundamental tenet of quantum mechanics, that a phase shift can occur in the absence of any classical force. A study of this effect using neutron interferometers has been conducted.[37] The two advantages of using atoms are the greater magnetic moment and the greater sensitivity. Care has to be taken to compensate for the large DC Stark phase shift which may limit the practical size of the fields which can be applied. The total phase measured in the neutron experiment was only 3 mrad.[37] In contrast we will be able to achieve up to 1 radian of Aharonov-Casher phase for sodium atoms in our interferometer. This should greatly reduce the statistical error and will also allow us to study the predicted dependence on the dipole orientation. Varying the velocity and the velocity distribution of our sodium beam will allow us to explicitly investigate the nondispersive character of the Aharonov-Casher effect and therefore give direct evidence of its topological nature.[38]

5 Conclusion

The future of atoms interferometers looks bright. Atom beams are relatively inexpensive and intense. Several techniques have now been demonstrated to make interferometers, and the atoms which may be used in them come with a wide range of parameters such as polarizability, mass, and magnetic moment. One can even imagine molecular interferometers. We predict the application of atom and molecular interferometers to a wide range of measurements of both fundamental and practical interest.

Our recent work on atom interferometers and atom optics is supported by the Army Research Office contracts DAAL03-89-K-0082, and ASSERT 29970-PH-AAS, the Office of Naval Research contract N00014-89-J-1207, and the Joint Services Electronics Program contract DAAL03-89-C-0001. J.S. was supported in part by an *Erwin Schrödinger Fellowship* of the *Fond zur Förderung der Wissenschaftilchen Forschung* in Austria. T.D.H. was supported by a National Science Foundation Graduate Fellowship.

REFERENCES

1. Th. Young, Phil. Trans. Roy. Soc., London XCII **12,** 387 (1802).
2. H. Fizeau, Ann. Phys. Chem. **3**, 457 (1853); Ann. Chem. Phys. **D66**, 429 (1862).
3. A.A. Michelson, Am. J. Sci., (3) **22**, 20 (1881); and E.W. Morley, Am. J. Sci. **31**, 377 (1886).
4. Lord Rayleigh, Proc. Roy. Soc., **59**, 148 (1896).
5. L. Marton, J. Arol Simpson, and J.A. Suddeth, Phys. Rev. **90**, 490 (1954), Rev. Sci. Inst. **25**, 1099 (1954).
6. G. Mollenstedt, and H. Duker Naturwissenschaften **42**, and Z. Phys. **145**, 377 (1956).
7. H. Rauch, W. Treimer, and U. Bonse, Phys. Lett. **A47**, 369 (1974).
8. Matter Wave Interferometry, a special review issue of Physica B&C **151**, (1988).
9. S. Altschuler and L.M. Franz, U.S. Patent #3,761,721.
10. R. Beach, S.R. Hartmann, and R. Friedberg, Phys. Rev. **A25**, 2658 (1982).
11. V.P. Chebotayev, B. Ya. Dubertsky, A.P. Kasantsev, and V.P. Yakovlev, J. Opt. Soc. Am. **2** (11), 1791 (1985).

12. Ch. J. Borde, Phys. Lett. **A140**, 10 (1989).
13. D.W. Keith, C.R. Ekstrom, Q.A. Turchette, and D.E. Pritchard, Phys. Rev. Lett. **66**, 2693 (1991); D.W. Keith, M.L. Schattenburg, Henry I. Smith, and D.E. Pritchard, Phys. Rev. Lett. **61**, 1580 (1988); P.L. Gould, G.A. Ruff, and D.E. Pritchard, Phys. Rev. Lett. **56**, 827 (1986).
14. O. Carnal, J. Mlynek, Phys. Rev. Lett. **66**, 2689 (1991).
15. F. Riehle, Th. Kisters, A. Witte, J. Helmcke, J. Borde, Phys. Rev. Lett. **67**, 177 (1991).
16. M. Kasevich, S. Chu, Phys. Rev. Lett. **67**, 181 (1991).
17. J.Robert et.al. Eur. Phys. Lett. **16,** 29 (1991).
18. F. Shimizu, K. Shimizu and H. Takuma, Phys. Rev. **A46**, 46 (1992).
19. U. Sterr, K. Sengstock, J.H. Muller, D. Bettermann and W. Ertmer, Appl. Phys. **B54**, 341 (1992).
20. D.E. Pritchard, *"Atom Interferometry"*, Proceedings of ICAP 13 (Munich 1992).
21. B. Levy, Physics Today, **17** (1991).
22. For an overview of recent demonstration experiments see: "*Optics and Interferometry with Atoms*" Special issue Applied Physics **B54** (1992), Plenum New York (1986) editors: J. Mlynek, V. Balykin, P. Meystre.
23. D.W. Keith, R.J. Soave, and M.J. Rooks, J. Vac. Sci. Technol. **9**, 2846 (1991); C.R. Ekstrom, D.W. Keith, and D.E. Pritchard, Appl. Phys. **B54** , 369 (1992).
24. B.J. Chang, R. Alverness, and E.N. Leith, Appl. Optics **14**, 1597 (1975).
25. R. Molof, H. Schwartz, T. Miller and B. Bederson, Phys. Rev. **A10**, 1131 (1974); W.D. Hall and J.C. Zorn, Phys. Rev **A10**, 1141 (1974).
26. F. Shimizu, K. Shimizu, and H. Takuma, J. Appl. Phys. **B31**, L436 (1992).
27. F. Riehle, A. Witte, Th. Kisters, and J. Helmcke, Appl. Phys. **B54**, 383 (1992).
28. J.D. Jackson, *"Classical Electrodynamics"*, John Wiley & Sons (1975).
29. V.F. Sears, *"Neutron Optics"*, Oxford Uinversity Press (1990).
30. V. Bagnato, D. Pritchard and D. Kleppner, Phys. Rev. **A35**, 4357 (1987); E. Tiesinga, A.J. Moerdijk, B.J. Verhaar, and H.T.C. Stoof, Phys. Rev. **A46**, R1167, (1992).
31. J.F. Clauser, Physica B&C **151**, 262 (1988).
32. M. Kasevich, S. Chu, App. Phys. **B54**, 321 (1992).
33. D. Weiss and S. Chu, Phys. Rev. Lett (in press) (1993).
34. A Stern, Phys. Rev. Lett. **68**, 1022 (1992); Y. Aharonov and A. Stern, Phys. Rev. Lett. **69**, 3593, (1992).
35. J. Anandan, Phys. Rev. Lett. **48**, 1660 (1982).
36. Y. Aharonov and A. Casher, Phys. Rev. Lett. **53**, 319 (1984).
37. A. Cimmino, G. I. Opat, A.G. Klein, H. Kaiser, S.A. Werner, M. Arif, and R. Clothier, Phys. Rev. Lett. **63**, 308 (1989).
38. A. Zeilinger in: *"Fundamental Aspects of Quantum Theory"* NATO ASI Series B, Vol.144, p. 311, editors: Vittorio Gorini and Alberto Frigerio, (Plenum, New York 1986).

CAVITY QED

MANIPULATION OF QUANTUM FIELDS IN A CAVITY BY ATOMIC INTERFEROMETRY

S. HAROCHE, J.M. RAIMOND and M. BRUNE
Laboratoire de Physique de l'Ecole Normale Supérieure,
24 rue Lhomond 75231, Paris Cedex 05, France.

ABSTRACT

We discuss atomic interference schemes involving Rydberg atoms crossing high Q cavities and show how they can be used to produce and detect various non classical radiation states. Quantum superpositions of small coherent fields differing by their phase, amplitude or spatial localization could be generated. These experiments, whose recent progresses are reported, will explore the frontiers between classical and quantum physics and the foundations of quantum measurement theory.

1. Introduction. General principles of field manipulation by atomic interferometry.

The behaviour of atoms and photons in small cavities illustrate some of the basic concepts of quantum physics and make possible the development of new, very sensitive field sensors. The study of these effects has opened a new field of quantum optics called Cavity Quantum Electrodynamics (CQED)[1]. In a first generation of CQED experiments, enhanced and inhibited spontaneous emission and cavity induced atomic level shifts have been demonstrated, in the optical as well as in the microwave domains[2]. Various applications of these phenomena have been proposed to improve the operation of micro-optical devices. In more recent developments, the interest has shifted to the study of strong atom-field correlation effects, observable on long lived atomic states resonantly coupled to very high Q cavity modes. Atom-cavity bound or antibound states, resulting from the exchange of a single photon between an atom and a cavity have been observed by very sensitive spectroscopic methods[3]. These atom-cavity states could be used, for example, to control with a single atom the transmission of a laser beam through an optical resonator.

Stimulated by the success of these experiments, many novel ideas have been proposed and discussed in order to manipulate quantum fields stored in a high Q

cavity[4, 5, 6]. Several research groups - including ours at Ecole Normale Supérieure (ENS) - are presently trying to reduce theory to practice. A variety of proposals are trying to exploit the non-resonant interaction between an atom and a cavity mode, resulting in light shifts for the atomic energy levels and in dispersive effects for the field in the cavity. These effects are particularly large if the atom is in a Rydberg state and if the cavity mode has a very narrow bandwidth, making it possible to bring the atomic transition very close to its frequency, without being quite resonant.

The energy level shifts being position dependent in the cavity, they correspond to tiny dispersive forces exerced on the atom[7], which could for example be exploited to measure weak field intensities in a non destructive way, providing a striking illustration of the measurement theory in quantum physics. The force being dependent on the number of photons stored in the cavity, a detection of the trajectories of atoms crossing the resonator should yield in principle a measurement of the field intensity[8]. The splitting of the atomic beam into beamlets associated to different photon numbers would be a kind of Stern-Gerlach effect, in which the particle position would be correlated not to its own internal state - as in the usual Stern-Gerlach experiment - but to the internal state of the deflecting system.

The dispersive interaction with the cavity field affects not only the atom's mean position, but even more importantly, it alters the phase of the atomic wave packet crossing the resonator. A retardation of the order of one tenth of the atom de Broglie wavelength - much less than a tenth of an Angström for thermal velocity atoms - corresponds to a phase shift of the order of one radian, easily detected by atomic interferometry methods. In order to measure the field intensity, one has to split the atom's wave function into two parts, experiencing different phase shifts, and to recombine these parts after the atom has left the cavity. Interferences should then be observed in the probability of detecting the atom in a given quantum state.

The splitting of the atom's wave packet is most readily obtained by preparing it in a superposition of two states experiencing different forces in the cavity field. The wave packet splitting and recombination are achieved by having the atom cross, before and after the cavity, two identical auxiliary field zones in which they experience $\pi/2$ pulses on a transition linking the two atomic levels of interest. This is the well known Ramsey method of separated oscillatory fields[4]. The probability of detecting the atom in either of the two levels behind the second zone exhibits fringes as the number of photons in the cavity is varied. This detection can be achieved, for Rydberg states, by the well known selective field ionization method. Two detectors operating with slightly different electric field levels successively ionize the two Rydberg states of interest and the resulting ions are counted in two separate counters. Recording the Ramsey fringes with an ensemble of atoms crossing the cavity provides a sensitive quantum non demolition (QND) method for measuring the photon number[4, 5]. The measurement reduces the field into a Fock state, in which the intensity fluctuations are totally suppressed. Once a measurement has achieved this reduction, any subsequent measurement will leave the field unperturbed and yield the same result. Several atoms are of course required to record a fringe pattern and to achieve the QND measurement. At the beginning of this reduction process, and after only one atom has crossed the cavity, intermediate states of the field are produced, which correspond to linear superpositions of fields with different phases, known as "Schrödinger Cats"[5]. A general sketch of the QND and "Schrödinger Cat" experiment is presented in Figure 1.

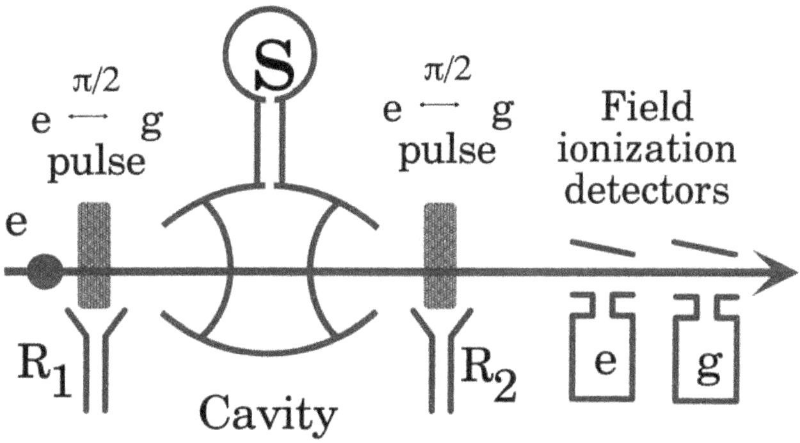

Figure 1: General sketch of the QND and "Schrödinger cat" experiment. An initial field is fed into the cavity by a source S. After being disconnected from the source, this field is manipulated by a beam of Rydberg atoms crossing the cavity in a superposition of two quantum states e and g. The atoms are detected downstream by two state-selective ionization counters. Interferences, essential for this manipulation, are produced by the Ramsey zones R_1 and R_2.

Since the theory of the generation and detection of Fock and Schrödinger Cat states has been presented in several articles by the ENS group, we have chosen to discuss in this paper one interesting aspect of these experiments, not yet emphasized, which is their connection with the "Einstein-Podolsky-Rosen" (EPR) problem[9] and the Quantum Measurement theory. In this discussion, we will concentrate on general concepts and refer the interested reader to more detailed papers recently published or in print.

2. Generation of Schrödinger Cats of the Field: A new EPR situation.

The build up of atom-field correlations is an essential feature of these experiment. The preparation of a "Schrödinger Cat" consists in injecting a coherent field in the cavity and having this field interact with a single atom prepared in a superposition of two internal states producing different phase shifts on the field[5]. The coherent field injection is readily achieved by coupling the atom to a microwave source during a finite time. The atom is then sent through the cavity, in a time short compared to the cavity damping. As the atom wave packet crosses the cavity, the field experiences a "back-action" dephasing which results from the "index of refraction effect" produced by the atom. The field dephasing is correlated to the atom's internal state. The final state of the combined system is quite similar to the entangled particle states of the Einstein-Podolsky-Rosen (EPR) experiment.

To be more specific, let us call α the complex amplitude of the coherent field initially present in the cavity, e and g the two atomic substates and let us assume that the atom in level e dephases the field in the cavity by an angle $\pi/2$ (transforming its amplitude into $\beta = i\alpha$) whereas the atom in level g dephases the field by the opposite angle $-\pi/2$ (transforming its amplitude into $-\beta$). Such large phase shifts are quite realistic with Rydberg atoms coupled to microwave resonators. The phase shift is proportional to the time spent by the atom in the cavity, i.e. inversely proportional to the atom's velocity which has thus to be carefully selected. Each state of the atom is associated to a state of the field with a different phase and the atom-field system is left after the atom exits the cavity in a linear superposition of these possible combined states:

$$|\Psi_{\text{Atom-Field}}\rangle = \frac{1}{\sqrt{2}}\left[|e,\beta\rangle + |g,-\beta\rangle\right] \tag{1}$$

(in each of the above kets the first symbol refer to the atom's state and the second one to the field's). There is a strong analogy with the two spins 1/2 particles of the EPR experiment, one "particle" being here the atom in one of the two substates e or g and the other one being the field in one of the two states $\pm\beta$. The strong correlation between the two subsystems subsists after the atom has left the cavity. As in the EPR experiment, detecting the atom in one of its quantum states results in the reduction of the field into the correlated radiation state. What is truly quantum in this correlation is that it can be expressed in different bases, depending upon the detectors which are actually being used. The choice of the detectors (and hence of the basis) can be made after the atom has left the cavity, as in the famous "delayed choice" EPR procedure. For example, Eq(1) can equally well be written as:

$$\begin{aligned}|\Psi_{\text{Atom-Field}}\rangle &= \frac{1}{2\sqrt{2}}\Big[\left(|e\rangle + |g\rangle\right)\left(|\beta\rangle + |-\beta\rangle\right) \\ &\quad + \left(|e\rangle - |g\rangle\right)\left(|\beta\rangle - |-\beta\rangle\right)\Big],\end{aligned} \tag{2}$$

which means that the detection of the atom in one of the two superpositions $|e\rangle \pm |g\rangle$ leaves the field in one of the two states $|\beta\rangle \pm |-\beta\rangle$. These states represent quantum superpositions of two classical fields with opposite phases, not to be confused with a classical superposition of two such fields which would correspond to the vacuum! This is clearly a non classical field, which we will call a "Phase Schrödinger Cat".

Of course, atomic superposition states of the form $|e\rangle \pm |g\rangle$ cannot be detected directly by a field ionization detector which is sensitive to the atom's energy and can discriminate $|e\rangle$ from $|g\rangle$ but not $|e\rangle + |g\rangle$ from $|e\rangle - |g\rangle$. The trick to detect these superpositions is to have the atom cross, after the cavity, a second microwave zone performing a $\pi/2$ pulse on the atom's state. Detecting $|e\rangle$ or $|g\rangle$ after this zone amounts to detecting $|e\rangle \pm |g\rangle$ directly after the cavity. This analyzis gives another interpretation to the Ramsey zones of the QND experiment of Fig(1). The first zone (before the cavity) prepares the atomic coherence leading to Eq.(1) and the second one defines the "delayed choice" of basis which leads to the "cat" state preparation.

3. Macroscopic coherence versus statistical mixture: the fast decoherence process.

The quantum coherence of the Schrödinger cat is very sensitive to dissipation. In a very short time, of the order of the cavity energy damping time t_{cav} divided by the number of photons n in the cavity, this coherence is washed out and the "cat" evolves into a mere statistical mixture of the two fields components[5]. As a result, one can realisticaly observe only "cats" consisting of relatively small "mesoscopic" fields, made of a handful of photons. In order to demonstrate the existence of these mesoscopic cats, one can send a second probe atom through the same apparatus after a small delay and measure the probability of detecting this atom in either one or the other of the two quantum states[10].

The cat's coherence produces interference effects in the atom's detection probability, which should thus be quite different for a cat and a statistical mixture. It can easily be shown that the probability of detecting the probe atom in the same state as the first one (whose detection has prepared the cat) is equal to one if no decoherence has occured betwen the two atoms, and only to 1/2 if the decoherence process has transformed the "cat" into a statistical mixture of two nearly orthogonal coherent field states. The complete experiment thus consists in sending two atoms through the cavity and correlating the outcomes of the detections of these two atoms, as a function of the delay between them. The double atom count signal should drastically change on a time scale of the order of t_{cav}/n, with the conditional probability of finding both atoms in the same state dropping from 1 to 1/2. This experiment would be a striking illustration of the transition between quantum physics (where probability amplitude superpositions play an important role) and classical physics (where probabilities merely add up without interfering). In this respect, one could say that "small cats" made of only a handful of photons are really quantum since their coherence decays basically on the same time scale as their energy, whereas "large cats", made of hundreds or thousands of photons are "classical" since, during most of the system life time, the coherence between their two parts is destroyed.

4. Quantum switches and "amplitude cats" in a cavity.

A different version of this experiment[11] can be performed in the following way: instead of injecting first a coherent field in the cavity and dephasing it afterwards by its interaction with an atom, one can couple the source S to a slightly detuned cavity, hereby preventing the field to build up inside, and then send one atom in a superposition of two states, one of which (e) results in the tuning of the cavity in resonance with the source, and the other (g) leaving the cavity mistuned. The atom behaves then as a switch, opening the cavity to the source in one of its states and leaving it closed in the other. It is indeed a quantum switch, in a coherent superposition of its open and closed states. The atom must again cross first a $\pi/2$ pulse zone in front of the cavity, and the atom-field state immediately after the atom exits the cavity is now:

$$|\Psi_{\text{Atom-Field}}\rangle = \frac{1}{\sqrt{2}}\left[|e,\alpha\rangle + |g,0\rangle\right]. \tag{3}$$

α is an amplitude proportional to the time spent by the atom inside the cavity (time of switch closing). Detecting the atom in $|e\rangle$ or $|g\rangle$ dowstream would merely project the field in one of the two states $|\alpha\rangle$ or $|0\rangle$ (vacuum), a classical result. If however the atom crosses a second $\pi/2$ zone downstream, the levels e and g are again mixed and in this way, the final atom state measurement cannot tell us in which state the atom has crossed the cavity. In this case, one shows easily that, after detecting the atom in level e or g behind this second zone, the field is left in one of the two states:

$$|\Psi_{\text{Field}}\rangle = \frac{1}{\sqrt{2}}\left[|\alpha\rangle \pm |0\rangle\right], \tag{4}$$

which are linear superpositions of two fields with different intensities (Amplitude Schrödinger Cats). The set up for this experiment is again of the Ramsey type (separated oscillatory zones sandwiching a cavity) and is the same as in Fig(1). Amplitude cats, as phase cats are very sensitive to dissipation and can be detected by performing double atom count experiments.

5. Non local field coherences

Other "Schrödinger cat" experiments could be performed with two identical cavities crossed in succession by a single atom prepared and detected in a linear superposition of its quantum states[11]. It should then become possible to prepare "non local cat states" in which a coherent field would belong either to one or the other cavity, again a quantum alternative associated to a mesoscopic coherence. The set up is shown in Fig(2). The two cavities, coupled by a symmetrical T-shaped wave guide to the slightly off resonant source are now crossed in succession by the same atom.

The cavities are sandwiched between two $\pi/2$ microwave pulse zones preparing and analyzing the atomic superposition states. A π pulse zone between the two cavities exchanges the e and g states and therefore the "open" and "closed" states of the quantum switch. If the field can penetrate in the first cavity (because it is crossed by the atom in level e), then it cannot enter the second one (which is crossed by the atom in level g) and reciprocally. As a result, it is easily shown that the detection of the atom after the last microwave zone in level e or g leaves the field in one of the two states:

$$|\Psi_{\text{Field}}\rangle = \frac{1}{\sqrt{2}}\left[|\alpha, 0\rangle \pm |0, \alpha\rangle\right]. \tag{5}$$

The first and second symbols in each ket represent now the field in the first and second cavity respectively. These superpositions are indeed non local states in which the field belongs to either one of two cavities with a quantum coherence between the two corresponding probability amplitudes. Such superpositions states could also be tested by detecting a second probe atom crossing the same apparatus after a short delay. The probability of detecting the second atom in e or g after it has crossed the apparatus presents interference terms which are sensitive to the coherence between the two parts of the field quantum state. This coherence evolves again on a time scale of the order of t_{cav}/n and this evolution should reflect on the detection probability of the second atom.

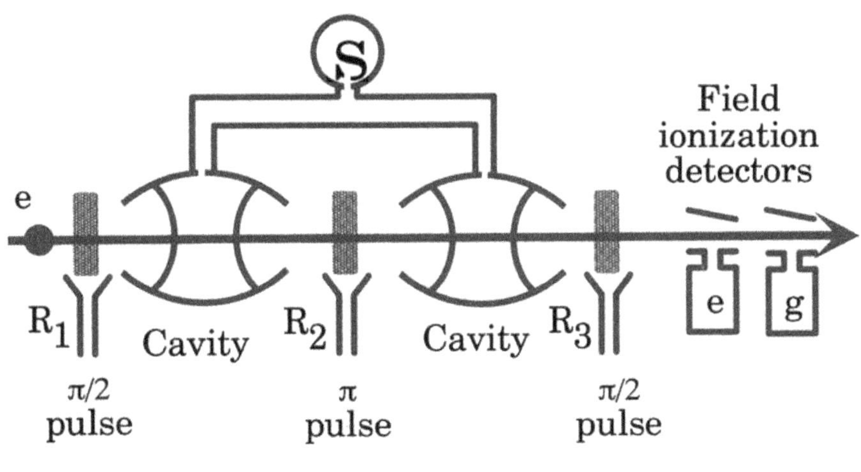

Figure 2: Set up for preparation of a "non local" field state.

6. Preliminary experiments.

These Cavity QED experiments require very weakly damped atomic states and cavity fields, so that atom-field correlations can survive to be detected long after the atom has interacted with the field. Circular Rydberg atoms coupled to very high Q microwave superconducting cavities at subKelvin temperatures constitute ideal systems for these studies. Atom and photon life times in the tenth to hundredth of a second range should be obtained, leaving enough time to perform the field QND measurements and the mesoscopic quantum coherences preparation and detection. In a preliminary set of experiments at ENS[12], we have been able to prepare efficiently circular Rydberg atoms with principal quantum number n around 50, whose radiative life time is 3. 10^{-2} s, which propagate along the 20 cm long apparatus without appreciable decay. We have also built open Fabry-Perot cavities made of superconducting niobium mirrors, which are resonant or quasi resonant with transitions betwen adjacing circular Rydberg atoms around 51 GHz. The Q of these cavities is in the 10^8 to 10^9 range, corresponding to a cavity damping time of the order of a millisecond. By applying a small electric field across the mirrors of these cavities, we have been able to propagate the circular atoms through them, without noticeable perturbation. Circular atoms are indeed notoriously fragile and require a small "directing" field normal to the orbit of the valence electron to preserve their circular character. Finally, we have applied resonant microwave pulses in two Ramsey zones sandwiching the cavity and observed fringes on the atomic detection signal as we varied the frequency of the microwave. These fringes demonstrate that coherence between adjacing circular levels can be propagated across the cavity. We have also observed a shift of the fringe pattern when a coherent field is injected in the cavity detuned from the atomic transition by 27 MHz. This shift, proportional to the cavity photon number, is still too small to be usable for the photon QND and "Schrödinger cat" experiments (a translation of one fringe required about 5000 photons in the cavity). By selecting a different cavity mode, more strongly coupled to the atom and by reducing the atom-cavity detuning to less than a MHz, we plan to achieve the same shift with about one photon only in the cavity, thus reaching the sensitivity required for the experiments described in this chapter. Experiments with two cavities to study non local field coherences will be attempted at a later stage.

7. Conclusion. Connection with quantum measurement theory.

At a fundamental level, the experiments described in this chapter illustrate some of the basic aspects of quantum mechanics, in one of its more mysterious aspects having to do with the quantum measurement and the postulate of state reduction[13]. The atomic beam crossing the cavity in Fig. 1 can indeed be considered as a "meter" used for the measurement of the photon number contained in the cavity. As the

Ramsey fringes are progressively revealed by the detection of successive atoms, the photon number distribution evolves towards a single n value, realizing so to speak the "collapse" of the field wave function. At the same time, the complementary "observable", the field phase, is randomized. This QND experiment would thus allow to observe the "direct" manifestation of the uncertainty principle in a real measurement process. This interesting aspect has been analyzed in details in previous papers.

Alternatively, one can consider in Fig.2 that the field left in the cavities after the atom has crossed them is a "meter" measuring the internal quantum state of the atom just before it enters the first cavity. If the field is found in the first cavity, the atom was in state e. It was in state g if the field is found in the second cavity. In the first stage of the experiment, just after the atom exits the second cavity, an entangled state of the observed system (the atom) and the meter (the field) is produced. This entangled state can be expressed in various bases, corresponding to different possible correlations between "observable" and "meter" states. Before an actual detection is made by observing where the field actually is, all these bases are equally legitimate and "meter" coherences of the form described by Eq(5) are present in the system. By measuring the field location, one reduces in effect its state to either one of two "classical" outcomes (field in first or second cavity) and the atom is left in one of the two correlated states (e or g). How this reduction occurs remains a puzzling aspect of quantum mechanics. Conceptually, it means that a coherent state such as (5) is transformed into a statistical mixture of two states $|\alpha, 0\rangle$ and $|0, \alpha\rangle$ with equal weights, since the two "meter" outcomes are equally probable. This non unitary transformation can be seen as due to the coupling of the apparatus to its environment which contains a great many states and provides a fast relaxation mechanism for the meter[14]. In this respect the decoherence process analyzed above appears as a direct observation of the "meter" collapse into one of its two classical positions. We can thus say that the observation of the Schrödinger cat states and of their fast decay will be a direct illustration of the most fascinating aspects of the measurement process in quantum mechanics.

References

1 . S.Haroche and J.M.Raimond, Scientific American, April 1993; S. Haroche in *Fundamental systems in Quantum Optics, Les Houches Summer School Proceedings, Session LIII*, J. Dalibard, J.M. Raimond and J. Zinn-Justin eds, North Holland, Amsterdam (1992).

2 . S.Haroche and D.Kleppner, Physics Today, January 1989; E.A. Hinds in *Advances in Atomic and Molecular Physics, Vol XX*, D.R. Bates and B. Bederson eds, Academic Press, New York (1990)

3 . R.J.Thompson, G. Rempe and H.J. Kimble, Phys.Rev.Lett. **68**, 1132 (1992); F.Bernardot, P. Nussenzveig, M. Brune, J.M. Raimond and S. Haroche, Euro.Phys.Lett. **17**, 33 (1991).

4 . M.Brune, S. Haroche, V. Lefèvre, J.M. Raimond and N. Zagury, Phys.Rev.Lett. **65**, 976 (1990).

5 . M.Brune, S. Haroche, J.M. Raimond, L. Davidovich and N. Zagury, Phys.Rev **A. 45**, 5193 (1992).

6 . M.J.Holland, D.F. Walls and P. Zoller, Phys.Rev.Lett. **67**,1716 (1991).

7 . S. Haroche, M. Brune and J.M. Raimond, Euro. Phys. Lett. **14**, 19 (1991); B.G. Englert, J. Schwinger, A.O. Barut and M.O. Scully, Euro. Phys. Lett. **14**,25 (1991).

8 . S.Haroche, M. Brune and J.M. Raimond in *Atomic Physics 13, Proc. 13th Intern.Conf on Atomic Physics*, H.Walther et al editors (1993); D.Ivanov et T.A.B.Kennedy, Phys.Rev. **A47**, 566 (1993).

9 . A. Einstein, B. Podolski and N. Rosen, Phys. Rev. **47**, 777 (1935)

10 . S.Haroche, M. Brune, J.M. Raimond and L. Davidovich in *Fundamentals of Quantum Optics III*, F.Ehlotzky editor, Springer Verlag (1993).

11 . L. Davidovich, A. Maali, M. Brune, J.M. Raimond and S. Haroche Phys.Rev.Lett. to be published.

12 . P. Nussenzveig, F. Bernardot, J. Hare, J.M. Raimond, S. Haroche and W. Gawlik, Phys. Rev. to be published (Nov 1993)

13 . J. von Neumann, *Die Matematische Grundlagen der Quantenmechanik*, Springer Verlag, Berlin (1932); K. Hepp, Helv. Phys. Acta **45**, 237 (1972); J.S. Bell, Helv. Phys. Acta **48**, 93 (1975); *Quantum Theory and Measurement*, J.A. Wheeler and W. Zurek ed., Princeton Univ. Press, Princeton (1983); W. Zurek, Phys. Rev. **D24**, 1516 (1986); F. Haake and D. Walls, Phys. Rev. **A36**, 730 (1987).

14 . W. Zurek, Physics Today, Oct 1991, p. 36.

One–atom maser: Parity states

Berthold-Georg Englert, Nicoletta Sterpi, and Herbert Walther[a]
Max-Planck-Institut für Quantenoptik
Ludwig-Prandtl-Strasse 10, D-85748 Garching, Germany

When the one–atom maser is to be used as a which–way detector for atoms, the change in the photon field resulting from the emission of one additional photon must be noticeable. To ensure this, an appropriate photon field has to be initialized prior to the emission. The quality of the which–way detector is determined by this initial photon state. An optimal choice is a state with a definite parity, that is: a state for which the photon number is known to be either even or odd. The additional photon alters the parity, so that the resulting final photon state is certainly distinct from the initial one, although the spread in photon number may be rather large. We show how the parity of the photon state in the one–atom maser can be measured, and thus how states of definite parity can be prepared. The parity measurement can be performed while the resonator is pumped in the standard one–atom maser fashion.

Introduction

When an atom, excited to a suitable Rydberg state, traverses a high–quality microwave resonator a resonant transition to a neighbouring Rydberg state can be induced whereby a photon is emitted into the privileged mode of the cavity. The left–behind photon can be used to obtain which–way information if such a resonator is put into an arm of an atom interferometer. Distinct from standard textbook examples,[b] a which–way detector of this kind does not fall prey to Heisenberg's position–momentum uncertainty relation, because the atomic center–of–mass motion is not affected by the emission of the microwave photon.[3,4] Based upon this observation various experiments[5–8] have been discussed which all involve atom interferometers supplemented with such quantum–optical which–way detectors. These experiments would demonstrate a fundamental consequence of the Principle of Complementarity:[7] one cannot observe the interference pattern built up by many atoms and, in addition, acquire which–way information for each individual atom. Inasmuch as complementarity is here enforced without invoking uncertainty relations as the mechanism at work, an important lesson is learnt, namely that complementarity is

[a]BGE and HW are also at Sektion Physik, Universität München, Am Coulombwall 1, D–85748 Garching; NS is on leave from Dipartimento di Fisica dell'Università, Via Celoria 16, I–20133 Milano, Italy.

[b]Einstein's recoiling–slit arrangement is discussed by Bohr in Ref. 1; Feynman's light–scattering scheme is described in Ref. 2.

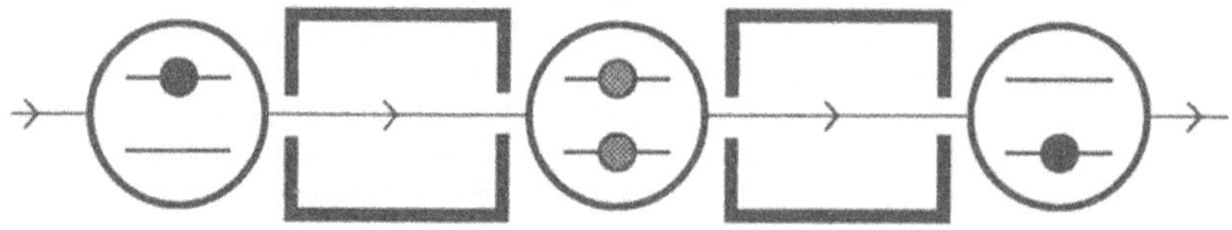

Figure 1: Ramsey fringe experiment. An atom in the upper one of two Rydberg states traverses two microwave cavities. The velocity, and therefore the interaction time, is chosen such that in each cavity alone the resonant photon emission happens with 50% probability. When the atom is finally detected in the lower Rydberg state, the transition could have occurred in either one of the two resonators. As long as these two 'ways' through the interferometer remain indistinguishable, the alternatives can interfere quantum mechanically and so produce Ramsey fringes in the probability for detecting the atom in the lower state. To exhibit these fringes, a static electric field of tunable strength (not indicated in the figure) is applied in the region between the two cavities, where the atom is in a coherent superposition of the two Rydberg states. The differential Stark shift experienced by these states introduces a relative phase, proportional to the strength of the electric field, and enables the experimenter to scan through the Ramsey fringes.

a more general concept than the 'wave–particle duality of matter' that is expressed by Heisenberg's uncertainty relations for the position (particle–like) and momentum (wave–like) attributes of a quantum mechanical object.

The two 'ways' to be distinguished by the which–way detector need not be two different paths in the three–dimensional physical space. In the Ramsey fringe set–up of Ref. 8, which is recalled in Fig. 1, the two ways refer to the more abstract alternatives of emitting a photon into either one of the two resonators. Ramsey fringes can only be observed if it is impossible to decide in which cavity the emission has occurred. This impossibility does not mean a lack of practical feasibility, but rather a principle absence of information stored in the resonators. The (abstract) way the atom took through the interferometer must be unknowable — not just unknown — for the Ramsey fringes to exist.

A functioning which–way detector has to make the atom's way through the interferometer knowable. For this purpose, the final photon field(s) in the resonator(s) must be distinguishable from the initial one(s), although they differ by one additional photon only. In the next section we shall employ the quantum mechanical overlap of the initial and final photon states to formulate a numerical criterion that distinguishes good from bad which–way detectors. This criterion then tells us that, for example, initial states with a definite photon number do the job. States of a definite parity, that is: states for which the photon number is known to be either even or odd, are suited equally well at first sight. A second look, however, speaks a lot in favor of the parity states, because — as shown below — they are prepared and detected much more easily than number states. In addition, the one–bit information, whether the atom took this way or the other through the interferometer, is directly reflected by

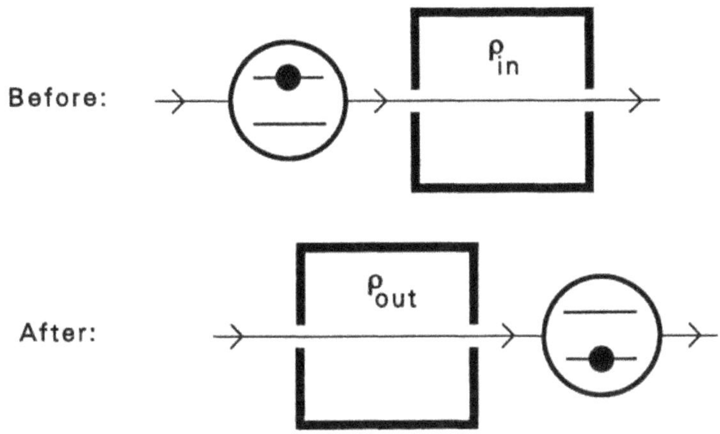

Figure 2: An atom enters the resonator in the upper Rydberg state. If it is found leaving in the lower one, the photon state has changed from ρ_{in} to ρ_{out}. The one additional photon in the resonator provides us with which–way information if ρ_{out} is noticeably different from ρ_{in}.

the one–bit choice of parity: even or odd. The costly preparation of a number state — though feasible in principle by either one of the schemes discussed by Brune *et al.* in Refs. 9 and 10 — is quite unnecessary for the purpose of which–way detection.

Good and bad which–way detectors

For simplicity, we shall consider the which–way information to get stored in one cavity only. Thus we face the situation of Fig. 2. Prior to the interaction the state of the photon field is $\rho = \rho_{\text{in}}(a^\dagger, a)$, which is a function of the photon ladder operators $a^\dagger$ and a. After the interaction we have

$$\rho = U_{\text{free}}^\dagger \, \rho_{\text{out}} \, U_{\text{free}} \,, \tag{1}$$

where U_{free} accounts for the free evolution of the photon state and

$$\rho_{\text{out}} = \frac{S^\dagger \rho_{\text{in}} S}{\operatorname{tr}\left\{S^\dagger \rho_{\text{in}} S\right\}} \tag{2}$$

is the final photon state in the interaction picture. The operator

$$S = \frac{\sin\left(\varphi\sqrt{aa^\dagger}\right)}{\sqrt{aa^\dagger}}\, a \tag{3}$$

summarizes the effect of the resonant interaction, whereby φ is the total Rabi angle corresponding to the specific experiment. For instance, one has $\varphi = \pi/4$ for the Ramsey fringe set-up of Fig. 1 and $\varphi = \pi/2$ for Young's double-slit arrangement discussed in Refs. 7 and 4.

As in Ref. 11 we now take the normalized quantum mechanical overlap of ρ_{in} and ρ_{out} as the numerical measure q of the quality of the which-way detector,

$$q \equiv \frac{\operatorname{tr}\{\rho_{\text{in}}\rho_{\text{out}}\}}{\sqrt{\operatorname{tr}\{\rho_{\text{in}}^2\}\operatorname{tr}\{\rho_{\text{out}}^2\}}}. \tag{4}$$

The normalizing denominator guarantees $q = 1$ for $\rho_{\text{in}} = \rho_{\text{out}}$ and $0 \le q < 1$ for $\rho_{\text{in}} \ne \rho_{\text{out}}$. Of course, both ρ_{in} and ρ_{out} are of unit trace. The extreme values of q and their significances are

$$\begin{aligned} q = 1 \ &: \ \text{no which-way information available,} \\ & \ \ \text{lousy which-way detector;} \\ q = 0 \ &: \ \text{full which-way information available,} \\ & \ \ \text{perfect which-way detector;} \end{aligned} \tag{5}$$

and intermediate values of q mean that partial which-way information is available.

Upon inserting Eq. 2 into Eq. 4 we get the more explicit expression

$$q = \frac{\operatorname{tr}\left\{S^\dagger\rho_{\text{in}}S\rho_{\text{in}}\right\}}{\sqrt{\operatorname{tr}\left\{SS^\dagger\rho_{\text{in}}SS^\dagger\rho_{\text{in}}\right\}\operatorname{tr}\{\rho_{\text{in}}^2\}}}. \tag{6}$$

If ρ_{in} projects to a classical coherent photon state, then the S and $S^\dagger$ operators can be equivalently replaced by their expectation values, that is: $S \to \operatorname{tr}\{S\rho_{\text{in}}\}$ everywhere, so that $q = 1$ here. In contrast, if ρ_{in} projects to a photon number state, then ρ_{out} projects to the number state with one photon more, with the consequence that the numerator in Eq. 6 vanishes, so that $q = 0$ now. In short: number states store which-way information perfectly, coherent states do not store it at all. These two extreme examples illustrate the folk wisdom that the one additional photon is only noticeable if the spread in the photon number is small enough. Although this statement is very plausible, it is false nevertheless. Consider the third example of a photon state with a definite parity,

$$(-1)^{a^\dagger a}\rho_{\text{in}} = \rho_{\text{in}}(-1)^{a^\dagger a} = \begin{cases} +\rho_{\text{in}} & : \ \text{even parity,} \\ & \ \ \text{or} \\ -\rho_{\text{in}} & : \ \text{odd parity.} \end{cases} \tag{7}$$

Because ρ_{out} contains an extra photon, it has the opposite parity, so that $q = 0$. Consequently, initial parity states provide us with a perfect which-way detector, no matter how large is the spread in the photon number.

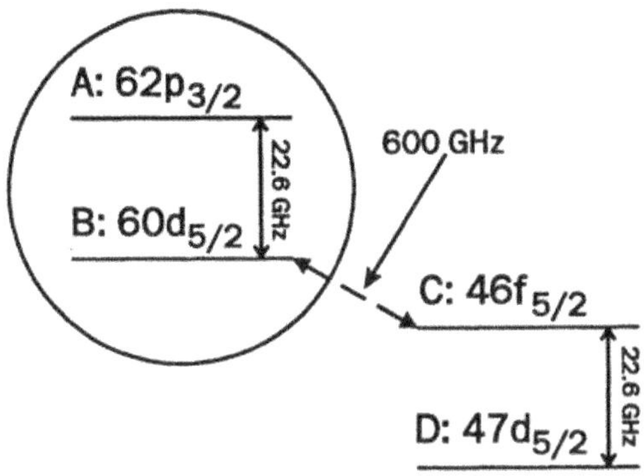

Figure 3: Relevant levels of ^{85}Rb. The transition from the $62p_{3/2}$ state A to the $60d_{5/2}$ state B is *not* in resonance with the cavity mode. It is used for the parity measurement. The symbolic two-level atom in Fig. 4 refers to these states. The transition from the $46f_{5/2}$ state C to the $47d_{5/2}$ state D *is* in resonance with the cavity mode. It can be used for the purposes of Figs. 1 and 2. Both transitions are near 22.565 GHz with the frequency of C→D being about 250 kHz less than that of A→B. The two pairs of levels are separated by about 600 GHz, so that the transitions do not interfere with each other.

How to measure the parity of the photon state

The ^{85}Rb atom possesses two Rydberg transitions in the convenient 20 GHz regime that differ only by a few hundred kHz in frequency, see Fig. 3. Since this relative detuning is nevertheless large compared to the kHz-range Rabi frequencies of the couplings to the cavity mode, one can use one transition for a resonant interaction, like those discussed in Figs. 1 and 2, and the other for the nonresonant interaction required for the parity measurement scheme that we shall now describe.

Based on the one-atom maser experiments performed in Garching,[12] we propose the set-up of Fig. 4 for the preparation of a state $\rho_{\rm in}$ of known parity for the photon field in the cavity. A ^{85}Rb probe atom is prepared in a coherent superposition of the two rather neighbouring levels A and B, specified in Fig. 3, by means of a classical Ramsey field. Then the atom is injected into the high-Q resonator, whose TE_{121} mode is tuned to the transition from state C to state D. As a result of this detuning, the nonresonant transitions from A to B are strongly suppressed. The atom emerging from the cavity is, therefore, still in a superposition of states A and B but with a change in the relative phase, a change that arises from the dynamical Stark shifts experienced by the states A and B. After leaving the resonator the atom passes through the second Ramsey field which oscillates at the same frequency as the first one with a controlled relative phase. The state of the atom is then changed into another superposition of A and B. Finally, by state-selective ionization and detection of the stripped-off electron

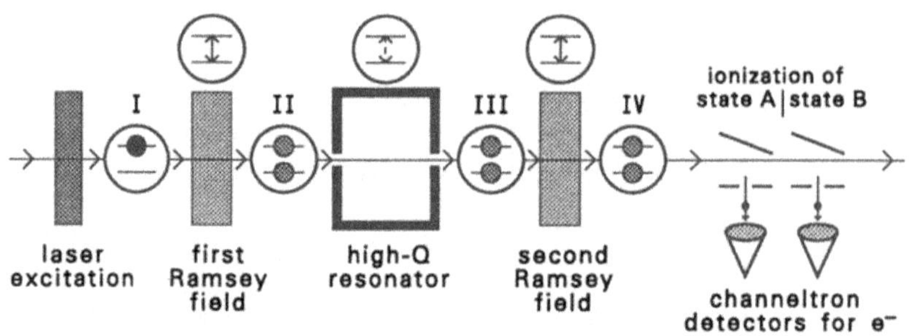

Figure 4: Schematic set-up of the parity measurement. At stage I the atom has been excited to the state A. The first Ramsey field induces a superposition of states A and B (stage II). After the passage through the high-Q resonator, at stage III, this superposition is changed. The second Ramsey field analyzes the superposition and gets the atom to stage IV. The experiment ends with a state-selective ionization and the detection of the stripped-off electrons. For a judicious choice of the parameters the parity of the photon state in the resonator is known to be odd when the final state A is detected, and even when B is found.

the final state of the atom is found to be either A or B.

When the parameters in the experiment are chosen fittingly one can ensure that the final state is A if the number of photons in the TE_{121} mode is odd, and B if it is even (or, of course, the other way round). In this way, the parity of this privileged mode of the quantized photon field is determined, and a state of known parity is prepared in the course.

Let us briefly report the essential features of the interaction. For simplicity, we choose all parameters appropriately and refer the reader to Ref. 13 for more details. Let us emphasize though that the required values of the parameters are within experimental reach.

After the laser excitation, at stage I in Fig. 4, the state of the atom is A, symbolically:

$$|I\rangle = |\text{A}\rangle. \tag{8}$$

The Ramsey fields effect the transitions between the stages I and II as well as III and IV according to:

$$|\text{A}\rangle \to \frac{1}{\sqrt{2}}\left(|\text{A}\rangle + e^{\mp i\pi/4}|\text{B}\rangle\right) \quad , \quad |\text{B}\rangle \to \frac{1}{\sqrt{2}}\left(|\text{B}\rangle - e^{\pm i\pi/4}|\text{A}\rangle\right), \tag{9}$$

where the upper sign holds for the first Ramsey field, the lower one for the second. Between the stages II and III the atom traverses the resonator. The nonresonant interaction with the photons produces dynamical Stark shifts of the two Rydberg levels. Owing to the relatively large detuning between the atomic transition and the cavity resonance these phase shifts are proportional to the number of photons n for state B and proportional to $n+1$ for state A. For appropriately chosen parameters the net effect is given by

$$|\mathrm{A}\rangle \rightarrow i^{-n-1}|\mathrm{A}\rangle \quad , \quad |\mathrm{B}\rangle \rightarrow i^{n}|\mathrm{B}\rangle. \tag{10}$$

Upon combining Eqs. 9 and 10 to propagate the initial state of Eq. 8 to the final state $|IV\rangle$ we find

$$|IV\rangle = i^{-n-1}\frac{1-(-1)^n}{2}|\mathrm{A}\rangle + i^{-n-1/2}\frac{1+(-1)^n}{2}|\mathrm{B}\rangle = \begin{cases} i^{-n-1}|\mathrm{A}\rangle & \text{for } n \text{ odd,} \\ i^{-n-1/2}|\mathrm{B}\rangle & \text{for } n \text{ even.} \end{cases} \tag{11}$$

Indeed, when the final state A is detected, the parity of the photon state is known to be odd, and when B is found it must be even.

Repeated parity measurements performed on the identically prepared photon state ρ determine the parity expectation value $\langle(-1)^{a^\dagger a}\rangle = \mathrm{tr}\{(-1)^{a^\dagger a}\rho\}$. It is worth mentioning that Wigner's phase space function $W(p',q')$ associated with $\rho(a^\dagger, a)$ is closely related to such parity expectation values, inasmuch as[14]

$$W(p',q') = \mathrm{tr}\left\{2(-1)^{a^\dagger a}\rho\left(a^\dagger + \frac{q'-ip'}{\sqrt{2}}, a + \frac{q'+ip'}{\sqrt{2}}\right)\right\} \tag{12}$$

holds. In particular, the value of the Wigner function at the origin in phase space equals twice the parity expectation value, $W(0,0) = 2\langle(-1)^{a^\dagger a}\rangle$. Consequently, the central value of the Wigner function is experimentally available as the measured mean value of the parity operator $(-1)^{a^\dagger a}$.

As implied by Eq. 11, the measurement process prepares the TE_{121} mode in a state ρ_{in} of known parity. In fact, due to the state reduction associated with the detection of the final atomic state A or B, the density operator ρ_{in} of the photon mode is

$$\rho_{\mathrm{in}} = \frac{\left[1 \mp (-1)^{a^\dagger a}\right]\rho_0}{1 \mp \langle(-1)^{a^\dagger a}\rangle_0}, \tag{13}$$

where the subscript '0' refers to the instant immediately before the probe atom enters the cavity and we have restricted the discussion to situations in which the density operator is a function of the photon number $a^\dagger a$ only and does not depend on a and $a^\dagger$ individually. The upper/lower sign applies when the parity is detected to be odd/even.

As an example, consider an initial thermal state,

$$\rho_0 = \frac{1}{\nu+1}\left(\frac{\nu}{\nu+1}\right)^{a^\dagger a}, \tag{14}$$

where ν is the number of thermal photons. The parity expectation value for this ρ_0 is $\langle(-1)^{a^\dagger a}\rangle_0 = 1/(2\nu+1)$, so that

$$\rho_{\text{in}} = \begin{cases} \frac{\nu+1/2}{\nu(\nu+1)}\left[\left(\frac{\nu}{\nu+1}\right)^{a^\dagger a} - \left(-\frac{\nu}{\nu+1}\right)^{a^\dagger a}\right] & \text{for odd parity,} \\ \frac{\nu+1/2}{(\nu+1)^2}\left[\left(\frac{\nu}{\nu+1}\right)^{a^\dagger a} + \left(-\frac{\nu}{\nu+1}\right)^{a^\dagger a}\right] & \text{for even parity.} \end{cases} \tag{15}$$

The probability for detecting even parity is $\frac{1}{2}\left[1+\langle(-1)^{a^\dagger a}\rangle_0\right] = \frac{\nu+1}{2\nu+1}$, that for odd parity is $\frac{1}{2}\left[1-\langle(-1)^{a^\dagger a}\rangle_0\right] = \frac{\nu}{2\nu+1}$.

Between the periods of interaction with the successive parity measuring atoms, the photon field then relaxes toward the thermal state of Eq. 14, if nothing else (pumping, which-way detection,...) is going on in addition. We model this 'free' evolution in the standard way with the aid of the master equation

$$\begin{aligned} \frac{\partial\rho}{\partial t} &= -\frac{A}{2}(\nu+1)\left(a^\dagger a\rho - 2a\rho a^\dagger + \rho a^\dagger a\right) - \frac{A}{2}\nu\left(aa^\dagger\rho - 2a^\dagger\rho a + \rho aa^\dagger\right) \\ &\equiv L\rho\,, \end{aligned} \tag{16}$$

where $1/A$ is the lifetime of the photons in the resonator.

Upon solving Eq. 16 with ρ_{in} of Eq. 15 as the initial state operator we find that the parity expectation value after the elapse of time t is given by

$$\langle(-1)^{a^\dagger a}\rangle_t = \frac{2\nu+1-(2\nu+1\pm 1)e^{-At}}{(2\nu+1)^2-4\nu(\nu+1)e^{-At}} \qquad \text{for}\begin{cases}\text{'odd,'}\\ \text{'even.'}\end{cases} \tag{17}$$

The short-time behavior

$$\langle(-1)^{a^\dagger a}\rangle_t = \begin{cases} -1+2(\nu+1)(2\nu+1)At & \text{for 'odd,'} \\ 1-2\nu(2\nu+1)At & \text{for 'even,'} \end{cases} \tag{18}$$

valid when $At \ll 1$, exhibits an intriguing quadratic dependence on ν, the number of thermal photons. As an illustration, Fig. 5 shows $\langle(-1)^{a^\dagger a}\rangle_t$ of Eq. 17 along with the approximation given in Eq. 18. An experimental verification of this time dependence could possibly test the validity of the theoretical models upon which Eq. 16 is based.

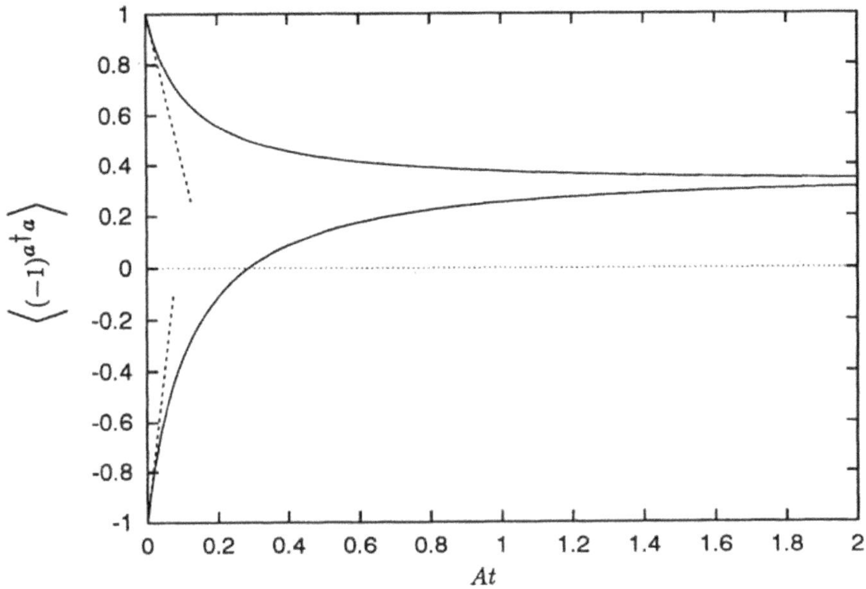

Figure 5: Time dependence of the parity expectation values according to Eq. 17, for $\nu = 1$. The dashed lines are the short–time approximations given in Eq. 18.

Statistics of emerging atoms

A measurement of the atomic beam statistics after the nonresonant interaction with the cavity photons would provide another valuable test. *Before* the interaction the beam is poissonian, that is: the probability that the time between two successive probe atoms is in the range $t \cdots t + \mathrm{d}t$ equals $P(t)\mathrm{d}t$ with

$$P(t) = re^{-rt}\,. \tag{19}$$

After the interaction one might determine the statistical properties of the partial beam composed only of the atoms emerging in either state A or B. For example, the temporal distribution of these atoms is described by

$$\left.\begin{matrix} P_{\mathrm{A}}(t) \\ P_{\mathrm{B}}(t) \end{matrix}\right\} = \frac{r}{2}\left[1 \mp \langle(-1)^{\;a^\dagger a}\;\rangle_t\right] \exp\left\{-\frac{r}{2}\int_0^t \mathrm{d}t' \left[1 \mp \langle(-1)^{\;a^\dagger a}\;\rangle_{t'}\right]\right\}. \tag{20}$$

But which is the actual value of $\langle(-1)^{\;a^\dagger a}\;\rangle_t$ that enters here and determines the statistics of the partial beam? The master equation governing the evolution between

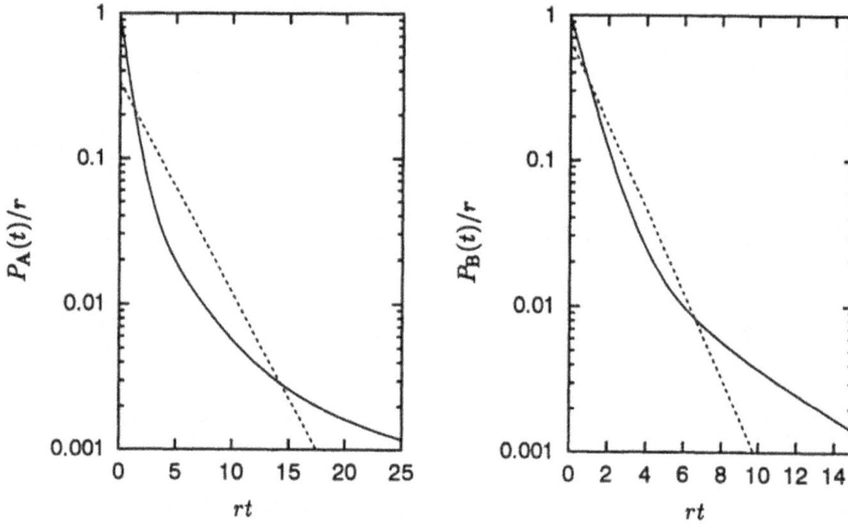

Figure 6: Statistics of the time intervals between successive atoms in the final state A (on the left) and in the final state B (on the right), for $\nu = 1$. The dashed lines show the corresponding poissonian statistics with the same effective rates.

two A atoms, say, is different from Eq. 16 because it has to account for all the B atoms that may traverse the resonator in the meantime. For a beam with an injection rate r, this master equation is, for $\rho = \rho(a^\dagger a, t)$:

$$\frac{\partial \rho}{\partial t} = \pm \frac{r}{2} \left[(-1)^{\,a^\dagger a} - \langle (-1)^{\,a^\dagger a} \rangle_t \right] \rho + L\rho \,, \tag{21}$$

where the upper/lower sign applies to A/B atoms.[c] The corresponding initial states are given on the right-hand side of Eq. 15. Note that, because of the occurrence of the parity expectation value $\langle (-1)^{\,a^\dagger a} \rangle_t$ in Eq. 21, this master equation is nonlinear in ρ.

In Fig. 6 we show plots of P_A and P_B, for $r/A = 20$ and $\nu = 1$, and we compare them with the corresponding poissonian distributions to the same effective rates. The effective rates are, of course, the reciprocals of the average times between two successive atoms in either state A or B. These average times are $(2\nu + 1)/(\nu r) = 3/r$

[c]Equation 21 is a special case of a more general master equation that governs the evolution between detector clicks. For its derivation and a discussion of other statistical properties of the emerging atoms in one–atom maser experiments, please consult Ref. 15.

in the A–beam and $(2\nu+1)/((\nu+1)\,r) = \frac{3}{2}/r$ in the B–beam, both considerably longer than the average time interval $1/r$ in the incoming poissonian beam.

Please note that the partial beams exhibit definitely nonpoissonian statistics. Both rather short and rather long time intervals are more probable than in a poissonian beam, and, as a compensation, waiting times of intermediate duration are less likely: the partial beams consist of superpoissonian bunches. This is as expected. If an atom has just emerged in the A state, say, the parity of the cavity field is odd; an atom, which follows immediately, encounters this odd–parity field and thus ends up in the state A as well. If, however, a long time has passed since the last A atom, a large number of B atoms have traversed the cavity in the meantime, thus pushing its field toward even parity; the probability that the next atom ends up in the state A is then rather small. In other words: if there has not been an A atom for a long while, then it is unlikely that the next one arrives shortly.

Summary

After showing that photon states of a definite parity enable one to conceive a perfect quantum optical which–way detector for atoms, we have discussed how the parity of the photon state can be measured. The proposed experimental scheme is based upon the relative dynamical Stark shift that affects a two–level atom interacting nonresonantly with the cavity photons.

The preparation of a state with a definite parity is achieved after a single atom is detected. This simplicity is quite distinct from the costly preparation of a number state, which requires many successive atoms whereby each atom must be prepared with a different velocity[9] or the velocity of each atom must be determined after its detection.[10] Further, a beam of parity measuring atoms can be used to monitor the parity quasi–continuously. The corresponding monitoring of the photon number appears to be impossible. Another advantage of the parity states is the possibility to pump the resonator with another, resonant, Rydberg transition — or to perform the which–way detection — while the parity measurements are going on.

We have also pointed out that measurements of the temporal evolution of the mean value of the parity can test the theoretical models used to describe the dissipative contributions to the dynamics of the photon state. For the same purpose, the statistics of the atoms emerging from the resonator after the interaction can be studied. The theoretical treatment predicts superpoissonian bunches of atoms in the same final Rydberg state.

Acknowledgments

NS would like to express her gratitude for the kind hospitality experienced at the Max–Planck–Institut für Quantenoptik. We thank W. Lange for useful discussions and, in particular, for his help with regard to the rubidium spectrum.

References

1. N. Bohr in *Albert Einstein: Philosopher–Scientist* (ed. P. A. Schilpp) pp. 200-241 (Library of Living Philosophers, Evanston, 1949); reprinted in *Quantum Theory and Measurement* (eds. J. A. Wheeler and W. H. Zurek) pp. 4-49 (Princeton University Press, 1983).
2. R. Feynman, R. Leighton, and M. Sands, *The Feynman Lectures on Physics* Vol. III (Addison Wesley, Reading, 1965).
3. B.–G. Englert, J. Schwinger, and M. O. Scully, in *New Frontiers in Quantum Electrodynamics and Quantum Optics* (ed. A. O. Barut) (Plenum, New York, 1990), pp. 513–519.
4. B.–G. Englert, M. O. Scully, and H. Walther in these Proceedings.
5. M. O. Scully, B.–G. Englert, and J. Schwinger, *Phys. Rev. A* **40**, 1775 (1989).
6. M. O. Scully and H. Walther, *Phys. Rev. A* **39**, 5229 (1989).
7. M. O. Scully, B.–G. Englert, and H. Walther, *Nature* **351**, 111 (1991).
8. B.–G. Englert, H. Walther, and M. O. Scully, *Appl. Phys. B* **54**, 366 (1992).
9. M. Brune, S. Haroche, V. Lefevre, J.M. Raymond, and N. Zagury, *Phys. Rev. Lett.* **65**, 976 (1990).
10. M. Brune, S. Haroche, J.M. Raymond, L. Davidovich, and N. Zagury, *Phys. Rev. A* **45**, 5193 (1992).
11. B.–G. Englert and M. O. Scully, in *New Frontiers in Quantum Electrodynamics and Quantum Optics* (ed. A. O. Barut) (Plenum, New York, 1990), pp. 507–512.
12. H. Walther, *Phys. Rep.* **219**, 263 (1992).
13. B.–G. Englert, N. Sterpi, and H. Walther, *Opt. Comm.*, in print.
14. B.–G. Englert, *J. Phys. A* **22**, 625 (1989).
15. H.–J. Briegel, B.–G. Englert, and N. Sterpi, in preparation.

The micromaser welcher–weg detector revisited

Berthold-Georg Englert, Heidi Fearn, Marlan O. Scully, and Herbert Walther
Max-Planck-Institut für Quantenoptik[a]
Ludwig-Prandtl-Strasse 10, D-85748 Garching, Germany

We extend our earlier theoretical work concerning a two–slit experiment with atoms in order to clarify a few points of importance. We show that the transverse–motion objections raised by Storey *et al.* are invalid for several reasons: they use an inappropriate set–up of the fields and cavities; they do not finish the calculation. The complete calculation confirms our original statement that it is the presence of which–way detectors, not Heisenberg's uncertainty relation, that enforces complementarity.

Introduction

In recent years we have discussed various experiments — of the gedanken type and also more realistic ones — that involve atomic interferometers with microwave resonators as which–way detectors: the Stern–Gerlach interferometer,[1] a quantum–beat experiment,[2] Young's double–slit arrangement,[3] Ramsey's two–field set–up,[4] and also an incomplete Stern–Gerlach interferometer.[5] The first four illustrate a fundamental consequence of the principle of complementarity, namely that both which–way information and interference pattern cannot be available simultaneously, whereas the fifth demonstrates the surrealistic nature of Bohm trajectories.

In each example it is essential that the which–way knowledge is acquired without changing the atomic center–of–mass wave function noticeably — in marked contrast to Einstein's recoiling–slit experiment[6] and Feynman's light scattering scheme.[7] This property of the quantum optical which–way detectors has been questioned by Matthew Collett in an interesting talk given at this conference and in a recent preprint.[8] We quote from Ref. 8:

> Taking the proposal of Scully, Englert and Walther[3] as an example, we show that in any path detection scheme involving a fixed double–slit, complementarity is enforced by means of momentum transfer in accordance with Heisenberg's uncertainty [relation].

---

[a]BGE and HW are also at Sektion Physik, Universität München, Am Coulombwall 1, D–85748 Garching; HF is permanently at Department of Physics, California State University Fullerton, Fullerton, CA 92634, USA; MOS is also at Department of Physics, Texas A&M University, College Station, TX 77843–4242, USA.

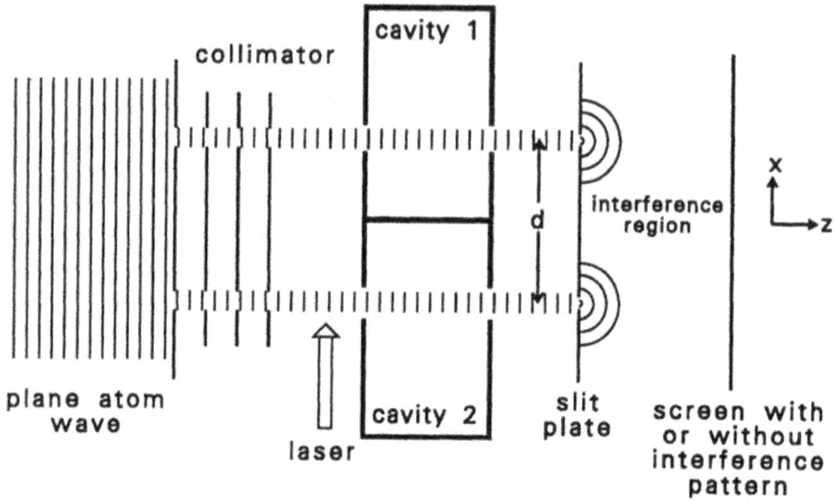

Figure 1: Young's double–slit interferometer for atoms with quantum optical which–way detectors. A set of wider slits collimates two atom beams which illuminate the narrow slits where the interference pattern originates. Each partial beam passes through a cavity. Prior to entering a laser excites the atoms to a Rydberg state. Inside the resonators the atoms perform resonant transitions to a neighbouring Rydberg state. The left–behind microwave photon provides the which–path information. Incidentally, please note that in Fig. 1 of Ref. 8 the cavities are inappropriately placed in the interference region.

Our response to Collett's critique of our work led to a spirited debate at the conference, and at the encouragement of the conference organizers, we here present our point of view. This takes the form of a supplement to our earlier theoretical treatment of the longitudinal center–of–mass motion of masing atoms in Ref. 9, by additional details concerning the transverse motion. This analysis confirms our position that it is the information in the which–way detectors which 'rubs out' the interference pattern and not Heisenberg's uncertainty relation.

We formulate the problem in the context of Young's double–slit interferometer as sketched in Fig. 1. The atoms approaching the slit plate from the left are collimated so that they are not scattered off the openings of the cavities. The cross sections of the two partial beams ($\sim$ 0.5mm in diameter) are considerably smaller than the openings ($\sim$ 3mm in diameter), yet much larger than the pin holes or 'slits' ($\sim$ 10nm in diameter and $\sim$ 5mm apart) at which the interference pattern originates. Thus the collimation does not affect the interference pattern, but it enables us to place the microwave resonators around the partial beams without disturbing them mechanically. The interferometer is operated with a very small atomic flux so that, at any instant, one atom at most is present in the apparatus. If this atom emits a microwave photon into one of the resonators, the information concerning which way the atom took to the screen is available, provided the initial state of the photon field is appropriate.

Please note that which-way information is obtained although a highly precise position measurement is not performed at all. We would only know in which partial beam the atom approached the slit plate, but this knowledge is quite sufficient, of course, to decide through which pin hole the atom went.

The problem is this. The total wave function in the interference region and on the screen is the sum of two contributions, one from each path through the double-slit:

$$\Psi = \psi_1(\vec{r})|\phi_1^f, \phi_2^i\rangle + \psi_2(\vec{r})|\phi_1^i, \phi_2^f\rangle , \tag{1}$$

where ψ_1 and ψ_2 are the two branches of the probability amplitude for the center-of-mass position $\vec{r}$, and $|\phi_1^f, \phi_2^i\rangle$, for example, denotes the state of the photon field where an atomic interaction has taken place in cavity 1 and none in cavity 2. (ϕ_j^i and ϕ_j^f represent the initial and the final state of the field in the cavity j.) Whereas the individual contributions are products of factors referring to the atom and photon degrees of freedom, their sum Ψ does not factorize in this way. This implies that these degrees of freedom have become entangled. The capability for interfering is lost as a result of this entanglement.

Consider the probability density for finding the atom somewhere on the screen; this is given by

$$\Psi^\dagger\Psi = |\psi_1|^2 + |\psi_2|^2 + \psi_1^*\psi_2\langle\phi_1^f, \phi_2^i|\phi_1^i, \phi_2^f\rangle + \psi_2^*\psi_1\langle\phi_1^i, \phi_2^f|\phi_1^f, \phi_2^i\rangle . \tag{2}$$

The first two terms produce the diffraction pattern of the single slits; the other two terms are the cross terms that potentially give rise to the two-slit interference fringes. Here, however, the photon overlaps $\langle\phi_1^f, \phi_2^i|\phi_1^i, \phi_2^f\rangle = \langle\phi_1^i, \phi_2^f|\phi_1^f, \phi_2^i\rangle^*$ may vanish, so that the fringes do not materialize. For example, if one has no photons initially and one photon in either cavity finally, then[b]

$$\langle\phi_1^f, \phi_2^i|\phi_1^i, \phi_2^f\rangle = \langle 1_1, 0_2|0_1, 1_2\rangle = 0 \tag{3}$$

vanishes, indeed, and there are no fringes. In contrast, if we had classical coherent states for the photons,

$$\langle\phi_1^f, \phi_2^i|\phi_1^i, \phi_2^f\rangle = \langle\alpha_1, \alpha_2|\alpha_1, \alpha_2\rangle = 1 \tag{4}$$

the overlap is perfect[c] and the interference fringes are there. In this situation, no entanglement is present and the total wave function Ψ of Eq. 1 factorizes.

This statement is true only if the interaction between the atom and the photon fields does not affect the branches ψ_1 and ψ_2 of the center-of-mass wave function

[b]The ket $|0_1, 1_2\rangle$, for instance, symbolizes the state with no photons in cavity 1 and one photon in cavity 2.

[c]In general, it is not equal to unity, but rather a pure phase factor.

to such an extent that their capability for interfering gets lost in the first place. This could come about by uncontrollable momentum transfers to the atom in the course of the interaction that supplies the which–way information. Indeed, this is the mechanism in standard textbook examples. Now, if the wave function branches can no longer interfere to begin with, the subject of entanglement becomes a side issue.

This is not the situation for the micromaser which–way detector. The emission of the resonant microwave photon is essentially recoil–free and leaves no noticeable trace on the center–of–mass wave function. This is the assertion questioned by Collett *et al.*[8] We shall rise to the occassion and present a clearcut demonstration of our case.

We proceed with a discussion of the arguments presented by Einstein and Bohr, Feynman and finally the Auckland group.[8] Then we consider both the set–up partially analysed by the Auckland group and our more fitting arrangement of the cavities and fields for the specific case of the double–slit interferometer. We take the transverse motion of the atom into account and show that one can, indeed, see fringes for both configurations.

In Appendix A we report the theory of an atom passing through a micromaser whereby it interacts resonantly with one photon mode. An alternative approach to the problem of recoil–free emission that does not make use of a specific size or shape of the resonator and does not refer to a particular experiment is presented in Appendix B.

Which–way detection and uncertainty relations

The double–slit experiment has born the brunt of many attacks on quantum mechanics. At the Fifth Solvay Congress at Brussels in 1927, Einstein[6] devised his own experiment in order to try and find an inconsistency in the then–new quantum theory. He considered photons incident on a moveable screen with a single slit. The photons then pass through a fixed double slit before hitting a fluorescent screen. According to Einstein photons moving toward one of the two slits impart a characteristic momentum to the moving screen. Thus by measuring the recoil of this moving slit Einstein hoped to gain which–way information, for each individual photon, whilst still observing the interference pattern built up by many photons. In this way he hoped to observe both the particle–like and wave–like nature of the photon and hence prove the principle of complementarity false. In defence Bohr[6] insisted that the recoiling screen must also be treated by the laws of quantum theory. In order to determine through which slit the photon passed one would need to know the initial momentum of the moveable screen to within an uncertainty of $\delta p \ll \Delta p = 2\pi\hbar d/\lambda D$ where Δp is the momentum transfer to be detected, λ is the photon wave length, d is the double slit separation, D is the distance between the moveable screen and the double slits (which is the same as that between the double slits and the fluorescent screen) and $\hbar$ is Planck's constant. Heisenberg's uncertainty relation $\delta x \delta p \geq \hbar/2$ then implies that the uncertainty in the position of the moveable slit obeys $\delta x \gg \lambda D/d$ which is exactly what is required to wash out the interference pattern, because $\Delta x = \lambda D/d$ is the separation of maxima in the fringe pattern. One should note that the uncertainties δx and δp refer to the moveable screen with the single slit and not to the photons.

A similar argument for electrons can be found in Feynman's lecture notes.[7] In this case a light source, positioned somewhere behind the double–slits, scatters light off the electrons after they have passed through the slits. Momentum is transfered to the electrons via the scattering event with a photon from the light source. Which–way information is obtained by 'seeing' a flash of light in the vicinity of either slit. From this one may deduce that the electron passed through slit 1 or 2. The scattering event gives an uncertainty in momentum to the electron which is just sufficient to wash out the interference pattern. If light of a larger wave length is used (to reduce the momentum kick) we find that we see fringes once more but we loose the which–way information. We see only a fuzzy flash of light not sufficiently localized to determine through which slit the electron passed. In this example, please note that the complementary uncertainties in position and momentum refer to the electrons that should compose the interference pattern.

However, Wootters and Zurek[10] have studied Einstein's modification of the double–slit experiment and come to a different conclusion. By modeling Einstein's moveable single slit by an harmonic oscillator in its ground state and giving each slit an aperture function, Wootters and Zurek calculate the wave function for the photons at the viewing screen and determine the interference pattern. They show that the situation is not so clean cut as it might first appear. It was their objective to find out to what extent one could obtain which–way information whilst still retaining an interference pattern. Their findings are quite remarkable. We quote from Ref. 10:

> ...out of 100 photons we expect 99 of them to pass through the more likely slit. An accuracy of 99% is certainly high, and one would expect that such a measurement should destroy the interference pattern. But a straightforward calculation shows that the crest–to–valley ratio ...equals approximately 3/2.

One should note that Heisenberg's uncertainty relation is, of course, observed throughout their calculation.

With this new insight we return to the micromaser problem outlined above.

The Auckland critique

We begin with a review of the Auckland critique[8] which follows along the same lines as the Bohr and Feynman arguments. They point out that when an atom emits a photon in either cavity, the atom will receive a recoil kick and thus an uncertainty in momentum. They suggest that the atom wave function is somehow altered by the interaction in the cavity in such a way as to preclude the possibility of obtaining which–way information and interference. We quote from Ref. 8:

> First we will argue that momentum transfers do occur in Scully's scheme, although their source is rather subtle. We then generalize the analysis to describe any detector which determines the path taken by the particle through a fixed double–slit. We show that the loss of interference in the presence of such a detector is due to momentum exchanges between the particle and the detector in the transverse direction ...This transverse momentm transfer accounts for the loss of interference on a distant screen ...

We disagree with their argument on the grounds that the atomic center–of–mass wave function itself carries no which–way information, all this information is stored in the fields of the cavities.

For the time being, we grant that Storey *et al.* give a reasonable wave function for the atoms at the slits in the form of an entangled state,

$$|\psi\rangle = \int dx\, \psi_i(x)[g_1(x)|1_1, 0_2\rangle + g_2(x)|0_1, 1_2\rangle + e(x)|0_1, 0_2\rangle] \otimes |x\rangle \tag{5}$$

where $\psi_i(x)$ is the probability amplitude before the interaction. This initial wave function has already passed through the collimators in Fig. 1 and is, therefore, narrowly peaked near the slit positions $x = \pm d/2$. More specifically $\psi_i(x)$ is the coherent sum of two branches, $\psi_i(x) = \psi_0(x - d/2) + \psi_0(x + d/2)$ where $\psi_0(x)$ is well localized near $x = 0$. We take $\psi_0(x)$ to be an even function, and then $\psi_i(x)$ is even too, consistent with the symmetry of the set–up in Fig. 1. The functions $g_1(x)$, $g_2(x)$ and $e(x)$ summarize the effect of the interaction and determine the probabilities that the atom emits a photon in cavity 1,2 or does not emit. They are given by

$$\begin{aligned} g_1(x) &= \sin(\beta \sin|2\pi x/\lambda|) \quad \text{for} \quad 0 \le x \le \lambda/2\,, \\ g_2(x) &= \sin(\beta \sin|2\pi x/\lambda|) \quad \text{for} \quad -\lambda/2 \le x \le 0\,, \\ e(x) &= \cos(\beta \sin|2\pi x/\lambda|) \quad \text{for} \quad -\lambda/2 \le x \le \lambda/2\,, \end{aligned} \tag{6}$$

and vanish outside the indicated ranges. Here λ (where $\lambda > d$) is the transverse wave length of the cavity mode and β is the product of the effective atom-field coupling g_0 and the interaction time τ. The kets represent numbers of photons present in the field inside the micromasers, subscripts 1 and 2 indicating the respective cavity. The above wave function follows from a similar treatment to the one given in Appendix A. In comparison with Wootters and Zurek we would need to consider the wave function at the screen and calculate the cross term there before making any statement about whether we do or do not see an interference pattern. The Auckland group failed to do this, so we shall continue where they left off.

The strategy of the Auckland group is to look at the Fourier transform $\tilde{g}_1(p)$ of $g_1(x)$, for example, and interpret that as the probability amplitude for a momentum transfer of amount p. Unfortunately, the Auckland group have been led astray by the intriguing structure possessed by $\tilde{g}_1(p)$. The product $\psi_i(x)g_1(x)$ is the quantity of interest. It can be computed by convolving $\tilde{g}_1(p)$ with the Fourier transform $\tilde{\psi}_i(p)$

of $\psi_i(x)$, followed by an inverse transform. Inasmuch as $\tilde{\psi}_i(p)$ is slowly varying, the convolution integral retains almost none of the structure in $\tilde{g}_1(p)$. Therefore, the conclusions drawn from the details of $\tilde{g}_1(p)$ are of no consequence.

The Garching rebuttal

In order to calculate the fringe pattern we must first find the wave function on the screen at $z = D$ in terms of what it is in the aperture at $z = 0$. This relation is given by,

$$\Psi(x,y,D) = \int\int G(x,y,D : x',y',0)\,[A(x'-d/2,y')\,|1_1,0_2\rangle\psi_i(x')g_1(x') + A(x'+d/2,y')|0_1,1_2\rangle\psi_i(x')g_2(x')]\,\mathrm{d}x'\mathrm{d}y' \tag{7}$$

where $G(x,y,z : x',y',z')$ is the appropriate Green's function and $A(x',y')$ is the aperture function which is nonzero only in the vicinity of $x' = 0$ and $y' = 0$. We have used the Auckland wave function of Eq. 5 in the $z' = 0$ plane and have picked out those terms that correspond to the atom being de–excited and one photon in either cavity. Note that their wave function does not display a y' dependence, although it should. Since $A(x',y')$ is well localized, no harm is done.

While we are at it, let us mention that in the context in which we shall use it, the Green's function for a wave with de Broglie wave length λ_{dB} is essentially given by,

$$G(x,y,D : x',y',0) = N\exp\left[\frac{i\pi}{D\lambda_{\text{dB}}}\left((x-x')^2+(y-y')^2\right)\right] \tag{8}$$

where the factor N contains all the constants that are irrelevant here. This Green's function applies in the Fraunhofer regime, when the screen is far away from the slits and the scattering is under small angles.

The Auckland result in Eq. 6 is obtained by considering a mode function whose transverse dependence is solely in the x direction, see Fig. 2a.[d] For reasons that will become obvious shortly, we prefer a mode which varies in the y direction only, see Fig. 2b. Thus, rather than the transverse mode function

$$\Omega(x,y) = \sin(2\pi|x|/\lambda) \tag{9}$$

we have

$$\Omega(x,y) = \cos(2\pi y/\lambda) \tag{10}$$

for the x, y dependence of the mode function. Equations 6 are thus particular cases of the general expressions

$$\begin{aligned} g_1(x,y) &= \sin\big(\beta\Omega(x,y)\big) \quad \text{for } x>0, \\ g_2(x,y) &= \sin\big(\beta\Omega(x,y)\big) \quad \text{for } x<0, \\ e(x,y) &= \cos\big(\beta\Omega(x,y)\big), \end{aligned} \tag{11}$$

[d] Consult also Fig. 1 of Ref. 8.

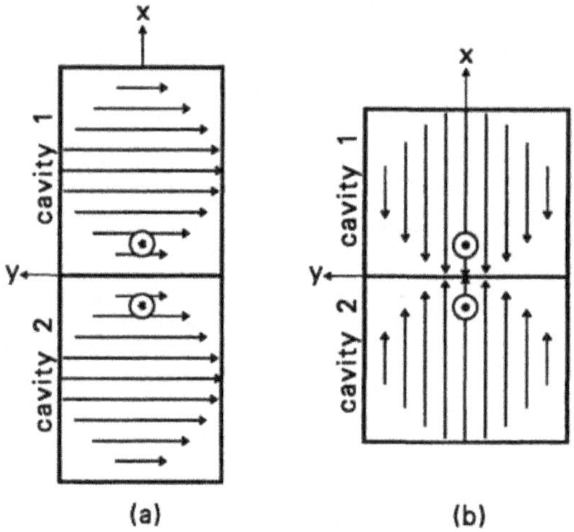

Figure 2: The cavity orientation and the direction of the electric vector of the microwave fields for (a) the Auckland scheme and (b) the Garching scheme. The cavities are shown as seen by somebody standing at the screen. The circles indicate the holes through which the atoms approach this observer; the shaded areas inside represent the branches of the atomic beam. The slits are not drawn. Note that in (a) there is a transverse gradient of the electric field. The atoms in the Auckland scheme suffer from a transverse net force that may potentially give rise to a splitting of the partial beams analogous to the Stern–Gerlach effect. This has been neglected by the Auckland group in arriving at Eq. 5. No such Stern–Gerlach splitting is present in the more fitting arrangement specified by Eq. 10.

obtained for the Auckland choice of Eq. 9. Our geometry in Eq. 10 produces

$$\begin{aligned} g_1(y) &= \sin\big(\beta\cos(2\pi y/\lambda)\big) \quad \text{for } x > 0\,, \\ g_2(y) &= \sin\big(\beta\cos(2\pi y/\lambda)\big) \quad \text{for } x < 0, \\ e(y) &= \cos\big(\beta\cos(2\pi y/\lambda)\big) \end{aligned} \tag{12}$$

instead. Since $\psi_i(x) = \psi_i(-x)$ and $g_1(x) = g_2(-x)$, any symmetric aperture function $A(x', y')$ — as we assume it — in conjunction with the $(x - x')^2$ dependence of the Green's function, ensures that the interference pattern is symmetric about the $x' = 0$ plane.

Realistically the initial wave function $\psi_i(x)$ is slowly varying over the aperture and can be taken to be constant (unity) over the width of the slit. We need $g_{1,2}(x, y)$ and $e(x, y)$ only in the apertures at $x = \pm d/2$, $y = 0$. To ensure that the no-emission probability is small, $e(x, y)$ must be zero there. This requires

$$\beta\sin(\pi d/\lambda) = \pi/2\,, \tag{13}$$

for the Auckland scheme, and

$$\beta = \pi/2 \tag{14}$$

for the Garching scheme.

In the Auckland scheme, see Eq. 11, we expand $g_1(x) = g_2(-x)$ around $x = d/2$ and find

$$g_1(x) = 1 - \frac{\pi^2}{2}\left[\frac{\pi d}{\lambda}\cot\left(\frac{\pi d}{\lambda}\right)\right]^2\left(\frac{x-d/2}{d}\right)^2 , \tag{15}$$

which simplifies to

$$g_1(x) = 1 - \frac{\pi^2}{2}\left(\frac{x-d/2}{d}\right)^2 \tag{16}$$

if $\pi d \ll \lambda$. Whereas in the Garching arrangement the corresponding approximation yields

$$g_1(y) = 1 - \frac{\pi^6}{2}\left(\frac{y}{\lambda}\right)^4 . \tag{17}$$

The difference $x - d/2$ in Eq. 16 and y in Eq. 17 range across the aperture, so they are both measured on the scale set by the slit width a ($\sim$ 5nm). Therefore, the deviation from unity in Eq. 16 is roughly given by $(a/d)^2 \sim (5\text{nm}/5\text{mm})^2 = 10^{-2}$, in contrast to $(a/\lambda)^4 \sim (5\text{nm}/50\text{mm})^4 = 10^{-28}$ in Eq. 17. The lesson taught by these numbers is two–fold. First, in both cases the deviation from unity is quite negligible even if the powers of π in Eqs. 16 and 17 are taken into account. Second, whereas it is true that the Garching scheme is an improvement over the Auckland geometry by many orders of magnitude, it is clear that the micromaser which–way set–up works for the Auckland scheme as well.

With the ingredients of Eq. 5 now at hand, we proceed to calculate the branches of the center–of–mass wave function at the screen and so demonstrate explicitly that the capability for interfering has not been lost as a result of the interaction. For the sake of simplicity, we choose a gaussian aperture function

$$A(x \mp d/2, y) = \exp\left(-\frac{1}{2a^2}\left[(x \mp d/2)^2 + y^2\right]\right) \tag{18}$$

The upper slit produces

$$\begin{aligned}
\Psi_1(x,y,D) &= \int \mathrm{d}x'\mathrm{d}y' N \exp\left(\frac{i\pi}{D\lambda_{\mathrm{dB}}}\left[(x-x')^2 + (y-y')^2\right]\right) \\
&\quad \times \exp\left(-\frac{1}{2a^2}\left[(x'-d/2)^2 + y'^2\right]\right)\left[1 - \frac{\pi^2}{2}\left(\frac{x'-d/2}{d}\right)^2\right] \\
&= \frac{2\pi a^2 N}{1 - ia/b}\underline{\underline{\left(1 - \frac{\pi^2}{2}\left(\frac{a}{d}\right)^2\left[1 - \left(\frac{x-d/2}{b}\right)^2\right]\right)}} \\
&\quad \times \exp\left(-(1 - ib/a)\frac{(x-d/2)^2 + y^2}{2b^2}\right)
\end{aligned} \tag{19}$$

in the Auckland case when Eq. 16 is to be used. The scale of the single–slit diffraction is set by $b = \lambda_{dB}D/(2\pi a)$ and terms of higher order in a/b have been discarded consistently. For $a \sim 5$nm, $\lambda_{dB} \sim 0.1$nm, $D \sim 1$m this characteristic distance is $b \sim 3$mm. The underlined factor represents the net change in ψ_1 due to the interaction of the atom with the cavity photons. Again we note that the deviation from the undisturbed unit value is of the order of 10^{-12} and perfectly irrelevant. When Eq. 17 is employed rather than Eq. 16, this factor is replaced by

$$1 - \frac{3\pi^6}{2}\left(\frac{a}{\lambda}\right)^4 \approx 1 - 1.5 \times 10^3 \times 10^{-28} \approx 1 - 10^{-25} , \tag{20}$$

which is even more irrelevant.

The two–slit interference pattern is produced by

$$\mathrm{Re}\,[\psi_2^*(x, y, D)\psi_1(x, y, D)] = \mathrm{Re}\,[\psi_1^*(-x, y, D)\psi_1(x, y, D)] \ . \tag{21}$$

Without the atom–photon interaction it is given by

$$(2\pi a^2|N|)^2 \exp\left(-\frac{x^2 + y^2 + (d/2)^2}{b^2}\right) \cos\left(\frac{xd}{ab}\right) , \tag{22}$$

which identifies the fringe spacing $2\pi ab/d = \lambda_{dB}D/d$. The underlined factor in Eq. 19, or the corresponding one in Eq. 20, obviously does not 'rub out' the interference pattern.

Conclusion

It is apparent that the Auckland calculations were incomplete; they were making premature statements about the interference pattern. An important observation is that the atoms are collimated and so confined to a small spatial region to begin with. The aperture samples a fraction of that region. Therefore, only a small portion of the atomic wave function contributes to the interference pattern, and this portion remains essentially unaffected by the which–way detection. We have shown that this is true even for the somewhat inappropriate geometrical arrangement chosen by the Auckland group, where the correction is found to be typically of order 10^{-12}. In practical terms this says: only a few atoms out 10^{12} will go astray and not contribute to the interference pattern. For the more fitting arrangement that we prefer, the corresponding number is 10^{-25}, virtually zero. In summary, we maintain that our earlier assertion is correct: The center–of–mass wave function does not carry which–way information.

Acknowledgments

We are grateful for insights provided by J. Schwinger and for the valuable, stimulating discussions we had with V. B. Braginsky, M. Collett, and D. Walls. MOS wishes to thank the ONR for its support of his work.

Appendix A: Theory of two-level atoms passing through the micromaser

Let us consider a beam of two-level atoms incident on two micromasers as in Fig. 1. A single atom can either pass through cavity 1 or 2 and may emit a photon in either of these cavities. The two branches of the center-of-mass wave function can be treated separately, and so we shall here concentrate on the interaction between one atom and the photon of one privileged mode in one resonator. We take the upper and lower Rydberg states of the atom to be A and B with an energy separation of $\hbar\omega$. This is also the photon energy in the privileged mode; the photon ladder operators are a and $a^\dagger$.[e] Initially the atom is in the excited state A and the photon state is ϕ^i, which we shall later specify to be the vacuum state or a classical coherent state. Distinct from Figs. 1 and 2, we now adopt a coordinate system such that the atom travels along the z axis, entering the resonator at $z = 0$ and leaving it at $z = L_z$.

We take for granted that the following reasonable conditions are obeyed:

(i) the atom's kinetic energy ($\sim 10^{-1}$ eV) is huge compared to the energy of the microwave photon ($\sim 10^{-11}$ eV), so that the center-of-mass motion can be treated semiclassically;

(ii) the natural spreading of the center-of-mass wave function is negligible during the time of interest, so that the transverse momentum values present in the center-of-mass wave function are negligibly small;

(iii) the atom moves with a well defined velocity along the z axis, so that one has a well defined de Broglie wave length λ_{dB};

(iv) the atom is well localized on the z axis, so that it probes only a small region in the transverse x, y plane;

(v) on the z axis there is no transverse gradient of the privileged eigenmode of the resonator, so that the atom is not exposed to a net force in the transverse direction.[f]

The atom-photon coupling vanishes outside the resonator, that is: for $z < 0$ and $z > L_z$, where L_z is the longitudinal extension of the cavity. The conditions (ii)-(v) also guarantee that no significant transverse motion happens. The Hamiltonian is then identical with that of Ref. 9 except that the transverse spatial dependence of

[e] We trust that this new use of the symbol a is not confused with the previous one, where a stood in for the slit width, as in Eq. 18, for example.

[f] Note that this condition is not obeyed by the Auckland mode function of Fig. 2a.

the mode function is now taken into account additionally.[g] As in Ref. 9 we introduce the coupling operator

$$\gamma \equiv |A\rangle\, a\, \langle B| + |B\rangle\, a^{\dagger} \langle A| \tag{23}$$

whose property

$$\gamma^2 = a^{\dagger} a + |A\rangle\,\langle A| \tag{24}$$

allows us to write the Hamilton operator compactly as,

$$H = \frac{1}{2m} p_z^2 + \hbar\omega\gamma^2 - \hbar g_0 \chi(z)\Omega(x,y)\gamma\,. \tag{25}$$

Here the microwave mode function is $\chi(z)\Omega(x,y)$ with the longitudinal dependence given by

$$\chi(z) = \begin{cases} \frac{\pi}{2}\sin(\pi z/L_z) & \text{for } 0 < z < L_z\,, \\ 0 & \text{elsewhere}\,, \end{cases} \tag{26}$$

and an example for the transverse mode function $\Omega(x,y)$ is provided by the Garching mode function of Eq. 10. If the joint atom–photon system is in an eigenstate of γ with eigenvalue γ', the atomic z motion is that of a one-dimensional particle exposed to a potential (energy) $-\hbar g_0\chi(z)\Omega(x,y)\gamma'$ whereby the x and y values are inert. Depending on the sign of γ' this potential may be attractive or repulsive which has the effect of either speeding up or slowing down the atom. Thus, when γ eigenvalues of both signs are present, one part of the total wave function will propagate in an attractive potential, the other in a repulsive one. The difference in the resulting dynamical behavior effects the resonant emission of the microwave photon.

The total wave function that we *would* have behind the cavity without the interaction is

$$\Psi_0(\vec{r}) = \psi_0(\vec{r})|A, \phi^i\rangle \tag{27}$$

where the ket stands for the internal state of the atom and the photon state. We think of it as a superposition of γ eigenstates,

$$|A, \phi^i\rangle = \sum_{\gamma'} |\gamma'\rangle\langle\gamma'|A, \phi^i\rangle \tag{28}$$

so that Eq. 27 is a multicomponent wave function in this respect.

With the interaction the atomic velocity is changed inside the resonator according to

$$\begin{aligned} v(\vec{r}) &= \left[v_0^2 + \frac{2}{m}\hbar g_0 \chi(z)\Omega(x,y)\gamma\right]^{1/2} \\ &= v_0 + \frac{\lambda_{\rm dB}}{2\pi} g_0 \chi(z)\Omega(x,y)\gamma \equiv v_0 + \Delta v \end{aligned} \tag{29}$$

[g] Note that in Ref. 9 the longitudinal motion is along the y axis, not along the z axis.

where $v_0 = 2\pi\hbar/(m\lambda_{dB})$ is the velocity outside the cavity. The second equality holds because of condition (i).[h] Note that the velocity change Δv depends on γ and in particular on the sign of the respective eigenvalue; this is the speeding up or slowing down mentioned above. The accumulated displacement equals

$$\Delta z = \int \frac{dz}{v_0}\, \Delta v = \frac{\lambda_{dB}}{2\pi}\, \beta\, \Omega(x,y)\gamma\ , \tag{30}$$

where $\beta = g_0 L_z/v_0 = g_0\tau$. In summary, the interaction alters $\Psi(\vec{r})$ of Eq. 27 according to

$$\Psi_0(x,y,z) \rightarrow \Psi(x,y,z) = \Psi_0(x,y,z-\Delta z) = \psi_0(x,y,z-\Delta z)|\mathrm{A},\phi^{\mathrm{i}}\rangle\ . \tag{31}$$

In view of condition (iii) the z dependence in $\psi_0(\vec{r})$ is a plane wave $\exp(2\pi i z/\lambda_{dB})$ with a very broad envelope, so that

$$\begin{aligned} \psi_0(x,y,z-\Delta z) &= \psi_0(x,y,z)\exp(-2\pi i\Delta z/\lambda_{dB}) \\ &= \psi_0(\vec{r})\exp\big(-i\beta\Omega(x,y)\gamma\big)\,. \end{aligned} \tag{32}$$

The net effect is a γ dependent phase factor. In conjunction with Eq. 28 we thus arrive at

$$\begin{aligned} \Psi(\vec{r}) &= \psi_0(\vec{r})\exp\big(-i\beta\Omega(x,y)\gamma\big)|\mathrm{A},\phi^{\mathrm{i}}\rangle \\ &= \psi_0(\vec{r})\sum_{\gamma'}|\gamma'\rangle\exp\big(-i\beta\Omega(x,y)\gamma'\big)\langle\gamma'|\mathrm{A},\phi^{\mathrm{i}}\rangle\,. \end{aligned} \tag{33}$$

We are now prepared to face two extreme cases: empty resonator initially and classical coherent field. If the resonator is initially empty we have

$$|\mathrm{A},\phi^{\mathrm{i}}\rangle = |\mathrm{A},0\rangle = \frac{1}{\sqrt{2}}\big(|\gamma_+\rangle + |\gamma_-\rangle\big) \tag{34}$$

with

$$|\gamma_\pm\rangle = \frac{1}{\sqrt{2}}\big(|\mathrm{A},0\rangle \pm |\mathrm{B},1\rangle\big) \tag{35}$$

which are the γ eigenstates with eigenvalues $\gamma' = \pm 1$,

$$\gamma|\gamma_\pm\rangle = \pm|\gamma_\pm\rangle\ . \tag{36}$$

For $\beta = \pi/2$, as in Eq. 14, Eq. 30 states that the two γ components are thus displaced, on the z axis, by a quarter of the de Broglie wave length λ_{dB} in either direction, the relative displacement being $\lambda_{dB}/2$. Here, $\langle\gamma_\pm|\mathrm{A},0\rangle = 1/\sqrt{2}$, so that Eq. 33 implies

$$\begin{aligned} \Psi(\vec{r}) &= \frac{1}{\sqrt{2}}\psi_0(\vec{r})\Big[|\gamma_+\rangle\exp\big(-i\beta\Omega(x,y)\big) + |\gamma_-\rangle\exp\big(i\beta\Omega(x,y)\big)\Big] \\ &= \psi_0(\vec{r})\Big[|\mathrm{A},0\rangle\cos\big(\beta\Omega(x,y)\big) - i|\mathrm{B},1\rangle\sin\big(\beta\Omega(x,y)\big)\Big]\,. \end{aligned} \tag{37}$$

[h] The situation is dramatically different for extremely slow atoms. These can be reflected from the cavity, as discussed in Refs. 11 and 12 for a lossless resonator. If dissipative effects are taken into account, as in Ref. 13, the reflection probability is, however, reduced to a few percent at most.

This is the basic ingredient in writing Eq. 5.

If, in contrast, one has a classical coherent state in the resonator initially, then one can replace the photon operators by their respective eigenvalues,

$$\begin{aligned} a \to \alpha &= \sqrt{n} e^{i\varphi} , \\ a^\dagger \to \alpha^* &= \sqrt{n} e^{-i\varphi} , \end{aligned} \tag{38}$$

where φ is the phase of the classical radiation and $\sqrt{n}$ is its amplitude expressed in terms of the (average) macroscopic photon number n. Consequently, we write

$$\begin{aligned} \gamma &\to \sqrt{n}\Big(|\mathrm{A}\rangle\, e^{i\varphi}\, \langle \mathrm{B}| + |\mathrm{B}\rangle\, e^{-i\varphi}\, \langle \mathrm{A}|\Big) , \\ \gamma^2 &\to n + |\mathrm{A}\rangle\, \langle \mathrm{A}| , \\ \gamma^2 |\mathrm{A}, \alpha\rangle &\to (n+1)|\mathrm{A}, \alpha\rangle , \end{aligned} \tag{39}$$

so that, using the identity

$$\exp(-i\beta\Omega\gamma) = \cos\left(\beta\Omega\sqrt{\gamma^2}\right) - i\gamma\frac{\sin\left(\beta\Omega\sqrt{\gamma^2}\right)}{\sqrt{\gamma^2}} \tag{40}$$

we find

$$\exp(-i\beta\Omega\gamma)|\mathrm{A},\alpha\rangle = |\mathrm{A},\alpha\rangle \cos\left(\beta\Omega\sqrt{n+1}\right) - ie^{-i\varphi}|\mathrm{B},\alpha\rangle \sin\left(\beta\Omega\sqrt{n+1}\right) . \tag{41}$$

Here, the total wave function

$$\Psi(\vec{r}) = \psi_0(\vec{r}) \left[|\mathrm{A}\rangle \cos\left(\beta\Omega(x,y)\sqrt{n+1}\right) - ie^{-i\varphi}|\mathrm{B}\rangle \sin\left(\beta\Omega(x,y)\sqrt{n+1}\right)\right] |\alpha\rangle \tag{42}$$

factorizes into an atomic and a photon part. Indeed, as stated at Eq. 4 there is no entanglement present.

Appendix B: A more general argument

We claim that quantum optical which-way detectors do not alter the atomic center-of-mass wave function.

The question is this. If $\psi_1(\vec{r})$ is the atomic center-of-mass wave function after the interaction with the photon field stored in the resonator and $\psi_0(\vec{r})$ is the wave function when there is no such interaction, is ψ_1 equal to ψ_0 or are the two different? The property 'equal' refers to the physical distinguishability. Accordingly, the quantitative criterion employs the quantum mechanical overlap of ψ_1 and ψ_0,

$$|\langle\psi_1|\psi_0\rangle|^2 \equiv \frac{\left|\int(\mathrm{d}\vec{r})\psi_1^*\psi_0\right|^2}{\int(\mathrm{d}\vec{r})\psi_1^*\psi_1 \int(\mathrm{d}\vec{r})\psi_0^*\psi_0} = \begin{cases} \simeq 1 & : \text{'equal,'} \\ \ll 1 & : \text{'not equal.'} \end{cases} \tag{43}$$

We begin by combining Eqs. 33 and 40 to arrive at

$$\begin{aligned}\Psi(\vec{r}) &= \psi_0(\vec{r})\Big[\cos\left(\beta\Omega\sqrt{a^\dagger a+1}\right)|\mathrm{A},\phi^{\mathrm{i}}\rangle \\ &\quad -i\sin\left(\beta\Omega\sqrt{a^\dagger a+1}\right)a^\dagger\frac{1}{\sqrt{a^\dagger a+1}}|\mathrm{B},\phi^{\mathrm{i}}\rangle\Big]\,. \end{aligned}\tag{44}$$

The probability that no emission has occurred is, therefore, given by

$$\int(\mathrm{d}\vec{r})\;|\psi_0(\vec{r})|^2\langle\phi^{\mathrm{i}}|\cos^2\left(\beta\Omega(x,y)\sqrt{a^\dagger a+1}\right)|\phi^{\mathrm{i}}\rangle\,. \tag{45}$$

For an initial zero–photon state this simplifies to the spatial expectation value

$$\int(\mathrm{d}\vec{r})\;|\psi_0(\vec{r})|^2\cos^2\big(\beta\Omega(x,y)\big)\equiv\Big\langle\cos^2\big(\beta\Omega(x,y)\big)\Big\rangle_0\,. \tag{46}$$

It is now time to make use of the fact that $\psi_0(\vec{r})$ is limited to the immediate vicinity of the z axis; this is condition (iv). Thus $\Omega(x,y)$ needs only to be known for small values of x and y. *On* the z axis, we normalize Ω to unity. *Near* the z axis $\Omega(x,y)$ is of the form

$$\Omega(x,y)=1-\frac{1}{2}(\pi x/L_x)^2-\frac{1}{2}(\pi y/L_y)^2 \tag{47}$$

with no linear terms as implied by condition (v). The Helmholtz equation obeyed by the complete mode function $\sim\sin(\pi z/L_z)\Omega(x,y)$ relates the constants L_x and L_y to ω and L_z through

$$\left(\frac{\omega/\pi}{c}\right)^2=\left(\frac{1}{L_z}\right)^2+\left(\frac{1}{L_x}\right)^2+\left(\frac{1}{L_y}\right)^2, \tag{48}$$

which involves the speed of light c. An example is $\Omega(x,y)=\cos(2\pi y/\lambda)$ as in Eq. 10, for which $L_x=\infty$, $L_y=\lambda/2$. In what follows, such precise knowledge of $\Omega(x,y)$ is not required, all we make use of is stated in Eqs. 47 and 48. Further details concerning the shape and size of the cavity are not revelant.

To be specific, we shall use the typical numbers of the ongoing Garching one–atom maser experiments,[i] in which the TE_{121} mode of a resonator with circular x,y cross–section is used. The transition frequency is $\omega\simeq2\pi\times$ 21.5 GHz; the cavity length is $L_z\simeq25$ mm; the Rabi frequency is $g_0\simeq2\pi\times$ 7kHz. Thus Eq. 48 supplies

$$\left(\frac{1}{L_x}\right)^2+\left(\frac{1}{L_y}\right)^2\simeq\left(\frac{1.4}{\mathrm{cm}}\right)^2. \tag{49}$$

This states that the mode function indeed changes very little across the atomic beam which has a diameter of $2s\simeq0.5$ mm.

[i]Ref. 14 is a recent review.

To obtain a small no–emission probability we obviously need $\beta = \pi/2$ in Eq. 46, see Eq. 14; then this probability is,

$$\left\langle \cos^2\left(\frac{\pi}{2}\Omega(x,y)\right)\right\rangle_0 = \frac{\pi^6}{16}\left\langle\left[(x/L_x)^2 + (y/L_y)^2\right]^2\right\rangle_0 . \tag{50}$$

The expectation values of the powers of x and y are — within factors of order unity — equal to the same powers of the beam radius s. Accordingly, with the numbers reported around Eq. 49, the no–emission probability

$$\left\langle \cos^2\left(\frac{\pi}{2}\Omega(x,y)\right)\right\rangle_0 \simeq \frac{\pi^6}{16}[0.025 \times 1.4]^4 \simeq 10^{-4} \tag{51}$$

is negligible as desired.

Since the emission happens practically with certainty, the center–of–mass wave function $\psi_1(\vec{r})$ after the interaction is given by what multiplies

$$a^\dagger \frac{1}{\sqrt{a^\dagger a + 1}}|\mathrm{B}, \phi^i\rangle = a^\dagger \frac{1}{\sqrt{a^\dagger a + 1}}|\mathrm{B}, 0\rangle = |\mathrm{B}, 1\rangle \tag{52}$$

in Eq. 44,

$$\psi_1(\vec{r}) = -i\psi_0(\vec{r}) \sin\left(\frac{\pi}{2}\Omega(x,y)\right) . \tag{53}$$

The decisive number is the overlap of Eq. 43; with Eq. 53 it reads

$$\begin{aligned} |\langle\psi_1|\psi_0\rangle|^2 &= \left\langle \sin\left(\frac{\pi}{2}\Omega(x,y)\right)\right\rangle_0^2 \Big/ \left\langle \sin^2\left(\frac{\pi}{2}\Omega(x,y)\right)\right\rangle_0 \\ &= 1 - \frac{\pi^{12}}{2^{10}}\left\{\left\langle\left[(x/L_x)^2 + (y/L_y)^2\right]^4\right\rangle_0 - \left\langle\left[(x/L_x)^2 + (y/L_y)^2\right]^2\right\rangle_0^2\right\} . \end{aligned} \tag{54}$$

The quantum mechanical variance in the curly brackets is of the order of $[0.025 \times 1.4]^8 = 2 \times 10^{-12}$ for the parameters in the Garching experiments. Accordingly, the overlap

$$|\langle\psi_1|\psi_0\rangle|^2 = 1 - 2 \times 10^{-9} \tag{55}$$

is so close to unity that ψ_0 and ψ_1 are physically equal. Indeed, the center–of–mass wave function does not carry which–way information.

As it stands, this calculation applies to the experimental situation in which the atom emits a photon with virtual certainty as in the double–slit set–up of Fig. 1, and also in the Stern–Gerlach interferometer[1] or the quantum beats experiment.[2] In other experiments, such as that exploiting Ramsey fringes,[4] the interaction with the quantized photon field puts the atom–photon system into a coherent superposition of $|\mathrm{A}, 0\rangle$ and $|\mathrm{B}, 1\rangle$. This is achieved by having $\beta = \pi/4$ rather than $\beta = \pi/2$. The decisive overlap is then between the exact final state

$$\Psi = \psi_0(\vec{r})\left[|\mathrm{A}, 0\rangle \cos\left(\frac{\pi}{4}\Omega(x,y)\right) - |\mathrm{B}, 1\rangle\, i \sin\left(\frac{\pi}{4}\Omega(x,y)\right)\right] , \tag{56}$$

in which the center-of-mass motion appears to be entangled with the other degrees of freedom, and the approximate one

$$\Psi' = \psi_0(\vec{r})\Big[|\mathrm{A},0\rangle\cos(\pi/4) - |\mathrm{B},1\rangle\, i\sin(\pi/4)\Big]\,, \tag{57}$$

in which no such entanglement is present. We find for this overlap

$$\begin{aligned}\frac{\left|\int(\mathrm{d}\vec{r})\Psi'^{\dagger}\Psi\right|^2}{\int(\mathrm{d}\vec{r})\Psi'^{\dagger}\Psi'\int(\mathrm{d}\vec{r})\Psi^{\dagger}\Psi} &= 1-\frac{\pi^6}{64}\left\langle\left[(x/L_x)^2+(y/L_y)^2\right]^2\right\rangle_0 \\ &\simeq 1-\frac{\pi^6}{64}[0.025\times 1.4]^4 = 1-2\times 10^{-5}\,;\end{aligned} \tag{58}$$

so here, too, the center-of-mass motion remains essentially unaffected.

References

1. M. O. Scully, B.-G. Englert, and J. Schwinger, *Phys. Rev. A* **40**, 1775 (1989).
2. M. O. Scully and H. Walther, *Phys. Rev. A* **39**, 5229 (1989).
3. M. O. Scully, B.-G. Englert, and H. Walther, *Nature* **351**, 111 (1991).
4. B.-G. Englert, H. Walther, and M. O. Scully, *Appl. Phys. B* **54**, 366 (1992).
5. B.-G. Englert, M. O. Scully, G. Süssmann, and H. Walther, *Z. Naturf.* **47a**, 1175 (1992).
6. See N. Bohr in *Albert Einstein: Philosopher–Scientist* (ed. P. A. Schilpp) pp. 200-241 (Library of Living Philosophers, Evanston, 1949); reprinted in *Quantum Theory and Measurement* (eds. J. A. Wheeler and W. H. Zurek) pp. 4-49 (Princeton University Press, 1983).
7. R. Feynman, R. Leighton, and M. Sands, *The Feynman Lectures on Physics* Vol. III (Addison Wesley, Reading, 1965).
8. P. Storey, S. Tan, M. Collett, and D. Walls, *Path Detection and the Uncertainty Principle* (preprint, University of Auckland, 1993; presumably submitted to *Nature*).
9. B.-G. Englert, J. Schwinger, and M. O. Scully, in *New Frontiers in Quantum Electrodynamics and Quantum Optics* (ed. A. O. Barut) (Plenum, New York, 1990).
10. W. K. Wootters and W. H. Zurek, *Phys. Rev.* **D19**, 473 (1979).
11. S. Haroche, M. Brune, and J. M. Raimond, *Europhys. Lett.* **14**, 19 (1991).
12. B.-G. Englert, J. Schwinger, A. O. Barut, and M. O. Scully, *Europhys. Lett.* **14**, 25 (1991).
13. M. Battocletti, *Reflexion langsamer Atome am gedämpften Mikromaser* (unpublished Diploma thesis, University of Munich, 1993).
14. H. Walther, *Phys. Rep.* **219**, 263 (1992).

PHYSICAL ORIGINS OF LOSS OF INTERFERENCE IN *WELCHER WEG* DETECTION

PIPPA STOREY, SZE TAN, MATTHEW COLLETT and DANIEL WALLS
Department of Physics, University of Auckland,
Private Bag 92019, Auckland, New Zealand

ABSTRACT

The standard examples of path detectors placed in a two-slit quantum interference device clearly involve the transfer of momentum to the particle whose path is being detected, in quantities sufficient to account for the loss of interference fringes that results from the path detection. A recently-proposed atomic path detector using micromaser cavities has been claimed not to involve any transfer of momentum between detector and particle; the fact that the interference fringes are nevertheless destroyed by the measurement suggests that the principle of complementarity must be accepted as an independent component of quantum mechanics, not a mere consequence of the uncertainty principle. We present a calculation showing that *any* efficient path detection scheme must have the possibility of transferring at least the amount of momentum required by a naive interpretation of the uncertainty principle, and hence sufficient to destroy the interference fringes. We illustrate this by showing that there is in fact such a momentum transfer in the micromaser path detector. The enforcement of the principle of complementarity in a two-slit interferometer is thus a direct consequence of the exchange of momentum between particle and detector, in an amount satisfying the requirements of the uncertainty principle.

1. Introduction

Young's double slit experiment with light has now also been performed for matter waves. Electrons, neutrons and even atoms have been shown to exhibit wavelike properties. In accordance with the wave particle duality of matter, any detector which localises the particle sufficiently to enable a determination of which path it took will wash out the interference fringes. This is an example of the principle of complementarity formulated by Niels Bohr. Such a description relates the results of the double-slit experiment in a very elegant way to a general theme of nature, but makes no reference to the physical mechanism by which the interference is lost.

The best known examples of *welcher Weg* (which-path) detectors include Einstein's recoiling slit[1] and Feynman's light scattering arrangement[2]. In these schemes and others[3] it can be shown that complementarity is enforced by means of momentum transfer in accordance with Heisenberg's uncertainty principle. This has prompted the question as to whether all *welcher Weg* schemes are subject to the position-momentum uncertainty relation. Scully, Englert and Walther[4] have responded in the negative, supporting their view with an ingenious scheme which they claim is a counterexample. On the other hand, it has been argued[5, 6] that this scheme is in principle no different from the Einstein recoiling slit: in each case there is a random phase correlated with a variable that is conjugate by an uncertainty relation to the one measured. In this paper, we shall show explicitly that the physical origin of this randomisation is momentum transfer.

We first analyse the general case of a detector which determines the path taken by a particle through a fixed double slit. We show that the loss of interference in the presence of such a detector is due to momentum exchanges between the particle and the detector in the transverse direction (that is, perpendicular to the primary direction of the beam's propagation). This transverse momentum transfer accounts for the loss of interference on a distant screen in the following way. In the far-field the double slit looks like a point object, and hence the position at which a particle strikes the screen indicates the transverse momentum that the particle had. The position distribution detected on the screen is therefore identical to the transverse momentum distribution of the beam, and may be altered if momentum is exchanged in the transverse direction. Our analysis shows that the amount of momentum transferred to the particle by a perfectly efficient detector (capable of resolving the path unambiguously) is related to the slit separation through the uncertainty principle. If only less momentum than this is ever transferred, the visibility cannot be reduced to zero, and the path detector cannot be perfectly efficient.

We then illustrate this result by showing that momentum transfers do occur in the scheme of Scully, Englert and Walther[4], although their source is rather subtle. Further insight into the situation is obtained by relating the transverse momentum transfer to longitudinal displacements of the wavefunction.

2. The General Argument

Consider an arbitrary *welcher Weg* detector which determines the path taken by a particle through a fixed double slit. The particle's position immediately after its passage through the double slit is encoded in the detector state $|D_x\rangle$, resulting in the entangled state

$$|\Psi\rangle = \int dx\, \psi_i(x)|x\rangle \otimes |D_x\rangle. \tag{1}$$

The wavefunction $\psi_i(x)$ is prepared by the double slit and consists of two narrow peaks at the slit positions x_1 and x_2.

The path is determined after the interaction by making a measurement on the detector. The measurement process may be described mathematically by projecting the detector into a state $|O_\xi\rangle$ of an orthonormal pointer basis, leaving the particle in a state described by the wavefunction

$$\psi_f(x) = N\psi_i(x)\langle O_\xi|D_x\rangle. \tag{2}$$

Thus the detection process multiplies the initial wavefunction by a distribution $\langle O_\xi|D_x\rangle$, which we may write more simply as $O_\xi(x)$.

Comparing the initial and final atomic wavefunctions $\psi_i(x)$ and $\psi_f(x)$, we see that the process of *welcher Weg* detection has multiplied the original wavefunction by a new distribution $O_\xi(x)$. In momentum space this is equivalent to convolving the initial atomic momentum distribution $\tilde{\psi}_i(p)$ by a "momentum kick distribution" $\tilde{O}_\xi(p)$, which is the Fourier transform of $O_\xi(x)$,

$$O_\xi(x) = \frac{1}{\sqrt{2\pi\hbar}} \int dp\, \tilde{O}_\xi(p) e^{ipx/\hbar}. \tag{3}$$

$\tilde{O}_\xi(p)$ can be interpreted very naturally as the probability amplitude for transferring momentum p from the detector (namely the cavity system) to the atom, since the convolution process takes every momentum component p_i of the initial distribution $\tilde{\psi}_i(p)$ and produces new momentum components $\tilde{O}_\xi(p + p_i)$ in the final distribution $\tilde{\psi}_f(p)$.

We now suppose that for every outcome of the detector measurement the momentum transfer is limited. That is, for every ξ the momentum kick distribution $\tilde{O}_\xi(p)$ is zero outside the range $|p| < p_m$, where p_m is the maximum transferred momentum. In the terminology of Fourier transform theory, the functions $O_\xi(x)$ are *band limited*. We shall show that this restriction defines a lower bound for the visibility of the total interference pattern, which is a measure of the efficiency of localisation.

The total interference pattern is the far-field distribution obtained by summing over all possible outcomes ξ of the detector measurement. Its visibility is equal to $\mathcal{V} = |\langle D_{x_1}|D_{x_2}\rangle|$, and may be regarded as a measure of the efficiency of localisation, with unity indicating no localisation, and zero signifying perfect localisation efficiency. This interpretation can be understood as follows. If $\mathcal{V} = 1$ the detector states D_{x_1} and D_{x_2} are equal (up to some phase factor) and there is no way to distinguish the path of the particle. If $0 < \mathcal{V} < 1$ the detector will provide an inefficient position measurement since, regardless of the pointer basis, there will always be at least one possible outcome ξ for which both $\langle O_\xi|D_{x_1}\rangle$ and $\langle O_\xi|D_{x_2}\rangle$ are nonzero, leaving the path ambiguous. If the visibility is zero then there will be at least one pointer basis for which every possible outcome determines the path of the particle unambiguously.

We now define the function $v(x) = \langle D_{x_1}|D_x\rangle$, which we shall use to derive a lower bound for the visibility $\mathcal{V} = |v(x_2)|$. Noting that $v(x)$ is a linear combination of the functions $O_\xi(x)$, and is therefore band limited to the same band, we can establish the following properties of $v(x)$:

1. $v(x_1) = \langle D_{x_1}|D_{x_1}\rangle = 1$,
2. $|v(x)| = |\langle D_{x_1}|D_x\rangle| \leq 1$ for all x,
3. $\tilde{v}(p) = 0$ for $|p| \geq p_m$.

Using these conditions we can apply Bernstein's theorem[7] to obtain a maximum value for the derivative of $v(x)$ for all x,

$$|v'(x)| \leq \frac{p_m}{\hbar}. \tag{4}$$

The mean value theorem now gives us the desired lower bound on the visibility as a function of the maximum momentum transfer p_m and the slit separation d,

$$1 \geq \mathcal{V} \geq 1 - \frac{p_m d}{\hbar}. \tag{5}$$

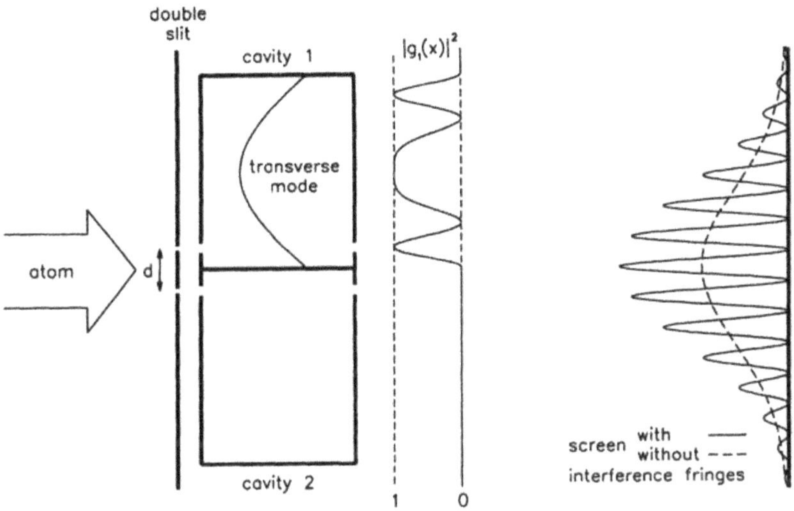

Figure 1: The micromaser cavity arrangement proposed by Scully, Englert and Walther[4] to detect the path of an atom passing through the double slit. Next to the cavities is plotted the distribution $|g_1(x)|^2$, which has a value of 1 at the upper slit and 0 at the lower slit. The chosen slit separation corresponds to a value for β of $3\pi/2$ through Eq. 11. The screen depicted on the right detects the far-field atomic position distribution (identical to the atomic momentum distribution).

Thus for a perfectly efficient *welcher Weg* detector (for which $\mathcal{V} = 0$), the maximum transferred momentum cannot be less than that required by the uncertainty principle for the given slit separation:

$$p_m d \geq \hbar. \tag{6}$$

3. The Micromaser Path Detector

The *welcher Weg* detector suggested by Scully, Englert and Walther[4] consists of two empty micromaser cavities, placed one at each slit of a double-slit arrangement. The interaction time and atom-field coupling are chosen so that an excited atom passing through the double slit will emit a photon into one of the cavities. The cavities may be examined afterwards to determine which slit the atom went through.

We claim that in this scheme, momentum in the transverse direction is transferred to the atom, due to uncertainty in the transverse momentum of the emitted photon. This uncertainty will exist even if the cavity mode structure is chosen so that the photon momentum is primarily longitudinal, and is due to the fact that the photon is localised within the width w of the cavity. Using the uncertainty principle, we find that the minimum momentum uncertainty is of the order of of $h/2w$ where h is Planck's constant.

For a slit separation d, the momentum separation between the maxima and minima of the double-slit interference pattern is $h/2d$, so a single photon emission can account for the loss of interference if the slits are located at the centres of their respective cavities. If the width of the cavities is greater than the slit separation, a single photon does not carry sufficient momentum to destroy the interference pattern, and a more complete analysis is required, including the virtual emission and reabsorption processes.

To this end we consider two identical cavities placed behind the double slit, as shown in Fig. 1. For convenience we shall assume that the cavities are rectangular boxes. The cavity modes then consist of standing waves in each of the three directions, and we can treat the transverse modes without regard to which longitudinal mode is excited. We shall assume that the atom may emit only into the lowest order transverse cavity mode, which is a single arc of a sine wave. This is the mode with the minimum transverse momentum variance, since it has the longest wavelength. The probability for emission is calculated using the Jaynes-Cummings model[8], and depends on the value of the mode function at the atom's transverse position x (which may be assumed not to change significantly during the interaction). This atomic position must be described by a probability amplitude distribution $\psi_i(x)$, which consists of two narrow peaks located at the slit positions $\pm d/2$. The interaction produces an entangled state between the atom and cavity system

$$|\Psi\rangle = \int dx\, \psi_i(x) \left[g_1(x)\,|1_1 0_2\rangle_g + g_2(x)\,|0_1 1_2\rangle_g + e(x)\,|0_1 0_2\rangle_e \right] \otimes |x\rangle \tag{7}$$

where

$$\begin{aligned} g_1(x) &= \sin\left(\beta \sin\left|\frac{2\pi x}{\lambda}\right|\right) \text{ for } 0 \le x \le \lambda/2 \text{ and 0 elsewhere,} \\ g_2(x) &= \sin\left(\beta \sin\left|\frac{2\pi x}{\lambda}\right|\right) \text{ for } -\lambda/2 \le x \le 0 \text{ and 0 elsewhere,} \\ e(x) &= \cos\left(\beta \sin\left|\frac{2\pi x}{\lambda}\right|\right) \text{ for } -\lambda/2 \le x \le \lambda/2 \text{ and 1 elsewhere} \end{aligned} \tag{8}$$

λ is the wavelength of the transverse mode and $\beta = gt$ is the product of atom-field coupling and the interaction time. The kets specify the number of photons in cavities 1 and 2, with a subscript indicating the final state of the atom. The atom will remain in the excited state (e) if no photon was emitted, or de-excite to the ground state (g) if a photon was emitted into either cavity 1 or cavity 2. The functions $e(x)$, $g_1(x)$ and $g_2(x)$ specify the probability amplitudes for each of these processes respectively.

Suppose that a photon is observed in cavity 1 after the interaction. The resulting atomic wavefunction $\psi_f(x)$, found by projecting out the state $|1_1 0_2\rangle_g$, is

$$\psi_f(x) = N\psi_i(x)g_1(x) \tag{9}$$

where N is a normalisation factor. The probability distribution describing the momentum kicks is independent of which cavity the photon is emitted into, and may be written as

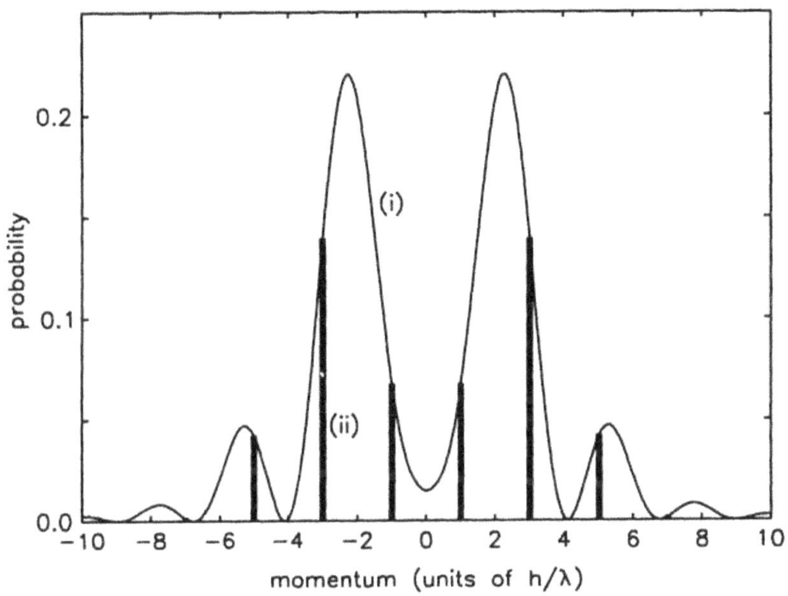

Figure 2: The probability distribution $|\tilde{g}(p)|^2$ describing the momentum kicks (i) for cavities of width $\lambda/2$. The underlying comb structure (ii) is the momentum kick distribution in the limit of very wide cavities. The slit separation is the same as was used in Fig. 1.

$|\tilde{g}(p)|^2$. The momentum variance of this distribution is

$$\mathrm{Var}(p) = \frac{1}{2}\left(\frac{h\beta}{\lambda}\right)^2 \left(\frac{1 + J_0(2\beta) + J_2(2\beta)}{1 - J_0(2\beta)}\right) . \tag{10}$$

This simple interpretation of $|\tilde{g}(p)|^2$ as the distribution of momentum kicks ignores quantum interference between kicks starting from different initial values of momentum. This is a reasonable approximation when, as here, the size of the typical momentum change is comparable to or larger than the momentum scale of structure in the initial atomic state (in this case the spacing of the interference fringes).

Given the slit separation d, the interaction time is chosen so that the parameter β satisfies the spin-flip condition

$$\sin\left(\beta \sin\left|\tfrac{\pi d}{\lambda}\right|\right) = 1 , \tag{11}$$

which ensures de-excitation of the atom. If the slit separation is $d = \lambda/2$, the slits are located at the centre of each cavity, where the value of mode function reaches its

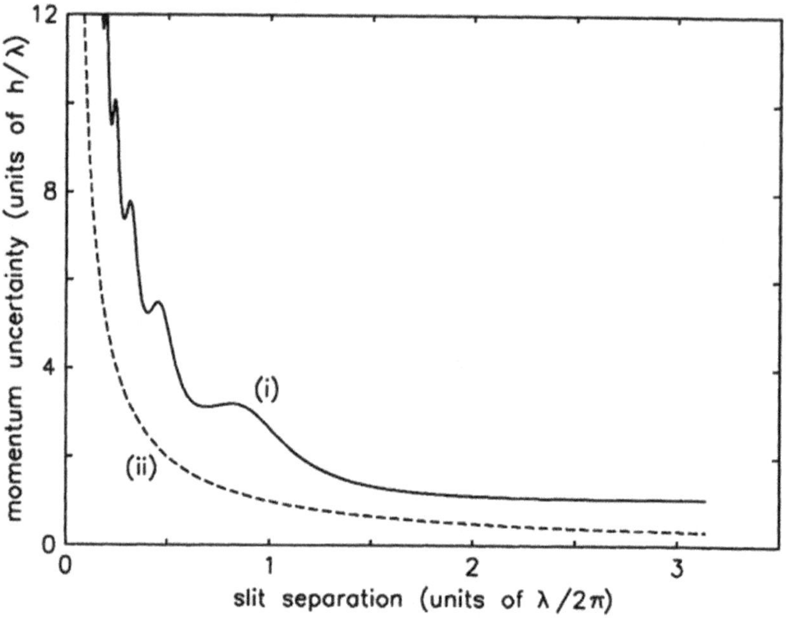

Figure 3: The uncertainty $\sqrt{\mathrm{Var}(p)}$ in the transferred momentum, plotted versus slit separation d for Scully's scheme (i). This lies above the minimum uncertainty curve $\hbar/d$ shown by the dashed line (ii).

maximum. At this point the value of β, and hence also the value of the interaction time, needed to satisfy the spin-flip condition reach a minimum. As the slit separation is reduced, the interaction time needed to satisfy the spin-flip condition increases.

Fig. 2 shows an example of the probability distribution $|\tilde{g}(p)|^2$ for a particular value of the slit separation. This distribution is smooth because it was calculated for cavities of width $\lambda/2$, which are so narrow that the photon momentum has a broad continuum of values. However, its underlying structure becomes evident if the cavities are extended outwards, keeping the wavelength constant, so that the transverse photon momentum becomes better defined. In the limit of very wide cavities the transverse photon momentum adopts the discrete values $\pm h/\lambda$. The spikes in the comb distribution occur at odd multiples of the photon momentum, indicating that virtual processes occur during the interaction in which the photon is emitted and then reabsorbed from the opposite direction, thereby exchanging twice the photon momentum between the atom and the cavity walls. During the time taken for a single Rabi flip, many such virtual processes can occur before the final emission leaves the atom in its ground state.

Using the relation between β and d given by Eq. 11, we can plot the uncertainty $\sqrt{\mathrm{Var}(p)}$ in the transferred momentum as a function of the slit separation (see curve (i) of Fig. 3). As the slit separation decreases from the value $d = \lambda/2$, the momentum uncertainty increases: since the atom must spend longer in the cavity in order to satisfy the spin-flip condition, it can scatter from the photon more often. In curve (i) of Fig. 3 there are clear edges as 3, 5, 7... photon exchanges each become the dominant process in turn. For every value of the slit separation, the momentum uncertainty for the micromaser scheme is greater than the value $\hbar/d$ determined by the uncertainty principle (curve (ii) of Fig. 3). This is consistent with the result of Section 2. that momentum transfers of at least $\hbar/d$ are required to resolve the path of the atom in any *welcher Weg* detection scheme.

The momentum uncertainty $\sqrt{\mathrm{Var}(p)}$ calculated in this way depends only on the wavelength of the transverse cavity mode and the value of the slit separation. For example, it does not depend on the width of the cavities, which may be extended outwards, keeping the wavelength constant, to include multiple arcs of the standing wave. Eq. 10 is however specific to the particular cavity mode structure chosen, namely that of a rectangular box cavity. It should therefore be taken as illustrative of the results of Section 2., rather than a general property of any micromaser arrangement, though of course the uncertainty relation Eq. 6 holds in all cases.

4. Longitudinal Displacements

An alternative analysis of the micromaser path detector, equivalent to the transverse momentum kick description, is in terms of longitudinal displacements of the atomic wavefunction. The displacement produced by the cavity has been calculated rigorously by Englert, Schwinger and Scully[9], but their result can be obtained using the following simple arguments.

The interaction energy between the atom and the cavity is either positive or negative depending on the atom's polarisation with respect to the cavity field. If the energy is negative, the cavity mode acts as a potential well, and the atom's longitudinal velocity during the interaction will be higher than in free space. The atom exits the cavity at the same longitudinal speed that it entered with (neglecting small corrections of the order of $\Delta v_\perp^2/v_\parallel$), but displaced forwards from the position it would have had in the absence of the cavity. If on the other hand the interaction energy is positive, the atom will travel slowly over the potential barrier formed by the cavity mode, and will be displaced backwards. The value of the displacement is determined by the spin-flip condition and, as we shall show below, turns out to be $\pm\lambda_{\mathrm{dB}}/4$ for a single Rabi flip, where λ_{dB} is the de Broglie wavelength of the atom. Although this displacement can be very small, it corresponds to a phase shift of $\mp\pi/2$ in the atomic wavefunction, which is exactly the phase shift needed to explain the changes in the interference pattern.

The principle of complementarity, as applied to the internal atomic states, prohibits us from knowing both the inversion of the atom and its polarisation. In fact the two possible polarisation states, which we may denote by $|+\rangle$ and $|-\rangle$, provide an alternative basis in which to express the state of the atom. In this basis the excited and ground states of the atom may be written as

$$
\begin{aligned}
|e\rangle_0 &= \frac{|+\rangle + |-\rangle}{\sqrt{2}} \\
|g\rangle_1 &= \frac{|+\rangle - |-\rangle}{\sqrt{2}}
\end{aligned}
\tag{12}
$$

where the subscript indicates the number of photons present. Since the atom enters the cavity in the excited state, there is an equal probability that the atom will be displaced backwards by $\lambda_{dB}/4$ (as for the $|+\rangle$ polarisation state) or forwards by $\lambda_{dB}/4$ (as for the $|-\rangle$ polarisation state). A displacement of $\lambda_{dB}/4$ corresponds to a phase shift of $-\pi/2$, which causes exactly one Rabi flip of the atom, since multiplication of the $|+\rangle$ component by $\exp(i\pi/2)$ and the $|-\rangle$ component by $\exp(-i\pi/2)$ transforms the excited state into the ground state and vice versa.

The important quantity in determining the interference pattern is the relative phase between the two peaks of the atomic wavefunction located behind the two slits. Each part of the atom interacts with a separate cavity, and since the longitudinal displacement induced on the atom by a single cavity is $\pm\lambda_{dB}/4$, the relative displacement by which the two parts of the atomic wavefunction are shifted can be 0 or $\pm\lambda_{dB}/2$. Unless some clever measurement is made on the cavities, the value of this relative displacement is unknown, and the interference pattern is washed out.

Such clever measurements were in fact discussed by Scully, Englert and Walther[4] and may be described as measurements of the relative phase between the fields in the two cavities. In describing such measurements we should not think of the photon as having been deposited in one cavity or the other, but as being shared between the two cavities with some relative phase. The principle of complementarity, in yet another guise, states that these two pictures are alternative descriptions of the system, each being equally valid until the time of measurement. At this time, the choice of picture is determined by the type of measurement made. Clearly the concept of relative phase does not make sense if it is known that one cavity contains a photon and the other a vacuum. Similarly, if a phase measurement is made, the photon number information is erased, and with it all information concerning the path of the atom. If the cavity fields are measured to be in the symmetric state, in which they have the same phase, then the two parts of the atomic wavefunction are displaced equally, and the interference pattern is unchanged. If on the other hand the fields are found to be in the antisymmetric state, that is, in antiphase, the two parts of the atomic wavefunction are displaced by $\lambda_{dB}/4$ in opposite directions, and the interference pattern is reversed.

The connection between a longitudinal displacement and a transverse momentum kick can be understood intuitively by visualising a rotation of the atomic wavefront. Since the atom's motion is primarily longitudinal, a transverse momentum kick will simply change its direction slightly. This implies a differential displacement of the atomic wavefront, the displacement being in the longitudinal direction by an amount which varies in proportion to the transverse position.

The mathematical equivalence between longitudinal displacements and transverse momentum kicks can be established by considering Eq. 3, which expands $g_1(x)$ as a sum of momentum kicks p, each with probability amplitude $\tilde{g}_1(p)$. The operator $e^{ipx/\hbar}$ can be interpreted as a transverse momentum kick operator, which transfers momentum p to the atom in the transverse direction. Alternatively, since the initial atomic state has a well-defined longitudinal momentum, it may be regarded as a longitudinal displacement operator. The initial atomic state is described by the wavefunction $\Psi(x,z) = e^{ip_z z/\hbar}\psi(x)$, where p_z is the longitudinal momentum of the atom. Acting with the operator $e^{ipx/\hbar}$ gives $e^{ip_z(z+px/p_z)/\hbar}\psi(x)$, which is displaced longitudinally relative to the original wavefunction by the amount $-px/p_z$. This is effectively the familiar phenomenon of refraction: a position-dependent change in the phase of a wave results in a change in the direction of propagation.

5. Conclusions

We have shown that the loss of interference from a double slit in the presence of a *welcher Weg* detector is physically caused by momentum kicks whose magnitude is determined by the uncertainty principle. We therefore conclude that the principle of complementarity is a consequence of the Heisenberg uncertainty relation. The source of these momentum kicks in the micromaser detector suggested by Scully, Englert and Walther[4] is the repeated emission and reabsorption of microwave photons by the atom.

This research has been supported by the Auckland University Research Committee and the New Zealand Lottery Grants Board.

1. N. Bohr in *Albert Einstein: Philosopher-Scientist*, ed. P. A. Schilpp (Library of Living Philosophers, Evaston,1949) p. 200-241.
2. R. Feynman, R. Leighton and M. Sands, *The Feynman Lectures on Physics* Vol. III (Addison Wesley, Reading, 1965).
3. Pippa Storey, Matthew Collett and Daniel Walls, *Phys. Rev.* **A47** (1993) 405.
4. M. O. Scully, B. Englert and H. Walther, *Nature* **351** (1991) 111.
5. S. M. Tan and D. F. Walls, *Phys. Rev.* **A47** (1993) 4663.
6. Rajendra Bhandari, *Phys.Rev.Lett.* **69** (1992) 3720.
7. J. Arsac, *Fourier transforms and the theory of distributions* (Englewood Cliffs, N. J. Prentice Hall, 1966).
8. E. T. Jaynes and F. W. Cummings, *Proc. IEEE* **51** (1963) 89.
9. B.-G. Englert, J. Schwinger and M. O. Scully in *New Frontiers in Quantum Electrodynamics and Quantum Optics*, ed. A. O. Barut (Plenum, New York, 1990) p. 513-519.

CORRELATED PHOTONS

COHERENCE, INDISTINGUISHABILITY, AND THE INTERPRETATION OF THE QUANTUM STATE

T.P. GRAYSON, X.Y. ZOU AND L. MANDEL

Department of Physics and Astronomy
University of Rochester
Rochester, NY 14627
U.S.A.

ABSTRACT

The relationship between coherence and indistinguishability has been studied in several quantum interference experiments with the help of the entangled photons produced by down conversions in non-linear crystals. In particular, we have measured the visibility of the interference when two signal photons and two idler photons come together, and when the two idlers follow a common trajectory. It is found that the degree of coherence between interfering signal photons can be controlled by an attenuator or by a differential time delay inserted in the idler paths. The results are consistent with the idea that a diagonal density operator reflects what is knowable in principle rather than what is known. These conclusions are shown to be applicable to the Schrodinger cat paradox.

1. Introduction

This is a review of some recent work on one photon and two-photon interference by the University of Rochester group in which various non-classical effects are observed. The effects are most readily understandable as the manifestation of the correspondence between mutual coherence and indistinguishability of two light beams. It can be shown, at least for the simple case in which the light from two sources comes together at some point in the far-field, and when we restrict ourselves to the emission of a single photon, that the degree of mutual coherence [1]$|\gamma_{12}|$ exactly equals the degree of intrinsic indistinguishability of the two sources.[2] It therefore follows that for two coherent sources the photon paths are indistinguishable by any measurement that does not disturb the interference experiment. But the converse is equally significant; if the two sources are mutually incoherent then, in principle, there exists some measurement scheme that allows the source of each detected photon to be identified and yet does not otherwise disturb the interference experiment.

Many of the experiments reported here depend on the process of parametric down-conversion, in which pump photons interacting parametrically with a non-linear crystal having a $c^{(2)}$ susceptibility split into an entangled pair of signal and idler photons which appear simultaneously.[3,4] The time separation between the two photons can be in the sub-picosecond range if the bandwidth of the down-conversion is not restricted deliberately.

2. Two-Photon Interference Experiments

Fig. 1 shows the outline of an experiment to study interference between down-converted light beams from two non-linear crystals NL1 and NL2.[5] Both crystals are optically pumped by

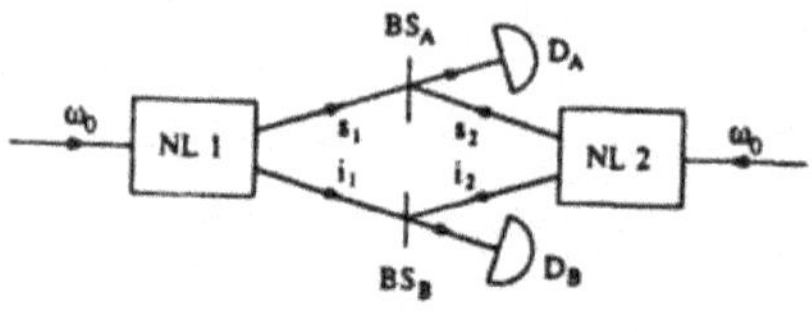

Fig 1. Outline of an experiment to measure interference between down-converted light beams from two non-linear crystals. [Reproduced from ref. 5]

light from the same argon ion laser that has been split into two mutually coherent beams. The 50%:50% beam splitter BS_A combines the s_1 and s_2 light beams from the two down-converters and lets the combination fall on detector D_A, and similarly for BS_B and D_B. We are interested in the question whether the mutual coherence between the two pump beams results in mutual coherence between the two signal beams s_1, s_2 and between the two idler beams i_1, i_2. If so then the photon counting rates registered by D_A and D_B oscillate as the optical path differences change. In fact it is found that both measured photon rates remain constant. However, if we measure and plot the two-photon coincidence rate for simultaneous photon detections by both D_A and D_B as a

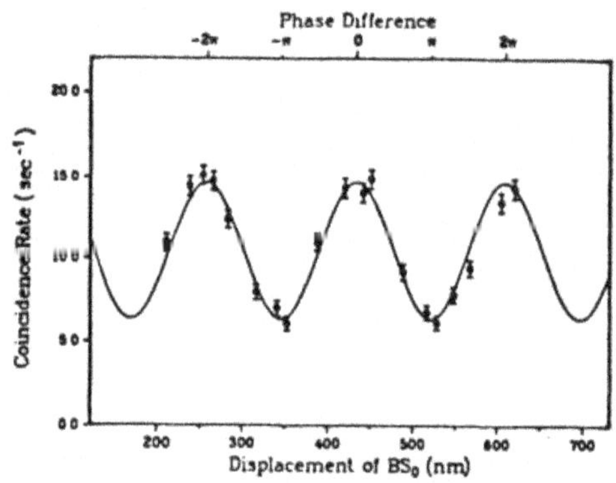

Fig 2. Plot of the measured 2-photon coincidence rate as a function of beam splitter displacement. [Reproduced from ref. 5]

function of optical path difference, then we obtain the results shown in Fig 2.[5] This time we evidently have an interference pattern.

These experimental results are readily understandable in terms of the indistinguishability of the photon paths through the interferometer. When a photon pair is registered by D_A and D_B together, there is no way to determine whether it originated in NL1 or in NL2 without disturbing the interference. However, if we are interested only in s_1, s_2 interference, we could insert two auxiliary, very efficient, photodetectors in the i_1, i_2 arms without disturbing the s_1, s_2 interference experiment. We could then make use of the fact that signal and idler photons are always emitted together to determine whether any given photon registered by D_A originated in NL1 or in NL2. This determination destroys all indistinguishability and therefore all mutual coherence between s_1 and s_2. We should note, however, that it is not necessary actually to carry out the auxiliary measurement; the mere possibility of being able to identify the source in this way is sufficient to wipe out all interference between s_1 and s_2, and similarly between i_1 and i_2.

3. One-photon Interference Experiments

Fig. 3 shows the outline of another interference experiment with two parametric down-converters. As before, both crystals NL1 and NL2 are optically pumped by light beams derived from the same laser with the help of the pump beam splitter BS_p, and down-conversion can occur

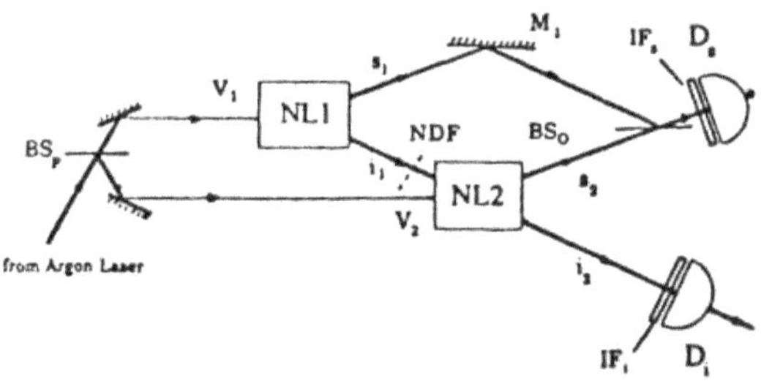

Fig 3. Outline of an experiment to measure interference between the signal beams s_1, s_2 from two parametric down-converters. [Reproduced from ref. 6]

at either crystal with the emission of s_1, i_1 or s_2, i_2 photons. This time the geometry has been altered so that the i_1 beam passes through NL2 and overlaps i_2. As before, s_1 and s_2 beams come together at beam splitter BS_0, which mixes them, and the resulting light beam falls on detector D_s. Once again we are interested to know whether s_1 and s_2 exhibit mutual coherence, so that the photon counting rate of D_s oscillates as the optical path difference is varied by translating BS_0. Whereas no one-photon interference was observed with one detector in the previous experiment, [5] this time an interference pattern is seen [6] so long as i_1 and i_2 are colinear (See Fig. 4, curve A). But when i_1 and i_2 are misaligned, or when i_1 is blocked and prevented from reaching NL2 by

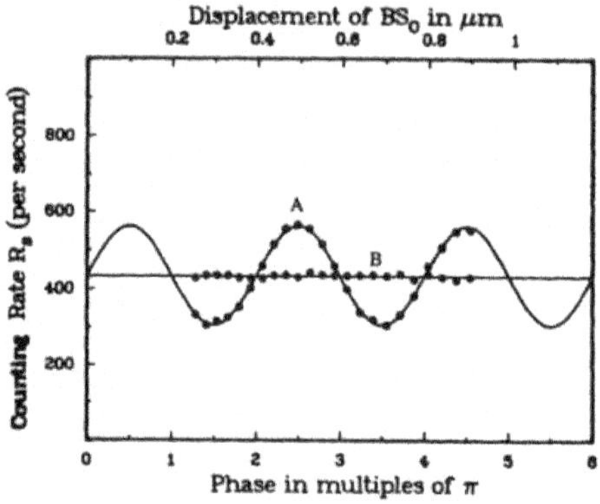

Fig 4. The observed counting rate of detector D_s as function of beam splitter BS_0 displacement; A with idlers i_1 and i_2 aligned; B with idler i_1 blocked. [Reproduced from ref. 6]

insertion of a beam stop, the interference disappears[6] (See Fig. 4, curve B). The average rate of photon counting is, however, the same in both cases, so that the down-conversions are evidently spontaneous. If, instead of blocking i_1, we place an attenuator or beam splitter of complex amplitude transmissivity T between NL1 and NL2, we find that the visibility of the interference pattern registered by D_s is proportional to $|T|$. This provides us with a new experimental technique for controlling the degree of mutual coherence of two light beams without affecting their intensities.

These observations are not understandable classically, because, if the light emission from NL2 is spontaneous, then it should not be influenced by the arrival of an i_1 photon from NL1. Yet evidently i_1 induces coherence between s_1 and s_2 in some sense, although without inducing any emission from NL2. Once again we find that the observations are most readily understood in terms of the potential indistinguishability of the s_1, s_2 paths. Let us suppose that an auxiliary detector D_i of near 100% efficiency is inserted in the idler beam at the same distance from NL2 as D_s, as shown in Fig 3. This new detector does not disturb the s_1, s_2 beams and their mutual coherence, but when i_1 is blocked it allows the source of each detected signal photon to be identified. For if a photon detection registered by D_s is accompanied by a simultaneous detection by D_i, with i_1 blocked, then the photons must have come from NL2. But if a detection by D_s is not simultaneously accompanied by a detection by D_i, with i_1 blocked, then that signal photon must have originated in NL1. The source of each detected signal photon is therefore identifiable, and this wipes out all indistinguishability and therefore all interference. Finally, we note, as before, that it is not necessary to carry out this auxiliary measurement; the mere possibility of being able to perform it without disturbing the interference experiment is sufficient to destroy the interference. When i_1 is blocked the density operator of the signal photons s_1, s_2 is diagonal in the photon numbers $|n_1\rangle_{s_1}, |n_2\rangle_{s_2}$ [6], and this state reflects not only what is known but what is knowable in principle. This emphasizes that what is observed depends on the state of the physical world rather than on human consciousness. We stress also that the disappearance of the interference pattern when i_1 is blocked is not the result of a large uncontrollable disturbance acting on the s_1, s_2 beams. Actually, the disturbance created by blocking i_1 is of a rather subtle kind; it affects only the possibility of obtaining path information.

It is well known that the visibility of an interference pattern can be strongly affected not only by attenuation of one of the light beams but also by insertion of a differential time delay. If the delay T is longer that the coherence time T_c of the interfering light, then interference effects exhibited by the total light intensity generally disappear. We have observed a similar phenomenon when the differential delay T is inserted not into the paths of the two interfering s_1, s_2 light beams, but in the idler i_1 path between the two non-linear crystals NL1 and NL2 in Fig 3.[7] Although the delay then appears to have no direct influence on the observed s_1, s_2 interference, it actually behaves much the same as if it were in one of the interferometer arms. For example, when T exceeds T_c the i_1, i_2 photons become distinguishable through the delay, and this means that the source of the detected signal photon becomes identifiable also. The interference is therefore destroyed. Fig. 5 shows the observed visibility V of the interference as a function of the delay T.[7]

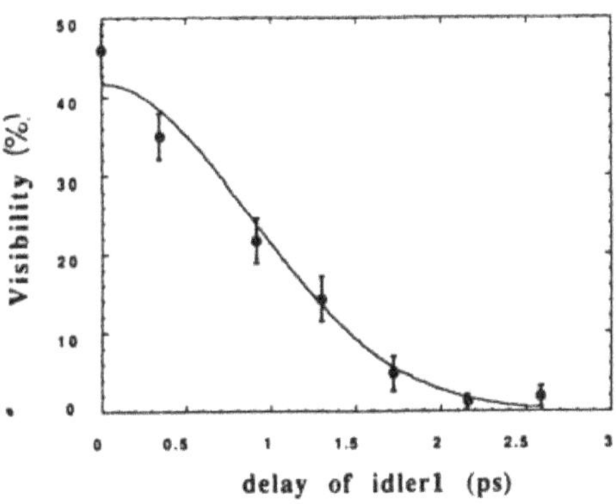

Fig 5. Variation of the observed visibility of the interference as function of the inserted time delay. The full curve is based on Eq. (1). [Reproducted from ref. 7]

The full curve is proportioned to the normalized overlap integral of the idler wave amplitudes $z_1(t), z_2(t)$ (with $|z_1(t)| = |z_2(t)| = |z(t)|$)

$$D_i \equiv \int_{-\infty}^{\infty} d\tau\, z_1^*(\tau) z_2(\tau + T) / \int_{-\infty}^{\infty} d\tau |z(\tau)|^2 \,, \qquad (1)$$

which measures the degree of indistinguishability of the two idlers. This appears to account for the variation of the observed visibility.

Although it may appear from Fig 5 that all interference affects disappear when $T >> T_c$, this is true only with respect to measurements of the light intensity. It can be shown that interference effects are present even when $T >> T_c$, but they show up in the spectrum of the interfering light. [8,9] We have therefore modified the experiment in Fig. 3 by inserting a scanning Fabry-Perot interferometer just before the detector D_s. The results of the interference experiment are shown in Fig 6, and they clearly exhibit the expected spectral modulation when the differential delay $T \sim 5\text{psec} = 5T_c$. The full curve is the theoretically predicted form , while the broken curve gives the

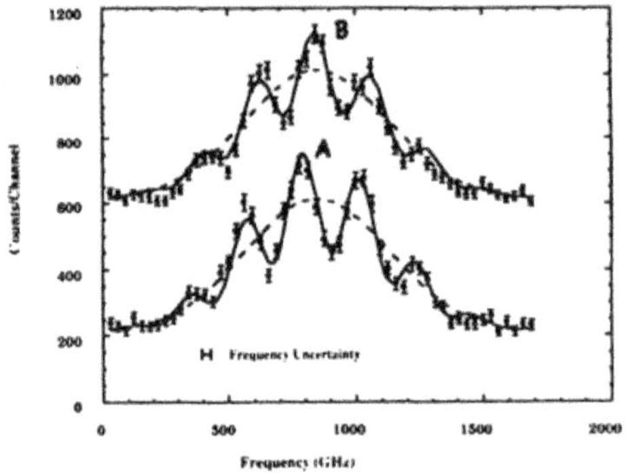

Fig 6. Variation of the observed spectral distribution as function of frequency when the time delay $T \sim 3T_c$ is inserted in s_1 (A) or in i_1(B). The broken curve shows the original, unmodulated spectrum. [Reproduced from ref. 8]

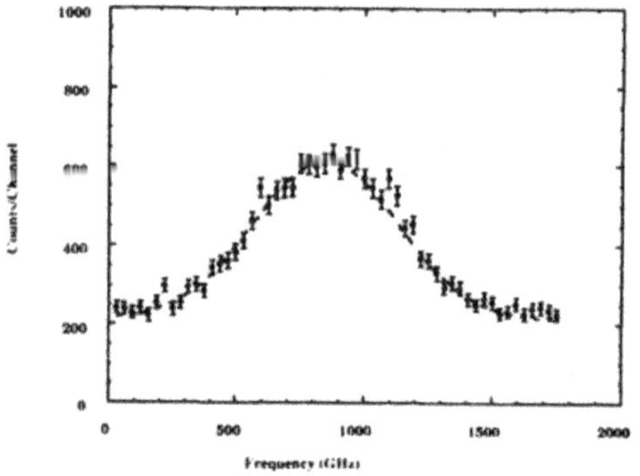

Fig 7. Variation of the observed spectral distribution as function of frequency when i_1 is blocked. [Reproduced from ref. 8]

original, unmodulated spectrum.[8] Fig 7 gives the observed spectral distribution when the idler i_1 beam is blocked. Once again we see that all interference effects then disappear and the original spectrum is recovered, because of the potential distinguishability of the photon path through the interferometer.

4. Application to the Schrödinger Cat Paradox

In this well-known paradox an enclosure contains a cat that is killed by some mechanism when a two-state quantum system makes a transition from one state to the other. One is then led to conclude that when the quantum system is in a superposition state, the cat is also in a superposition state of being alive and dead. Although most physicists have long been reconciled to accept the reality of quantum superposition states, they tend to baulk at the idea of a superposition state for a macroscopic object like a cat.

We have learned from experiments of the type shown in Fig. 3 that the concept of indistinguishability is central to the interference phenomenon. Whenever second order interference effects exist then the photon paths are intrinsically indistinguishability if the interference experiment is not disturbed. But the converse is also true. When there is no interference, and the density operator is diagonal in terms of photon numbers, then there exists some auxiliary observation that makes the photon paths distinguishable without disturbing the interference experiment.

It therefore follows that, whenever one out of a possible set of quantum states can be identified in principle, without the introduction of an uncontrollable disturbance, then the density operator is diagonal in that set of states. Then superposition states and interference effects disappear. If we apply this argument to the Schrödinger cat we are led to the conclusion that if it is possible to determine whether the cat is alive or dead without disturbing it, perhaps from the expression on its face, then there can be no superposition of live and dead states. The feline ghost is therefore exorcised by an operational argument, rather than by explicit consideration of the mechanism by which off-diagonal matrix elements disappear.

5. Acknowledgements

This work was supported by the National Science Foundation and by the U.S. Office of Naval Research.

6. References

1. See for example M. Born and E. Wolf Principles of Optics, 6th Ed, (Pergamon, Oxford, 1980)
2. L. Mandel, Opt. Lett. **16**, 1882 (1991).
3. B. Ya. Zeldovich and D.N. Klyshko, J.E.T.P. Lett. **9**, 40 (1969).
4. D.C. Burnham and D.L. Weinberg, Phys. Rev. Lett. **25**, 84 (1970).
5. Z.Y. Ou, L.J. Wang, X.Y. Zou and L. Mandel, Phys. Rev. A 41, 566 (1990).
6. X.Y. Zou, L.J. Wang and L. Mandel, Phys. Rev. Lett. **67**, 318 (1991).
7. X.Y. Zou, T.P. Grayson, G.A. Barbosa and L. Mandel, Phys. Rev. A; **47**, 2293 (1993).
8. X.Y. Zou, T.P. Grayson and L. Mandel, Phys. Rev. Lett. **69**, 3041(1992).
9. L. Mandel, J. Opt. Soc. Am. **52**, 1335 (1962).

INTERFERENCE AND INDISTINGUISHABILITY OF PHOTONS.

J G RARITY AND P R TAPSTER

DRA Malvern
St Andrews Rd, Malvern
Worcs., UK, WR14 3PS

ABSTRACT

We discuss recent experiments featuring one- and two-photon interference at extended distances noting applications to secure key sharing.

1 One-photon interference

1.1 Introduction

In 1909 G I Taylor[1] set up a simple Young's slit interference apparatus using a sensitive photographic plate as a screen. He attenuated his light source until the probability of more than one photon being between source and screen at any one time was extremely low. After waiting a period of some weeks he developed the photographic plate and discovered interference fringes. This was the first experimental evidence supporting Dirac's famous statement 'Each photon then interferes only with itself'.[2] Of course this experiment could in principle be explained by semi-classical means without recourse to a fully quantised field. Stronger support comes from a later experiment[3] using an atomic cascade source to create two simultaneous photons. Detection of one photon of the pair allows electronic gating of the detection of its partner. In the experiment this gated 'one-photon' state passed through a Mach-Zehnder interferometer and collection of many one-photon events at various interferometer settings showed the expected interference fringes. In principle this experiment could also be performed using the more recently favoured parametric down-converion sources where optical gating of 'one-photon' states has also been demonstrated.[4] In both cases resulting interference effects can only be explained by using a fully quantised electromagnetic field.

1.2 Experiments

Here we dicuss a variant of these one photon interference experiments adapted for use as a secure key sharing system with a range of 10km. The apparatus is shown schematically in figure 1. Two identical unbalanced interferometers are connected in series. A light source emitting short pulses of light is attenuated until the average

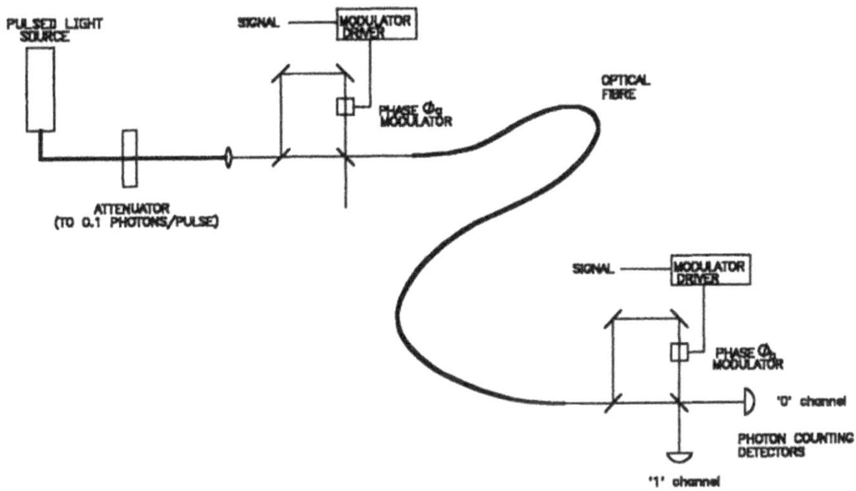

Figure 1: 10km long faint pulse interferometer using two time delay interferometers. Interference is seen between the two equal path-length routes through the system (either through the long then short path OR the short then long path). The phases can be adjusted using phase-plates ϕ_a, ϕ_b in the long arms of the interferometers.

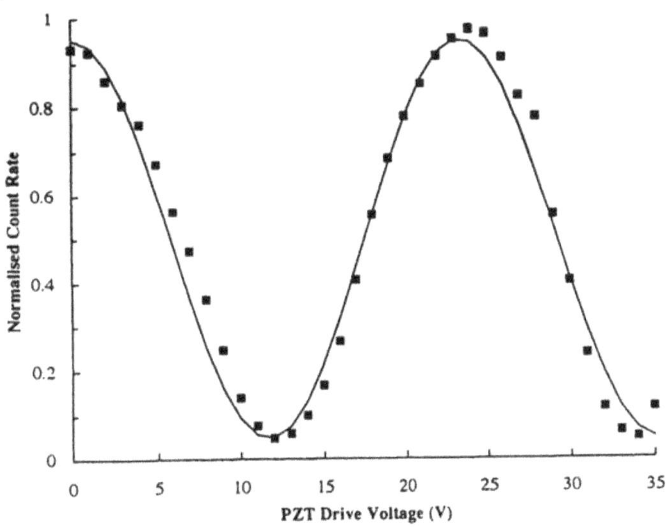

Figure 2: Singles rate as a function of phase (driven by a piezo-electric transducer) in a 10km long faint pulse interferometer (figure taken from ref. 5) .

photon number per pulse is about 0.1. The interferometer system works by separating the pulse in time in the first interferometer which has a path length difference much longer than the pulse width. The two pulses so formed propagate to the receiving interferometer which has an identical path difference. When the leading pulse takes the longer path and the trailing pulse takes the shorter path the pulses arrive simultaneously at the final beamsplitter and interference is seen. Events where the pulse takes the long (or short) path in both interferometers will not show interference but can be gated out as they will arrive later (or earlier) when compared to a timing reference originating in the source. Having gated out the non-interfering events the probability of seeing a photon in the '1' or '0' outputs for a single pulse is given by

$$P_{1/0} = \overline{m}\alpha\eta/4\,[1 \pm cos\,(\phi_a - \phi_b)] \qquad (1)$$

where ϕ_a is the transmitter phase, ϕ_b the receiver phase, α the lumped transmission, η the detection efficiency and $\overline{m}$ the mean number of photons per pulse. In the experimental realisations of this time-division interferometry we have seen interference visibilities greater than 0.95 (see figure 2) with detection rates of a few kilo-counts per second (Kcps) at ranges of 10km.[5,6] These experiments were carried out using a fibre coupled laser source and all fibre interferometers. The wavelength chosen was 1.3μm to allow use of low loss standard communications fibre for the 10km link between interferometers. Detectors were Germanium APD's operated in Geiger mode[7,8] with efficiency about 6% and dark counts of order 10Kcps.

This experiment can again be described by classical models because the source used is simply an attenuated laser source. To show truly non-classical behaviour we need to use one-photon states as created in pair photon gating experiments.[4] An alternative means of generating low n-states in future could be suitably engineered high efficiency LED's.[9,10] However the low photon number does ensure that very few pulses contain more than one photon

If we identify the outputs of the interferometer with 0's and 1's we can use this system to send a secure key.[11,12] A simple coding scheme randomly switches the transmitter phase between 0 (coding for '0') and π (coding for '1') with receiver phase set to zero. The receiver will obviously receive only a small fraction of the sent bits when $\alpha\eta$ is small but the valid bits can be selected on sharing time of arrival information of the detected photons. Absolute security can be guaranteed by randomly switching between non-orthogonal measurement bases. For this case two suitable receiver bases are $\phi_b = 0, \pi/2$. Transmitter phases can be randomly selected from $\phi_a = 0, \pi$ and $\phi_a = \pi/2, 3\pi/2$ coding '0's and '1's. Information on measurement basis is shared using conventional communication means and all non valid measurements (using the wrong receiver phase) are discarded to leave the transmitter and receiver with a near identical random bit string to be used as a key.

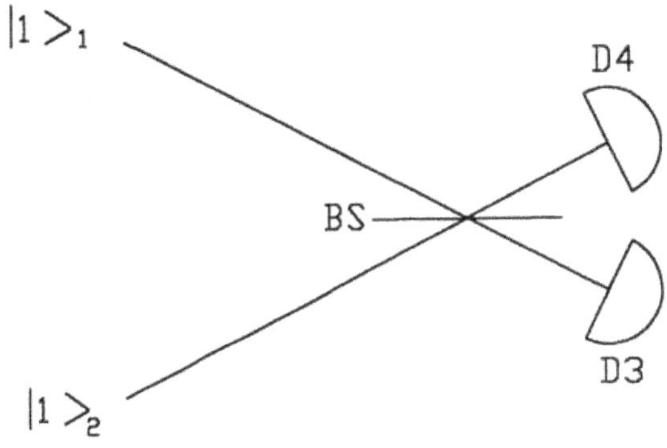

Figure 3: The interference of two one-photon states at a beamsplitter (BS). Coincidences are measured between photon detectors D3 and D4

2 Two-Photon Interference

2.1 Introduction

Non-classical two photon interference effects were first predicted for a two photon field falling on a screen.[13] A more striking example is the case where two 'one-photon' states are input into opposite ports of a beamsplitter[14,15] (see figure 3).

In a simple particle picture we expect there to be no correlation in the output taken by each photon of the pair. A quantum mechanical picture includes the one-photon probability amplitudes $|1\rangle_1, |1\rangle_2$ and the phase shift between transmission amplitude t and reflection ir. If the input photons are indistinguishable from measurements at the beamsplitter outputs the beamsplitter output state can be represented by

$$|\Psi\rangle = 2^{-1/2}\left[|t^2|1\rangle_3|1\rangle_4 - r^2|1\rangle_3|1\rangle_4 + irt|2\rangle_3 + irt|2\rangle_4\right] \tag{2}$$

We see that a destructive interference effect can occur in the probability for seeing photons simultaneously in both beamsplitter outputs.

$$P_{34} = t^4 + r^4 - 2r^2t^2 \tag{3}$$

In fact when $r^2 = t^2 = 1/2$ (a 50/50 beamsplitter) $P_{34} = 0$ and the photons always appear as **pairs** in random outputs of the beamsplitter.

Indistinguishable photon pairs can be obtained by using a parametric down-conversion crystal pumped by a short wavelength laser beam and arranging apertures and crystal tilt to satisfy symmetric phase and energy matching conditions. Pair beams can then be directed to opposite input ports of a beamsplitter allowing the effect to be experimentally demonstrated[16,17] as a reduction of the pair detection rate at the beamsplitter outputs. By scanning the path length difference from the crystal

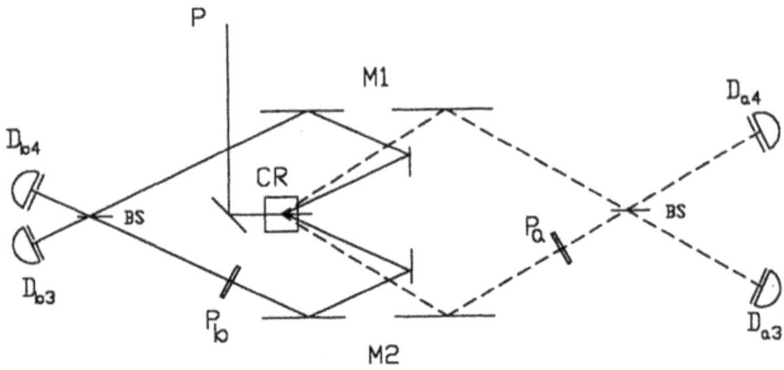

Figure 4: a. Schematic of the momentum state superposition Bell inequality. A pump beam (P) illuminates a non-linear crystal (CR). Double apertures (A) are used to select different energy photon pairs. a-colour (dashed) and b-colour photons (dashed) are emitted simultaneously EITHER into the upper a-mode and the lower b-mode OR into the lower a-mode and upper b-mode. These are refflected at mirrors M1, M2 and recombined at beamsplitters BS. The two possibile routes are indistinguishable when coincident detections are seen between b-detectors $D_{b3,b4}$ and a-detectors $D_{a3,a4}$. Phase plates P_a and P_b placed in the lower a- and b-modes can be used to alter the phase between these two routes hence producing an interference effect which redistributes the coincidences between upper and lower detectors without altering the singles rates.

to the beamsplitter inputs one can show that the pairs are coincident to within their coherence length which can be much less than 100fs. This time is several orders of magnitude less than the fastest resolving time of typical photon counting detectors.

In an extension of this experiment the fact that the pairs need not be localised objects was explored.[18] A representation of the experiment emphasising the non-locality is shown in (figure 4). We see that the interference arises because there are two indistinguishable routes for the photon pairs to propagate from the crystal to the detectors. Either the a-colour photon takes the lower and its partner the b-colour photon takes the upper route or vice-versa. This interference can be demonstrated by varying phase plates in the lower paths of either the a- or b-colour photons as shown in figure 4. The non-locality is emphasised when we look at the lower a-detector and calculate the probability of detecting partner photons in the b-detectors. Essentially the partners can be switched between upper and lower b-detectors by altering the local phase ϕ_a and in principle ϕ_a can be set after the pair has left the crystal. In the experiment the visibility of this non-local interference effect was 79% which was higher than the maximum possible using any classical model or hidden variable theory.

This experiment demonstrates a violation of Bell's inequality[19] using a superposition of momentum states. and was the first non-spin based experiment to be performed. The effect was limited to small (millimeter) spatial separations due to the geometry of the experiment. For communications applications of non-local correlations,

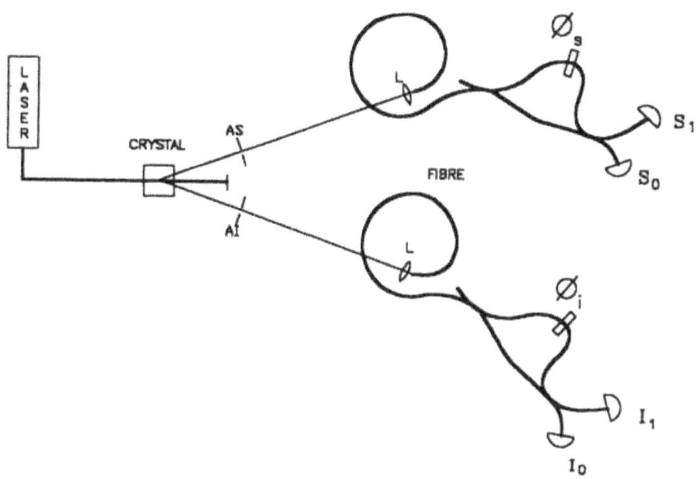

Figure 5: Schematic of the emission time superposition apparatus. Filters A_s and A_i select energy (and phase) matched photon pairs. They are launched into optical fibres and propagate through identical out of balance fibre-optic interferometers to signal $S_{0,1}$ and idler $I_{0,1}$ photon counting detectors. Coincidences between signal and idler detectors are measured as a function of phase shift ϕ_i and ϕ_s.

schemes suited to fibre optic implementation would be favoured. Experiments involving superpositions of emission times using out of balance interferometers[22–24] have recently been demonstrated in single mode fibre[25] and we discuss progress towards violation of Bell's inequality at kilometer separations in the following. In this form these strong quantum correlations can also be used in secure key sharing schemes.[26]

2.2 Recent Experiments

Time superposition interference experiments were first suggested by Franson[20] using atomic cascade pair photon sources. A schematic of a parametric downconversion based version using fibre-optic interferometers is shown in figure 5. A downconversion crystal cut for non-degenerate operation is pumped by a single frequency short wavelength laser. Signal and idler photons satisfying energy conservation are emitted in a broad band cone behind the crystal and pairs are selected by placing apertures in the cone satisfying phase matching conditions. The pair beams are launched into single mode fibres and propagate through identical fibre interferometers.

There are thus two routes from the crystal to the detector for each photon of the pair. A simplistic wavefunction for the two photons viewed from the detectors (but normalised at the crystal) assuming 50/50 beamsplitters is thus

$$\Psi_{si} = \frac{1}{4}[|1_{long}>_i |1_{long}>_s e^{i(\phi_i+\phi_s)} e^{i\Delta T(\omega_i+\omega_s)} + |1_{short}>_i |1_{short}>_s \tag{4}$$

$$+|1_{short}>_i |1_{long}>_s e^{i\phi_s} e^{i\Delta T\omega_s} + |1_{long}>_i |1_{short}>_s e^{i\phi_i} e^{i\Delta T\omega_i}]$$

where $|1>_{s,i}$ denotes a one-photon state in the signal/idler arm of the apparatus, the subscripts *long* and *short* denote propagation via the long and short paths through the interferometer and the small phase shifts $\phi_{i,s}$ incorporate any phase changes occuring on reflection while ΔT represents the mean time delay incurred on taking the long path in either interferometer. Containing a superposition of a *long, long* two-photon with a *short, short* two-photon this wavefunction directly expresses the uncertainty in time of emission of the photon pair. Coherence between the two possible emission times is ensured when the pump laser has a constant frequency ω_0 (ie a long coherence length as compared to ΔT) and when energy conservation is strictly applied in the relationship $\omega_i + \omega_s \equiv \omega_0$. Near coincident detections between *short, long* and *long, short* do not interfere because they are distinguishable. If detector time resolution is better than ΔT we can exclude these non-interfering events and evaluate the probability $P_{si}(j,k)$ of coincident detection in detectors indexed $j,k = 0,1$ with efficiencies $\eta_{s,i}$

$$P_{si}(j,k) = \frac{1}{8}\eta_s\eta_i[1 + (-1)^{j+k}V\cos(\phi_i + \phi_s + \omega_0\Delta T)]. \tag{5}$$

The interference has a maximum visibility of $V = 1$ and is clearly non-locally controlled by the phases ϕ_s, ϕ_i. In a realistic experiment the visibility is reduced from unity by dispersion effects in the fibres, changes of polarisation state in the interferometers and differences in the path length difference between the interferometers.

In an initial experimental realisation[25] these effects were partially avoided by using a single interferometer, with pairs entering both ports as shown in figure 6. A suitable delay fibre in the signal beam ensures the photons pass through the interferometer separated in time and a single phase ϕ replaces the independent phases in equation 5. The experiment uses a 406.7nm single frequency laser and a lithium iodate crystal cut to produce 813.4nm pairs in a cone of half angle 5° around the pump beam. Pair beams are selected and launched into a single mode fibre interferometer as shown and coincidences are measured at the interferometer outputs.

The coincidence rate as a function of phase shift obtained by selecting only the interfering events is shown in figure 7. The oscillation is fitted by a sinusoidal function of visibility 90.7%. Reduction from 100% visibility is due primarily to the high dispersion in the near infra-red fibres and imperfect polarisation state matching at the second coupler.

To use these effects in a key sharing system we need to demonstrate these correlations at kilometer separations. For long distance experiments it is best to work in the low loss 1.3-1.6μm wavelength region where we may expect improvement

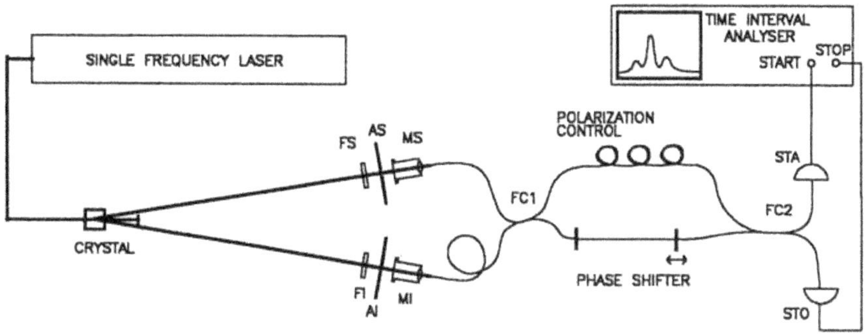

Figure 6: Schematic Fibre Apparatus: Degenerate photon pairs are selected by filters (FS, FI) and apertures (AS, AI). Microscope objectives (MS, MI; ×10, 0.2 NA) launch this light into a single mode fibre coupler (FC1) with 50/50 split ratio. A second coupler (FC2) to forms the interferometer with 25cm path length difference. A polarization controller matches polarisation states at FC2. A stepper motor driven fibre stretcher provides about 20μm adjustment of interferometer path length difference. Silicon APD photon counting detectors view the interferometer outputs and coincidences are measured in a time-interval analyser.

in visibility from lowered dispersion. However, detector technology in this wavelength region is a limiting factor. The germanium avalanche diodes described earlier[8] suffer from low efficiency (<8%) and high dark counts (>10Kcps) which coupled with launch and propagation losses can reduce the effective channel efficiency to well below 1% making background coincidence rates comparable to true coincidence rates. In present work we have adopted a hybrid approach using non-degenerate wavelengths of 820nm and 1.3μm for the seperate beams. The 1.3μm beam then passes through a 4Km coil of low loss communication fibre and through an all fibre interferometer. The 820nm light is detected locally using high efficiency silicon detectors after passing through an interferometer with an air gap path-length difference to reduce dispersion effects. The latest results from this system show interference visibilities above 87% allowing us to infer a violation of Bell's inequality at 4km spatial separation.[27] Application of this apparatus to key sharing[26] awaits improved polarisation matching as the present visibility represents too great an error rate.

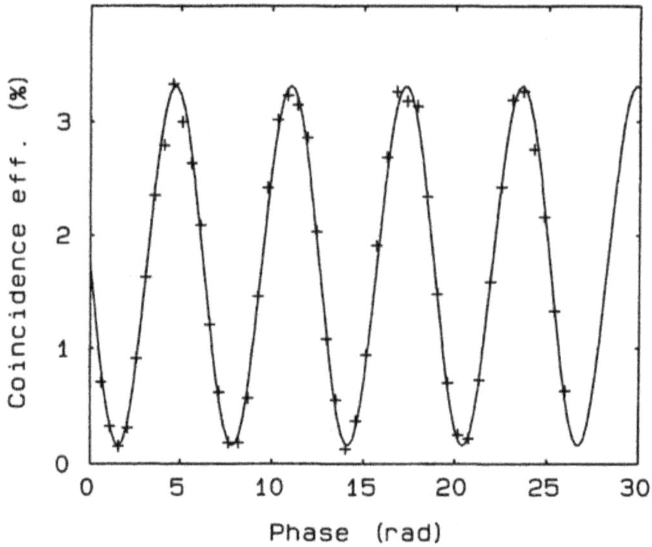

Figure 7: Coincidence rate (measured as an effective detector efficiency) as a function of phase shift ϕ. Phase shift calibrated by the sinusoidal fitting function (solid line) which has a visibility of 90.7%

3 Conclusion

In the introduction to this paper we quote the Dirac statement implying that a photon only interferes with itself. The work that has been carried out initially supported this statement but the more recent two-photon interference experiments fall outside this limited definition of interference. We have in the past extended this statement to include self interference of 2-photons and in principle this could be extended to n-photons. However the key factor in any interference phenomemena is the existence of multiple indistinguishable paths, the coherent addition of the probability amplitudes associated with each path and the subsequent evaluation of the square modulus to evaluate a detection probability. We feel that this principle provides a clearer intuitive understanding of these higher order interference effects.

Although this work initially has been an search for esoteric interference phenomena the recent discovery of possible applications has provided renewed interest in these long distance quantum correlations.

References

1. G.I.Taylor, Proc. Cambridge Philos. Soc. **15**, 114 (1909).
2. P A M Dirac, *The Principles of Quantum Mechanics*, Oxford University Press, Oxford (1930), p9.

3. P.Grangier, G.Roger and A.Aspect, Europhys. Lett. **1** 173 (1986).
4. J.G.Rarity, P.R.Tapster and E Jakeman, Opt. Commun. **62** 201 (1987).
5. P.D.Townsend, J.G.Rarity and P.R.Tapster, Electron. Lett. **29**, 634, (1993)
6. P.D.Townsend, J.G.Rarity and P.R.Tapster, Electron. Lett. **29**, 1291, (1993)
7. R.G.W.Brown, J.G.Rarity and K.D.Ridley, Applied Optics **25**, 4122 (1986)
8. P.C.M.Owens, J.G.Rarity, P.R.Tapster, D.Knight and P.D.Townsend, Applied Optics **XX**, XXX, 1994.
9. P.R.Tapster , J.G.Rarity and J.Satchell, Europhys. Lett. **4**, 293 (1993).
10. I.Schnitzer, E.Yablonovitch, C.Caneau, T.J.Gmitter and A.Scherer, Appl. Phys. Letts. **63**, 2174 (1993).
11. C H Bennett, Phys. Rev. Letts. **68**, 3121 (1992).
12. J.G.Rarity, P.R.Tapster and P.C.M.Owens, J. Modern Opt. **XX**, XXX (1994)
13. R.Ghosh and L.Mandel, Phys. Rev. Lett. **59**, 1903 (1987).
14. H. Fearn and R. Loudon, Opt. Commun. **64**, 485-490 (1987).
15. Z.Y.Ou, C.K.Hong, and L.Mandel Opt. Commun. **63**, 118-122 (1987).
16. C.K Hong, Z.Y.Ou and L.Mandel, Phys. Rev. Lett., **59**, 2044 (1987)
17. J G Rarity and P R Tapster, in 'Photons and Quantum Flutuations', E R Pike and H Walther eds., Adam Hilger, p122. (1988); J Opt Soc Am B, **6**, 1221 (1989).
18. J.G.Rarity and P.R.Tapster, Phys. Rev. Letts. **64**, 2495 (1990).
19. J.S.Bell, Physics **1**, 19, (1964)
20. J.D.Franson, Phys. Rev. Lett. **62**, 2205 (1989).
21. J.G.Rarity, P.R.Tapster, E.Jakeman, T.Larchuk, R.A.Campos, M.C.Teich and B.E.A Saleh, Phys. Rev. Letts. **65**, 1348 (1990).
22. J.G.Rarity and P.R.Tapster, Phys. Rev. A **45**, 2054 (1992).
23. J.Brendel, E.Mohler, and W.Martiennsen, Europhys. Letts. **20**, 575 (1992)
24. P.G.Kwiat, A.M.Steinberg and R.Y.Chiao, Phys. Rev. A **48**, R867 (1993).
25. J G Rarity, J Burnett, P R Tapster and R Paschotta, Europhys. Lett. **22**, 95 (1993).
26. A.K.Ekert, J.G.Rarity, P.R.Tapster and M.Palma Phys. Rev. Letts. **69**, 1293 (1992).
27. P.R.Tapster, J.G.Rarity and P.C.M.Owens, in preparation.

Acknowledgements

We would like to acknowledge the contribution of Dr P.D.Townshend to the single photon interference experiments. This work is supported in part by Esprit grant no. 6934 Quintec.

REALIZATIONS OF THE EPR PARADOX USING HIGH-INTENSITY FIELDS

J.D. FRANSON

Applied Physics Lab, The Johns Hopkins U.
Laurel, Maryland 20723 U.S.A.

ABSTRACT

The Einstein, Podolsky, Rosen paradox is usually limited to situations involving a small number of entangled particles. Several examples are given here in which these nonlocal effects are generalized to macroscopic systems containing a large number of photons. These effects include a nonlocal collapse of the wave function due to quantum phase measurements and nonlocal interferometry at high field intensities.

1. Introduction

A quantum-mechanical measurement at one location can collapse or reduce the wave function at a second, distant location, which gives rise to the Einstein, Podolsky, Rosen (EPR) paradox[1] as well as violations of Bell's inequality[2]. In most cases the initial system consists of two entangled particles whose correlations are responsible for the nonlocal effects obtained. More recent examples[3] involving N particles are not truly macroscopic in the sense that a quantum measurement on each of the individual particles is still required.

Several examples are considered here in which these nonlocal effects are generalized to macroscopic measurements involving a large number of photons. One example is that of quantum phase measurements made on two correlated but separated beams of light. A second example generalizes two-photon interferometry to nonlocal interferometry with high-intensity fields. A recent experimental demonstration of the EPR paradox based on photon timing measurements is also described.

2. Quantum Phase Measurements

Classically, a monochromatic beam of light can always be characterized by a unique but possibly unknown phase. If such a beam of light is split into a reflected and transmitted beam as illustrated in Fig. 1, then a measurement of the phase of the reflected beam will immediately determine the phase of the transmitted beam as well.

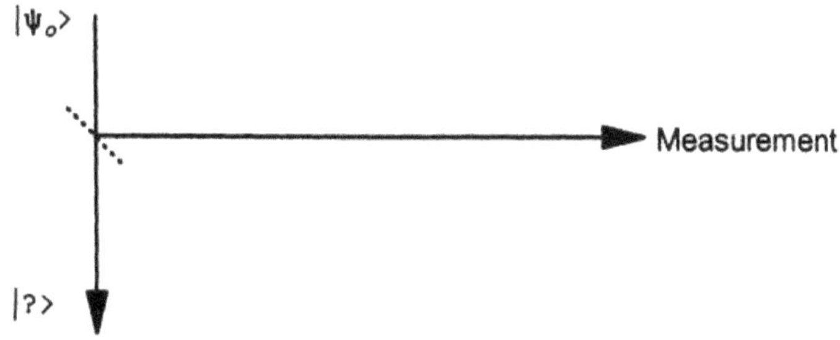

Fig. 1 Reduction of the wave function due to a measurement of the phase of the reflected beam.

The corresponding situation for a quantum state with indeterminate phase is more complicated as illustrated, for example, by a state of the form

$$|\psi_0\rangle = e^{-\alpha^*\alpha/2}\sum_{N=0}^{\infty}\frac{|\alpha|^N e^{i\phi_N}|N\rangle}{(N!)^{1/2}} \tag{1}$$

where each ϕ_N is a randomly chosen parameter and α is a complex constant. Such a state is characterized by a large number of nonequivalent phase parameters and it is not entirely obvious whether or not a measurement of the phase of the reflected beam will immediately determine the phase of the transmitted beam. If so, a nonlocal reduction of the wave function would be required, since both phases were indeterminate initially, which would provide a further[4] example of the EPR paradox for continuous variables.

The reduction of the wave function by quantum phase measurements has been investigated[5] based upon the sequence of measurements shown in Fig. 2. Classically, a small fraction of the energy in a beam of light can be split off and its phase measured by mixing it with a local oscillator in a coherent state $|\alpha\rangle$ with phase θ_i. In analogy with the classical result, the operator for the cosine of the measured phase ϕ is defined here as

$$\hat{\cos}(\phi-\theta_i) = \frac{\hat{N}_{1i}-\hat{N}_{2i}}{\langle\hat{N}_{1i}\rangle + \langle\hat{N}_{2i}\rangle} \tag{2}$$

where N_{1i} and N_{2i} are the number of photons counted in each of the detectors. This is equivalent to the operational phase measurement introduced by Noh, Fougeres, and Mandel[6] except for the denominator, which they replace with an operator expression.

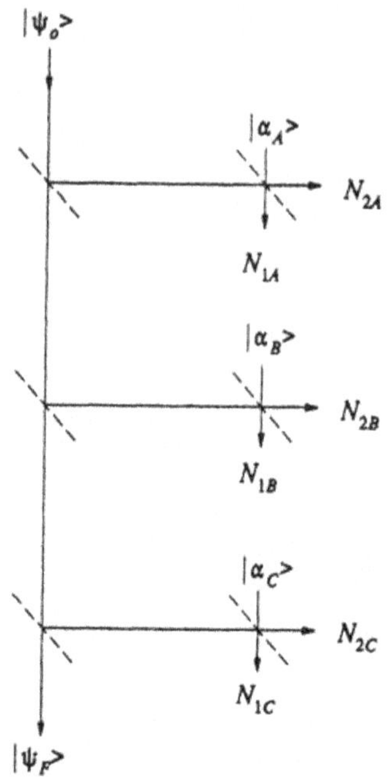

Fig. 2 A sequence of quantum phase measurements.

The consistency of two successive phase measurements can be investigated by setting the first two local oscillator phases to zero and calculating the mean squared difference Δ between those two measurements:

$$\Delta^2 = \langle[\hat{\cos}_A(\phi) - \hat{\cos}_B(\phi)]^2\rangle. \tag{3}$$

For an initial state with indeterminate phase, it can be shown[5] that

$$\Delta = \frac{(\langle\hat{N}_i\rangle + \langle\hat{N}_{\alpha A}\rangle + \langle\hat{N}_{\alpha B}\rangle)^{1/2}}{2\langle N\rangle} = \langle N\rangle^{-1/2}. \tag{4}$$

Here $\hat{N}$ is the photon number operator for the incident beam while $\hat{N}_{\alpha A}$, $\hat{N}_{\alpha B}$, and $\hat{N}_i$ are the corresponding operators for the coherent local oscillators and their input signals.

Thus a phase measurement in one beam determines the phase in the other, distant beam to within an uncertainty that vanishes for large N. This establishes the phase as an "element of reality" and provides an example of the EPR paradox for continuous, macroscopic variables. It can also be shown that successive phase measurements satisfy the usual trigonometric identities[5].

The reduction of the wave function after the first measurement (path A) can be investigated in more detail by expanding the initial state $|\psi_0\rangle$ and final state $|\psi_{TA}\rangle$ as

$$|\psi_0\rangle = \left(\sum_{N_i=0}^{\infty} \gamma_{N_i} |N_i\rangle\right) |\alpha_A\rangle |0\rangle$$

$$|\psi_{TA}\rangle = c\left(\sum_{N_{TA}=0}^{\infty} \delta_{N_{TA}} |N_{TA}\rangle\right) |N_{1A}, N_{2A}\rangle \tag{5}$$

where the γ and δ are complex coefficients. In general, the effect of a beam splitter is equivalent to a unitary transformation R:

$$|\psi_{out}\rangle = R\,|\psi_{in}\rangle. \tag{6}$$

The coefficients in the reduced wave function can then be shown to be given by

$$\delta_{N_{TA}} = \sum_{N_i = N_{TA}}^{N_{TA}+N_{1A}+N_{2A}} \gamma_{N_i} R^{T}_{(N_{\alpha A}, N_{RA}),(N_{1A}, N_{2A})} (-1)^{N_i - N_{TA}}$$
$$\times \left[\frac{N_i!}{N_{RA}!\,N_{TA}!} (\tau_s^{N_{TA}} \rho_s^{N_{RA}})\right]^{1/2} \frac{e^{-\alpha_A^{*}\alpha_A/2}\,\alpha_A^{N_{\alpha A}}}{(N_{\alpha A}!)^{1/2}} \tag{7}$$

where

$$N_{RA} = N_i - N_{TA}$$

$$N_\alpha A = N_{1A} + N_{2A} - N_{RA}. \tag{8}$$

The matrix elements $R_{(N1,N1),(n1,n2)}$ can be calculated using the recursion relations given by Campos, Saleh, and Teich[7] while ρ_s and τ_s are the reflection and transmission coefficients of the sampling beam splitter.

The reduced wave function of Eq. 7 forms the input to the next measurement and the process can then be repeated for each successive measurement. The reduction of the wave function obtained in this way is illustrated in Fig. 3 for the case in which an incident Fock (number) state containing 150 photons was subjected to both a sine and cosine measurement. The coefficient A_N is defined in the figure and corresponds to the contribution of each pair of terms to the positive frequency component of the electric field. It can be seen that the approximate result of two such measurements is to reduce the initial number state with indeterminate phase into a coherent state.

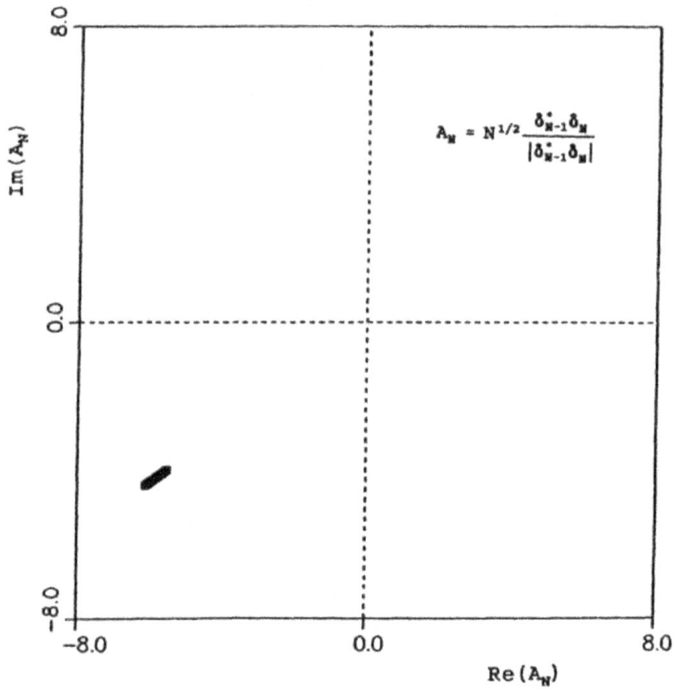

Fig. 3. Complex field amplitudes of the reduced wave function after a sine and cosine measurement.

The reduction of the wave function produces a number of nonclassical effects, such as the probability distribution shown in Fig. 4 for the measured cosine of a number state containing 50 photons. For larger values of N, the oscillations near $\phi = 0$ maintain the same amplitude but are confined to a smaller region while the total number of oscillations increases.

3. Nonlocal Interferometry

It was previously shown[8] that a pair of entangled photons incident upon two identical interferometers as shown in Fig. 5 can produce violations of Bell's inequality. Each interferometer contains a longer and a shorter path with a difference ΔT in transit times and with phase shifts θ_1 and θ_2 inserted in the longer paths. When $\theta_1 = \theta_2$ the photons are totally correlated with regard to whether they are detected in the primed or unprimed set of detectors.

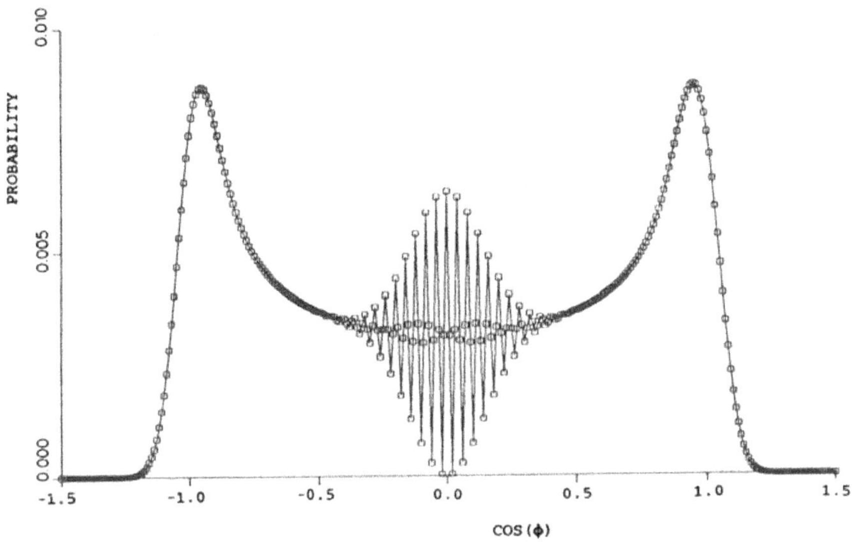

Fig. 4. Probability distribution of the estimated cosine for an incident number state.

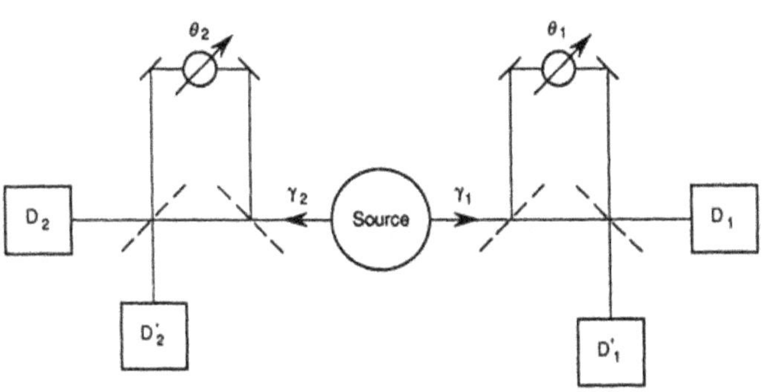

Fig. 5 Nonlocal interferometer.

These effects have recently been generalized[9] to high-intensity fields containing a large number of photons. In that case there are coincident bursts of energy that appear in the corresponding output ports of the two interferometers. The effects are truly macroscopic in that no single-particle measurements are required for their observation.

The quantum state producing such effects is given by

$$|\Psi> = \gamma\sum_n \frac{(\alpha c^\dagger)^n}{n!}|0> = \gamma e^{\alpha c^\dagger}|0> \tag{9}$$

where γ is a normalization constant and

$$c^\dagger = \sum_k f_k a_k^\dagger b_{k_0-k}^\dagger. \tag{10}$$

The operators $\hat{a}^\dagger$ and $\hat{b}^\dagger$ create photons of momentum k and k_0 - k in the two paths to the inteferometers, $c^\dagger$ creates entangled pairs of photons, k_0 is the momentum of the pump photons, and f_k is determined by any filters inserted into the two beams. There is currently no practical method for producing such states, but they nevertheless illustrate the kind of macroscopic effects that may occur.

This state can be written in the more useful form

$$|\Psi> = \gamma\prod_k e^{\alpha f_k a_k^\dagger b_{k_0-k}^\dagger}|0>. \tag{11}$$

The usual property of a coherent state

$$c|\Psi> = \alpha|\Psi> \tag{12}$$

is not obtained because $[c, c^\dagger] \neq 1$. However, the state does have the useful property that

$$a_k|\Psi> = \alpha f_k b_{k_0-k}^\dagger|\Psi>$$

$$b_k|\Psi> = \alpha f_k a_{k_0-k}^\dagger|\Psi>. \tag{13}$$

These two properties can be used to calculate[9] all of the desired results in the limit of $\alpha f << 1$, which corresponds to a non-degenerate but high-intensity field. In that limit, it can be shown that

$$E_A^+(x,t_A)E_B^+(x,t_B)|\Psi> = 2\pi\rho\gamma'\alpha f e^{i(k_0x-\omega_0 t_A)}\delta_A(t_A-t_B)|\Psi>. \tag{14}$$

where ρ is the density of states and ω_0 is the frequency of the pump laser. Eq. 14 shows that coincident events still dominate the coincident counting rate for sufficiently short counting intervals.

It follows that these high-intensity fields also violate the following classical inequality[8]:

$$
\begin{aligned}
&< |E_A^*(t)E_B^*(t)E_B(t+\Delta t)E_A(t+\Delta t)| > \\
&\leq \frac{1}{2} < E_A^*(t)E_B^*(t+\Delta t)E_B(t+\Delta t)E_A(t) > \\
&+ \frac{1}{2} < E_A^*(t+\Delta t)E_B^*(t)E_B(t)E_A(t+\Delta t) > .
\end{aligned}
\tag{15}
$$

Coincidence measurements using two detectors can be shown to give

$$
R_c = \eta \cos^2\left[\frac{\theta_A + \theta_B + \omega_0 \Delta t}{2}\right]
\tag{16}
$$

where η is the detection efficiency, which is the same as that obtained earlier for weak fields.

Although Eq. 16 violates Bell's inequality, the macroscopic nature of these effects can only be seen by considering higher-order events. A more lengthy calculation shows that the probability P_N of detecting N coincident pairs of photons in the corresponding output ports of the two interferometers is given by

$$
P_N = N! \, \eta^N \cos^{2N}\left[\frac{\theta_A + \theta_B + \omega_0 \Delta t}{2}\right] = N! \, P_1^N .
\tag{17}
$$

It can be seen that the probability of obtaining N coincident pairs of photons is enhanced by a factor of N! over what would be expected from N independent events. This corresponds to large bursts of energy that violate Bell's inequality. The origin of the N! enhancement can be seen for the case of N = 2, in which case photon A_1 can be coincident with photon B_1 while A_2 is coincident with B_2,or A_1 can be coincident with B_2 while A_2 is coincident with B_1.

The time at which the photons were emitted is totally uncertain in the quantum-mechanical sense, but the fact that they are coincident means that a measurement of the arrival time of a photon at one detector must determine the arrival time of the photons at a second detector. This in itself requires a nonlocal reduction of the wave function and provides an additional example of the EPR paradox for continuous variables. Such an effect was recently demonstrated[10] using high-speed detectors separated by a 100 meter distance in air and photon pairs produced by parametric down-conversion. The reduction of the wave function was unaffected to at least one part in 10^4 by the large distance over which the collapse must have occurred, which provides strong evidence against Furry's hypothesis.

4. Summary

Several examples have been given of situations in which the EPR paradox or Bell's inequality can be generalized to macroscopic measurements involving large numbers of particles. These effects suggest that no clear division between the macroscopic and quantum worlds exists. An experimental investigations of systems somewhere in between those two limits would be of interest.

This work was supported by the Office of Naval Research.

5. References

1. A. Einstein, B. Podolsky, and N. Rosen, *Phys. Rev.* **47** (1935) 777.
2. J.S. Bell, *Physics* **1** (1964) 195.
3. D.M. Greenberger, M.A. Horne, A. Shimony, and A. Zeilinger, *Am. J. Phys.* **58** (1990) 1131.
4. Z.Y. Ou, S.F. Pereira, H.J. Kimble, and K.C. Peng, *Phys. Rev. Lett.* **68** (1992) 3663.
5. J.D. Franson, submitted to *Phys. Rev. A*.
6. J.W. Noh, A. Fougeres, and L. Mandel, *Phys. Rev. Lett.* **67** (1991) 1426; *Phys. Rev.* **A45** (1992) 424; *Phys. Rev.* **A46** (1992) 2840.
7. R.A. Campos, B.E.A. Saleh, and M.C. Teich, *Phys. Rev.* **A40** (1989) 1371.
8. J.D. Franson, *Phys. Rev. Lett* **62** (1989) 2205; **67** (1991) 290.
9. J.D. Franson, submitted to *Phys. Rev. A*
10. J.D. Franson, to be submitted to *Phys. Rev. A*.

EINSTEIN-PODOLSKY-ROSEN CORRELATIONS IN HIGHER DIMENSIONS

A. ZEILINGER, M. ZUKOWSKI*
Institut für Experimentalphysik, Universität Innsbruck
A-6020 Innsbruck, Austria

M.A. HORNE
Stonehill College, North Easton, MA 02357, USA

H.J. BERNSTEIN
Hampshire College, Amherst, MA 01002, USA

D.M. GREENBERGER
City College of the City University of New York, New York, NY 10031, USA

ABSTRACT

Using multiport beam splitters it will be possible to study Einstein-Podolsky-Rosen correlations in higher dimensional Hilbert space. As an explicit example we present the design and theory of a tritter, which is a multiport beam splitter with three input ports and three output ports, such that any amplitude incident at one input port is distributed equally over the output ports. We will then show the results for a two-photon, two-tritter experiment, where novel Einstein-Podolsky-Rosen correlations occur.

1. Introduction

All experimental work concerning the Einstein-Podolsky-Rosen Paradox[1] and Bell's theorem[2] thus far is restricted to two-particle (in most cases two-photon) entangled states where the correlations can effectively be described by restricting the analysis to a Hilbert space of dimension 2 for each particle. These states can be two polarization states as proposed initially by Bohm[3] and first employed in an experiment by Freedman and Clauser[4], they can be two momentum eigenstates as in the experiment proposed by Horne and Zeilinger[5] and performed first by Rarity and Tapster[6], or, they can be two states which took beam paths of markedly different length on their way from the source to the detector as proposed by Franson[7]. This latter experiment has now been performed by various groups[8], the most conclusive experiment which showed a striking violation of a Bell-type inequality is due to Kwiat, Steinberg and Chiao[9].

There are two obvious routes for generalization. One is to consider more than two particles, the other is to analyze the case of more than two states available to each particle. The generalization to more than two particles has led to some new insight into the difference

---

* Permanent address: *Institute of Theoretical Physics and Astrophysics, University of Gdansk, PL-80952 Gdansk, Poland*

between quantum mechanics and local realistic theories[10]. But, due to the unavailability of coherent multi-particle sources this has not as yet resulted in an experiment.

In the present paper we would like to focus our analysis on another generalization. This is the case where each particle has more than two states available. The correlations are then defined in Hilbert spaces of higher dimension[11]. It is obvious that a possible route to generalizing EPR correlations to systems of higher dimension would be to investigate spin correlations between particles with spin-1 or higher (with the obvious and notable exception of the photon or other massless particles which have only 2 polarisation states.) Again, since at present there exist no sources for correlated particles of higher spin, such investigations based on spin correlations are purely theoretical to date[12].

This paper shows how to obtain such EPR correlations in more than two dimensions in real experiments. Such experiments are based on both the availability of parametric down-conversion as a source for highly correlated two-photon states[13] and on the use of multi-port devices[14]. Finally, we present some theoretical predictions for the novel correlations expected.

2. The Beam Splitter as a Four-Port Device

The beam splitter is a central element of many experiments in quantum optics. A general beam splitter has two input ports and two output ports (Fig. 1). Formally it may be described by a unitary operator in a two-dimensional Hilbert space. We should note here that for the present paper we deliberately adopt an explicit Hilbert space formalism because it is equally well suited for describing a beam splitter operating for any type of particle, be it electrons, photons, atoms or neutrons, to name just those types of radiation for which quantum interference experiments with beam splitters have been performed so far.

Fig. 1: A general beam splitter has two input ports and two output ports

The general beam splitter pure input state is a superposition

$$|\psi\rangle = \psi_a|a\rangle + \psi_b|b\rangle \tag{1}$$

where $|a\rangle$ and $|b\rangle$ describe a particle in beam a or b (see Fig. 1) respectively. We assume the normalization $\psi_a\,\psi_a{}^* + \psi_b\,\psi_b{}^* = 1$. Likewise the general output state is the superposition

$$|\psi'\rangle = \psi'_a|a'\rangle + \psi'_b|b'\rangle \tag{2}$$

in obvious notation. Input and output states may equally well be written in matrix notation as

$$\psi = \begin{pmatrix} \psi_a \\ \psi_b \end{pmatrix}, \quad \psi' = \begin{pmatrix} \psi'_a \\ \psi'_b \end{pmatrix}. \tag{3}$$

The general beam splitter operator U then couples ψ' to ψ, $\psi' = U\psi$ with $U^{+}U = I$.

We restrict ourselves now to 50-50 beam splitters. This means that a particle incident at any of the two input ports of a symmetric beam splitter has the same probability $p = 1/2$ to be found in any of the two output ports. It is well known that such a beam splitter is defined only up to arbitrary phase factors in the input and output ports[15].

Two possible 50-50 beam splitter operators are for example

$$U_t = \frac{1}{\sqrt{2}}\begin{pmatrix} 1 & 1 \\ 1 & -1 \end{pmatrix} \quad \text{or} \quad U_s = \frac{1}{\sqrt{2}}\begin{pmatrix} i & 1 \\ 1 & i \end{pmatrix} \tag{4}$$

where U_t represents a time-symmetric beam splitter and U_s represents a spatially symmetric one. The two beam splitters can be converted into each other using π phase shifts in one input and one output port, i.e.

$$U_t = \begin{pmatrix} -i & 0 \\ 0 & 1 \end{pmatrix} U_s \begin{pmatrix} 1 & 0 \\ 0 & i \end{pmatrix}. \tag{5}$$

The two beam splitter operators imply different transition rules for incident beams. These are

$$\begin{aligned} |a\rangle &\Rightarrow \frac{1}{\sqrt{2}}\{|a'\rangle + |b'\rangle\} \qquad |b\rangle \Rightarrow \frac{1}{\sqrt{2}}\{|a'\rangle - |b'\rangle\} \text{ for } U_t, \\ |a\rangle &\Rightarrow \frac{1}{\sqrt{2}}\{i|a'\rangle + |b'\rangle\} \qquad |b\rangle \Rightarrow \frac{1}{\sqrt{2}}\{|a'\rangle + i|b'\rangle\} \text{ for } U_s. \end{aligned} \tag{6}$$

The first beam splitter implies no phase change upon reflection from one side while reflection from the other side implies a phase change of π. The second beam splitter operator implies that both reflected beams acquire a phase shift of $\pi/2$ upon reflection.

We note here that beam splitters are just special cases of 4-port devices. Another example of a 4-port device would be a Mach-Zehnder interferometer.

3. Two-Particle Two-State Systems

Using these rules it is now easily possible to calculate the results of a two-particle two-state EPR-Bell experiment as shown in Fig. 2. A source emits two particles in the state

$$|\psi\rangle = \frac{1}{\sqrt{2}}\{|a\rangle|c\rangle + |b\rangle|d\rangle\}. \tag{7}$$

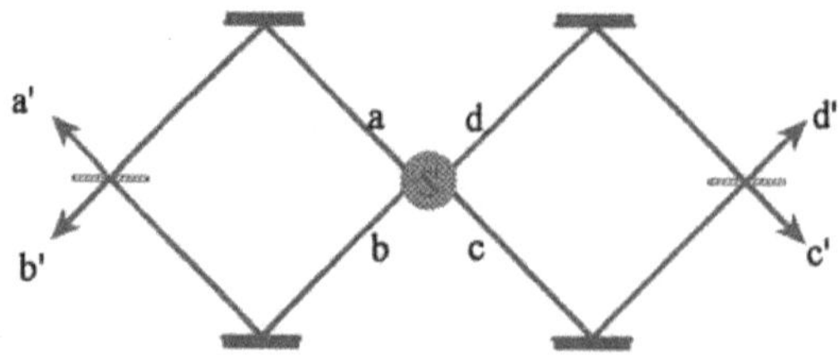

Fig. 2: Principle of a two-particle, two-state EPR-Bell experiment using beam splitters.

Here and below the first ket in a product always refers to particle 1 and the second to particle 2. Also, e.g., $|a\rangle|c\rangle$ implies the tensor product $|a\rangle\otimes|c\rangle$ etc. The beams a, b, c, d may then be subject to the phase shifts α, β, χ, δ such that the state becomes

$$|\psi\rangle=\frac{1}{\sqrt{2}}e^{i(\alpha+\gamma)}\{|a\rangle|c\rangle+e^{i\chi}|b\rangle|d\rangle\} \tag{8}$$

with $\chi=\beta+\delta-\alpha-\gamma$. Applying now the beam splitter rules (6) and, analogously,

$$|c\rangle\Rightarrow\frac{1}{\sqrt{2}}\{|c'\rangle+|d'\rangle\} \qquad |d\rangle\Rightarrow\frac{1}{\sqrt{2}}\{|c'\rangle-|d'\rangle\} \tag{9}$$

one obtains for the joint probabilities for two detectors to register the particles in coincidence

$$\begin{aligned} p(a',c')&=p(b',d')=\frac{1}{2}\cos^2(\chi/2)\\ p(a',d')&=p(b',c')=\frac{1}{2}\sin^2(\chi/2). \end{aligned} \tag{10}$$

Thus, perfect correlations arise for

$$\chi=n\pi. \tag{11}$$

For odd n detector a' fires in coincidence with detector d' and detector b' fires in coincidence with detector c' while for even n the coincidences are $a'-c'$ and $b'-d'$. These two different types of coincidences are represented in Fig. 3. In other words, for these parameter settings the path taken by a particle after its beam splitter is an Einstein-Podolsky-Rosen element of reality, i.e. firing of any one individual detector for one particle allows one to predict with certainty which detector will register the other particle.

These perfect correlations can be characterized via a value-assignment procedure introduced by Bell. The possible results obtained on either side are named A and B, and they are assigned the values ±1. It then follows that the two cases of perfect correlation are signified by $AB=+1$ and $AB=-1$ respectively. We call these values Bell numbers. We notice that one of the beam splitter operator representations (U_t) just contains Bell numbers (+1 and -1 for the two dimensional case). It will be seen later that for multiports the generalization of

Bell's value assignment procedure is quite interesting. Furthermore, in any dimension there are always multiports whose unitary representation contains only Bell numbers.

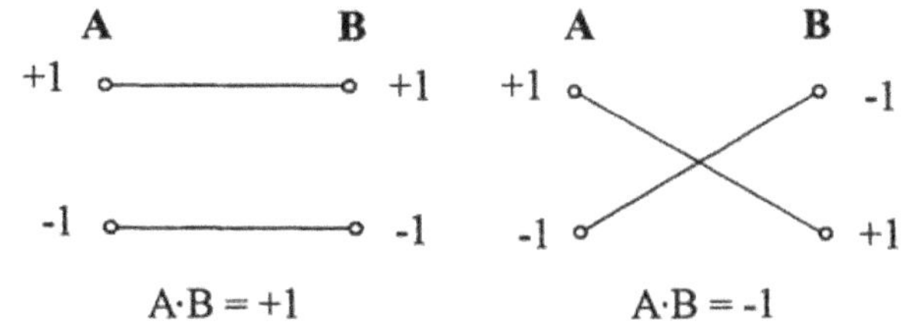

Fig. 3: Possible perfect correlations for the case of an experiment as shown in Fig. 2. The results A, B on either side can be +1 or -1, depending on which detector in which outgoing beam registers a particle. The perfect correlations can be signified by either $A \cdot B = +1$ or $A \cdot B = -1$.

We should mention that the results of this section are basically known. They were repeated here in order to prepare the reader for the less familiar situations in the following sections. An experimentally available source which prepares the two particles in the entangled state of Eq. (7) is a non-linear crystal where through the process of spontaneous parametric down-conversion an incident photon may split into 2 photons of lower energy.

4. The Tritter as an Example of a Multiport Device

In this section we first introduce the general concept of multiports and then we give some explicit examples. A general multiport has L input ports and M output ports* and is called N-port $(N = M + L)$. For simplicity we restrict our considerations to symmetric N-ports which are defined as having an equal number of input ports and output ports $(L = M = N/2)$ and, furthermore, which operate such that a single particle incident on any individual input port has equal probability (i.e. $p = 1/M = 2/N$) to be found in any specific output port. This is the generalization of the generic beam splitter discussed in section 2 above. We propose to call symmetric multiports "Critters" and specifically a critter with $L = M = 3$ is called a Tritter, one with $L = M = 4$ is a Quitter[16] etc.

Lossless symmetric multiports (critters) can be represented by unitary operators in an M-dimensional Hilbert space. Again, as was the case for the conventional beam splitter, there are many physically possible critters, but, as opposed to the beam splitter case, it is not always possible to transform all types of a specific critter (i.e. symmetric N-port with a given N) into each other by merely supplying external phase shifters or relabelling output ports[17].

Let us consider explicitly the tritter. The general input and output states are (Fig. 4)

$$
\begin{aligned}
|\psi\rangle &= \psi_a|a\rangle + \psi_b|b\rangle + \psi_c|c\rangle \\
|\psi'\rangle &= \psi'_a|a'\rangle + \psi'_b|b'\rangle + \psi'_c|c'\rangle.
\end{aligned}
\tag{12}
$$

or, in matrix notation,

---

*In general some physical ports can work both as input and output ports (viz. the Michelson interferometer).

$$\psi = \begin{pmatrix} \psi_a \\ \psi_b \\ \psi_c \end{pmatrix} \quad \text{and} \quad \psi' = \begin{pmatrix} \psi'_a \\ \psi'_b \\ \psi'_c \end{pmatrix}. \tag{13}$$

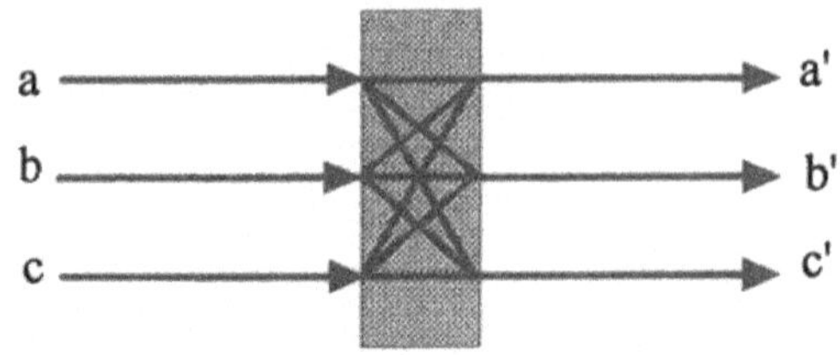

Fig. 4: A generic tritter is devised with three input ports and three output ports such that an amplitude incident on any one of the input port excites any of the output ports equally.

Again, a unitary operator couples the output state to the input state,

$$\psi' = U\psi. \tag{14}$$

This unitary operator can now be represented by a 3 x 3 matrix where the modulus of each matrix element is $1/\sqrt{3}$. Here again and for all critters it is possible to absorb any phase factors of the first row into phases of the input beams and to absorb any phase factors of the first column into phases of the output beams. Such a representation of a multiport only contains "1" in both the first column and the first row. We will call such a representation canonical. Thus, the general tritter operator can be written as

$$U = \frac{1}{\sqrt{3}} \begin{pmatrix} 1 & 1 & 1 \\ 1 & \varphi & \varphi^* \\ 1 & \varphi^* & \varphi \end{pmatrix} \tag{15}$$

with $|\varphi| = 1$ and $\varphi + \varphi^* = -1$. The only *two* possible choices for φ are $\varphi = \alpha$ and $\varphi = \alpha^2$ with $\alpha = e^{2\pi i/3}$.

Thus the tritter operator has two canonical representations, either

$$U_T = \frac{1}{\sqrt{3}} \begin{pmatrix} 1 & 1 & 1 \\ 1 & \alpha & \alpha^2 \\ 1 & \alpha^2 & \alpha \end{pmatrix} \quad \text{or} \quad U'_T = \frac{1}{\sqrt{3}} \begin{pmatrix} 1 & 1 & 1 \\ 1 & \alpha^2 & \alpha \\ 1 & \alpha & \alpha^2 \end{pmatrix}. \tag{16}$$

The transition rules for incident beams therefore are

$$
\begin{aligned}
|a\rangle &\Rightarrow \frac{1}{\sqrt{3}}\{|a'\rangle + |b'\rangle + |c'\rangle\} \\
|b\rangle &\Rightarrow \frac{1}{\sqrt{3}}\{|a'\rangle + \alpha|b'\rangle + \alpha^2|c'\rangle\} \\
|c\rangle &\Rightarrow \frac{1}{\sqrt{3}}\{|a'\rangle + \alpha^2|b'\rangle + \alpha|c'\rangle\}
\end{aligned}
\tag{17}
$$

for the tritter rule U_T. For U_T' the roles of α and α^2 are just interchanged. Note also that $U_T^{-1} = U_T'$ and that the two different types of tritter can be converted into each other by an odd number of permutations of rows or columns, e.g.

$$
U_T' = \begin{pmatrix} 1 & 0 & 0 \\ 0 & 0 & 1 \\ 0 & 1 & 0 \end{pmatrix} U_T. \tag{18}
$$

These results imply that sequential arrangement of tritters does not lead to new nontrivial tritters. In other words, given some tritter one can obtain any tritter by changing external phases and by a permutation of input and/or output ports, which may simply be achieved for example by flipping two output ports. Physically, there are many different possibilities of realising a tritter. A specific type with parallel input beams and parallel output beams is shown in Fig. 5. One can easily see that a tritter has more adjustable parameters than a beam splitter. These are the reflectivities of the partially reflecting mirrors and the nontrivial phase in the internal loop of the tritter.

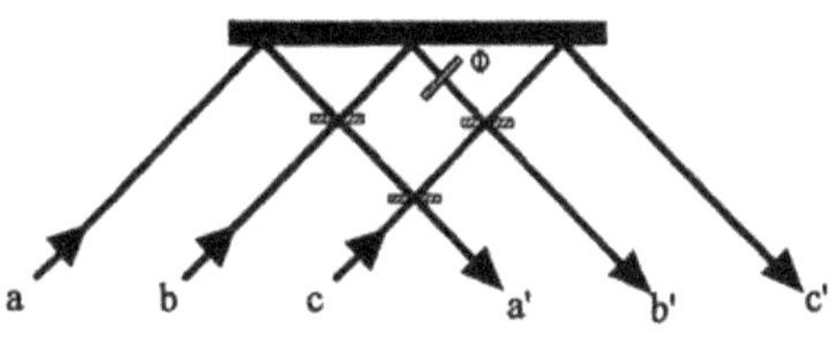

Fig. 5: Possible realization of a tritter using partially reflecting mirrors and a nontrivial internal phase $\phi = 0, \pi$.

Turning to higher multiports the number of experimentally adjustable nontrivial parameters grows quadratically with the number of ports. One of the most interesting results for higher multiports is the existence of distinct classes which cannot be transformed into each other by just changing external phases and by permutation of input and/or output ports. We leave a detailed discussion to a forthcoming paper.

5. Two-Particle Three-State Systems

It is evident that with multiports a large number of novel experiments in quantum optics become possible. Because of the availability of down-conversion photon sources, we only discuss here the case where a two-particle source is employed. Assume that such a source emits two particles in the state

$$|\psi>=\frac{1}{\sqrt{3}}\{|a\rangle|d\rangle+|b\rangle|e\rangle+|c\rangle|f\rangle\}. \tag{19}$$

Again, the first ket in a product state refers to particle 1, and the second to particle 2. The beams a,b,c,d,e,f are subject to the phase shifts $\alpha,\beta,\gamma,\delta,\varepsilon,\zeta$, respectively, and thus the state evolves into

$$|\psi\rangle=\frac{1}{\sqrt{3}}e^{i(\alpha+\delta)}\{|a\rangle|d\rangle+e^{i\varphi}|b\rangle|e\rangle+e^{i\chi}|c\rangle|f\rangle\} \tag{20}$$

with $\chi=\beta+\varepsilon-\alpha-\delta$ and $\varphi=\gamma+\zeta-\alpha-\delta$.

Suppose now that the three beams excited by particle 1 are fed into a tritter and likewise the three beams excited by particle 2 are fed into another tritter (Fig. 6). Clearly the final state is then obtained by applying the appropriate tritter operator Eq. (16) to state (20). Instead of writing down the final state explicitly, we focus on the count rates and on the correlations to be expected.

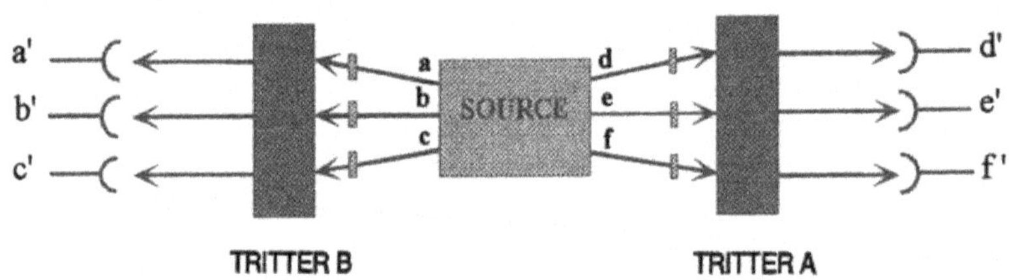

Fig. 6: Principle of a two-tritter, two-photon EPR experiment. In a practical realization the source can be parametric down-conversion.

The unconditional probability to detect a particle in any of the detectors is a constant, e.g. $p(a')=1/3$. The independence of any of the phases inserted between source and tritters is a consequence of the initial entanglement. Certainly this does not hold anymore for the various joint probabilities of detecting a particle in a given detector on one side together with detecting the other particle on the other side. These joint probabilities are:

$$p(a',d')=p(b',f')=p(c',e')=\frac{1}{27}[3+2\cos\chi+2\cos\varphi+2\cos(\varphi-\chi)],$$

$$p(a',e')=p(b',d')=p(b',f')=\frac{1}{27}[3+2\cos\chi'+2\cos\varphi'+2\cos(\varphi'-\chi')]$$

$$\text{with } \chi'=\chi+2\pi/3,\ \varphi'=\varphi-2\pi/3$$

$$p(a',f')=p(b',e')=p(c',d')=\frac{1}{27}[3+2\cos\chi''+2\cos\varphi''+2\cos(\varphi''-\chi'')]$$

$$\text{with } \chi''=\chi-2\pi/3,\ \varphi''=\varphi+2\pi/3 \tag{21}$$

and where, e.g., $p(a',e')$ is the probability to simultaneously detect a particle in detector a' and a particle in detector e'.

The joint probabilities of Eqs. (21) have a number of remarkable features. It is easy to show that all these probabilities are nonnegative and their maximum value is 1/3. This may be understood by analyzing for example the case where the first equation attains its maximum value which occurs when $\chi,\varphi = 2n\pi$. Then $p(a',d') = p(b',f') = p(c',e') = 1/3$ and all other joint probabilities vanish. This implies that if the phases in the two-tritter two-particle interferometer are set to these values then perfect correlations arise, and thus Einstein-Podolsky-Rosen elements of reality may be introduced. Explicitly, if, say, detector a' fires and the phases are set to the parameters just mentioned we can predict with certainty that the other particle will be registered by detector d'. Likewise, if particle 1 is registered by detector $b'(c')$, particle 2 will be registered by detector $f'(e')$. Thus, while it is always maximally uncertain which detector will register either of the particles, it is known with certainty which detector will register the second particle once the first particle has been observed, as long as the phases are set according to the above condition.

Another set of similar perfect correlations arises if the phases are set such that $\chi',\varphi' = 2n\pi$. Then the joint probabilities are $p(a',e') = p(b',d') = p(e',f') = 1/3$ with all others being zero. Here again perfect correlations occur but now between different detectors than before. Finally, a third set of perfect correlations arises for $\chi'',\varphi'' = 2n\pi$, then $p(a',f') = p(b',e') = p(c',d') = 1/3$ with all other joint probabilities vanishing. Fig. 7 shows these three possible ways of perfect correlations. Note that of the six possible one-to-one combinations between detectors on either side only three combinations are realized for perfect correlations. Here we should note the fact that these types of perfect correlations arise whenever we use the same tritter on each side (either the one represented by U_T or the one represented by U_T'). In case we chose to use different types of tritters on the two sides, the other three types of perfect correlations occur, with the original three now being excluded.

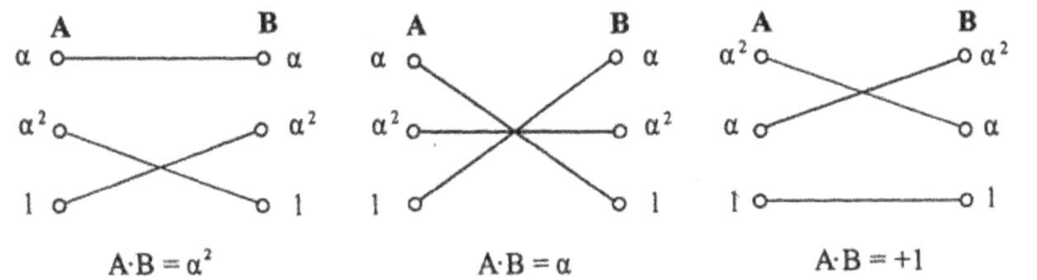

Fig. 7: Perfect correlations occurring in an experiment of the type of Fig. 6. The results on either side are best characterized by assigning them the value $A, B = \alpha, \alpha^2, 1$, where $\alpha = e^{2\pi i/3}$. The three cases of perfect correlations occurring are then signified by $A \cdot B = \alpha, \alpha^2, 1$.

The three types of perfect three-state correlations may be signified in the same way by value assignment as it was done originally by Bell for two-state correlations. One might be tempted to assign the values +1,0,-1 to the three possible outcomes on each side. Such a procedure does not succeed because, when calculating the product AB, if A is again the result for one particle and B the result for the other, appearance of a "0"-result always leads to $AB = 0$ independent of which type of correlation occurs and thus information is lost. A rather elegant procedure of value assignment is to choose α, α^2, α^3 (with $\alpha = e^{2\pi i/3}$) for the three possible outcomes on either side. It then follows that the three cases of perfect correlations are signified by $AB = \alpha, \alpha^2, 1$ (see Fig. 7). These numbers are now the Bell numbers for a three-dimensional Hilbert space.

In general, for the case of correlations between two particles, where each one is defined in an M-dimensional Hilbert space, at most M cases of perfect correlations (where EPR-elements of reality may be introduced) occur with a given set of multiports. It is thus natural to generalize the procedure just given by assigning the values $A, B = e^{2\pi i n/M}$ $n = 1,2,\ldots M$ to the results in order to signify the cases of perfect correlations by $AB = e^{2\pi i n'/M}$. As we will show in a forthcoming paper there is always at least one case of a specific multiport for any M where this procedure succeeds. But, we should point out, for $M\rangle 3$ these are also cases where this procedure fails. Obviously the case $M = 2$ as analyzed originally by Bell is just the most simple nontrivial case. This is the reason why we propose to call these general numbers used in value assignment Bell-numbers.

Concluding this section we note that besides introducing EPR elements of reality the way just given, one can also apply a generalized Bell inequality to the two-tritter correlations[18] thus providing the first feasible test for Bell's theorem for pairs of spins higher than 1/2 via an optical analog.

6. Concluding Comments

In general, an experiment using multiports which are fed the two correlated photons created in the process of parametric down-conversion provides a generalization of EPR correlations to Hilbert spaces of higher dimensions. These correlations are fully analogous to those between two particles with higher spin. Thus they are expected to give new interesting results going beyond those realizable in spin correlations between two spin-1/2 particles or two photons. A specific example are those correlations which are necessary to establish the Bell-Kochen-Specker paradox[19]. Using two correlated particles each defined in a higher-dimensional Hilbert space it is possible to establish the results for each individual measurement utilized in the Kochen-Specker argument as Einstein-Podolsky-Rosen elements of reality[20]. It is evident that using multiports together with a down-conversion photon source can provide immediate experimental realization of such correlations.

This work was supported by the Austrian Science Foundation FWF, project No. S6502 (Schwerpunkt Quantenoptik), and the US National Science Foundation, grant No. PHY92-13964.

7. References

1. A. Einstein, B. Podolsky, and N. Rosen, *Phys. Rev.* **47** (1935) 771.
2. J.S.Bell, *Physics* (NY) **1** (1964) 195.
3. D. Bohm, *Quantum Theory*, (Prentice-Hall, Englewood Cliffs, 1951).
4. S.J. Freedman and J.F.Clauser, *Phys. Rev. Lett.* **28** (1972) 938.

5. M.A. Horne and A. Zeilinger, in *Proc. of the Symposium on the Foundations of Modern Physics*, eds. P. Lahti and P. Mittelstaedt (World Scientific, 1985); M. Zukowski and J. Pykacz, *Phys. Lett.* **A127** (1988) 1; M.A. Horne, A. Shimony, and A. Zeilinger, *Phys Rev. Lett.* **62** (1989) 2209.
6. J.G. Rarity and P.R. Tapster, *Phys. Rev. Lett.* **64** (1990) 2495.
7. J.D. Franson, *Phys. Rev. Lett.* **62** (1989) 2205.
8. P.G. Kwiat, W.A. Vareka, C.K. Hong, H. Nathel, and R.Y. Chiao; *Phys. Rev.* **A41** (1990) 2910; Z.Y. Ou, X.Y. Zou, L.J. Wang, and L. Mandel, *Phys. Rev. Lett.* **64** (1990) 321; J.G. Rarity, P.R. Tapster, E. Jakeman, T. Larchuk, R.A. Campos, M.C. Teich, and B.E.A. Saleh, *Phys. Rev. Lett.* **65** (1993) 1348; J. Brendel, E. Mohler, and W. Martiensen, *Phys. Rev. Lett.* **66** (1991) 1142; *Europhys. Lett.* **20** (1992) 575.
9. P.G. Kwiat, A.M. Steinberg, and R.Y. Chiao, *Phys. Rev.* **A47** (1993) R 4272.
10. D.M. Greenberger, M. Horne, and A. Zeilinger, in *Bell's Theorem, Quantum Theory and Conceptions of the Universe*, ed. M. Kafatos (Kluwer Academic, Dordrecht, 1989), p. 73; D.M. Greenberger, M.A. Horne, A. Shimony, and A. Zeilinger, *Am. J. Phys.* **58** (1990) 1131; N.D. Mermin, *Phys. Rev. Lett.* **65** (1990) 1838.
11. L. Hardy (*Phys. Rev. Lett.* **68** (1992) 2981) has recently made a proposal employing two particles, each one defined in more than two states with the attempt at arriving at a GHZ-type contradiction between local realism and quantum mechanics. In this case and in a related proposal (H.J. Bernstein, D.M. Greenberger, M.A. Horne, and A. Zeilinger, *Phys. Rev.* **A47** (1993) 78) that goal is only achieved with some additional assumptions.
12. N.D. Mermin, *Phys. Rev.* **D22** (1980) 356; A. Garg and N.D. Mermin, *Phys. Rev Lett.* **49** (1982) 901; N.D. Mermin and G.M. Schwarz, *Found. Phys.* **12** (1982) 101.
M. Ardehali, *Phys. Rev.* **D44** (1991) 3336; G.S. Agarwal, *Phys. Rev.* **A47** (1993) 4608.
13. D.C. Burnham and D.L. Weinberg, *Phys. Rev. Lett.* **25** (1970) 84.
14. A. Zeilinger, H.J. Bernstein, D.M. Greenberger, M.A. Horne, and M. Zukowski, in *Quantum Control and Measurement*, eds. H. Ezawa and Y. Murayama (Elsevier, 1993).
15. A. Zeilinger, *Am. J. Phys.* **49** (1981) 88; S. Prasad, M.O. Scully, and W. Martienssen, *Opt. Comm.* **62** (1987) 139; Z.Y. Ou, C.K. Hong, and L. Mandel, *Opt. Comm.* **63** (1987) 118; H. Fearn and R. Loudon, *Opt. Comm.* **64** (1987) 485.
16. N.G. Walker and J.E. Carroll, *Optical and Quantum Electronics* **18** (1986) 355; N.G. Walker, *J. Mod. Opt.* **34** (1987) 15.
17. H.J. Bernstein, *J. Math. Phys.* **15** (1974) 1677.
18. The method is to be found in M. Zukowski, *Phys. Lett.* **A177** (1993) 290.
19. S. Kochen and E.P. Specker, *J. Math. Mech.* **17** (1967) 59; J.S. Bell, *Rev. Mod. Phys.* **38** (1966) 447.
20. M. Redhead, *Incompleteness, Nonlocality and Realism* (Clarendon, Oxford; 1987).

QUANTUM PHASE TRACING OF CORRELATED PHOTONS IN OPTICAL MULTIPORTS

MICHAEL RECK AND ANTON ZEILINGER
Institut für Experimentalphysik,
Universität Innsbruck, Technikerstr. 25
A–6020 Innsbruck, Austria

ABSTRACT

Multiport-beamsplitters permit experiments with photons in higher-dimensional Hilbert spaces. The classification and description of these experiments, which typically use beamsplitters, mirrors, and phase shifters, is achieved by methods borrowed from graph theory. An algorithmic approach to the calculation of the unitary matrix for an arbitrary arrangement of these components is introduced. As an example, the two-photon correlations in a generalized Franson type experiment using three-way beamsplitters are calculated.

1 Introduction

A multiport-beamsplitter can be viewed as a black box transforming n input states into n output states (see Figure 1). Thus multiports operate on photons in a higher-dimensional Hilbert space. Inside the black box we find the basic building components of a multiport experiment: beamsplitters, mirrors, and phase shifters. It seems convenient to find an abstract description of the workings of the black box as function of the parameters that can be set (internal phase shifts). A lossless experimental setup will be described by a unitary matrix.

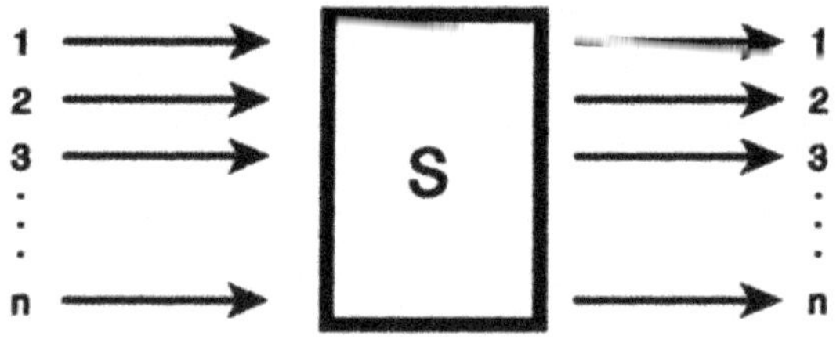

Figure 1: Any experiment with beamsplitters, mirrors, and phase shifters can be viewed as a black box. It will be described by a unitary matrix transforming n input states into n output states.

2 Formal Description of Multiport Experiments

Graph theory offers a convenient way of describing the basic topological setup of an experiment. A graph consists of vertices connected by edges. In the drawing of a graph a vertex is usually represented by a point and the (directed) edges by arrows connecting the points.[1] This drawing is an abstract description of an experiment with photons (see Figure 2):

1. The experimental setup begins after the photon sources and ends before the detectors.
2. Each optical component is represented by a vertex of a directed graph.
3. All possible paths through the experimental setup to the detector are described by the graph's edges. The graph may contain loops or multiple edges. (A loop is an edge connecting a point with itself.)
4. The degree of a vertex (optical component) is the number of edges incident. Thus the degree of a mirror or phase shifter is two (one input port and one output port), the degree of a beamsplitter is four (two input ports and two output ports).
5. The properties of all optical components having only one input and one output port (mirrors, phase shifters, etc.) are taken into the graphs edges.
6. The basic building block is the *beamsplitter*. The properties of the beamsplitters (complex reflectivity and transmittivity) are attached to the vertices of the graph.
7. Each beamsplitter must have four ports, therefore all "dangling" edges (i.e. external input and output ports of the system) can be left away in the drawing. Thus the number edges incident on each beamsplitter represented by a dot in the graph must be thought supplemented by edges not connected to other beamsplitters, making a total of four ports (see Figure 2).

Once the graph of an experiment is drawn it is easy to formalize its description. The experimental setup can be described by a matrix S transforming the probability amplitudes at the n input ports into the amplitude at the n output ports. The matrix is unitary if the setup is lossless.

3 Quantum Phase Tracing

The calculation of the unitary transform matrix is accomplished by "quantum phase tracing". All possible paths from the sources to the detectors are traced through the graph. The phases along the edges are summed up and the resulting probability amplitudes of indistinguishable alternative paths inside the experiment are added up.[2]

The detection probability at a given output is proportional to the absolute square of the probability amplitude.

The graph is the fingerprint of the experimental setup. Isomorphic graphs belong to equivalent experimental setups. Graph theory provides a list of all possible graphs with a given number of edges and vertices. Thus we can list all non-equivalent experimental setups with a given number of beamsplitters (see Figure 2).

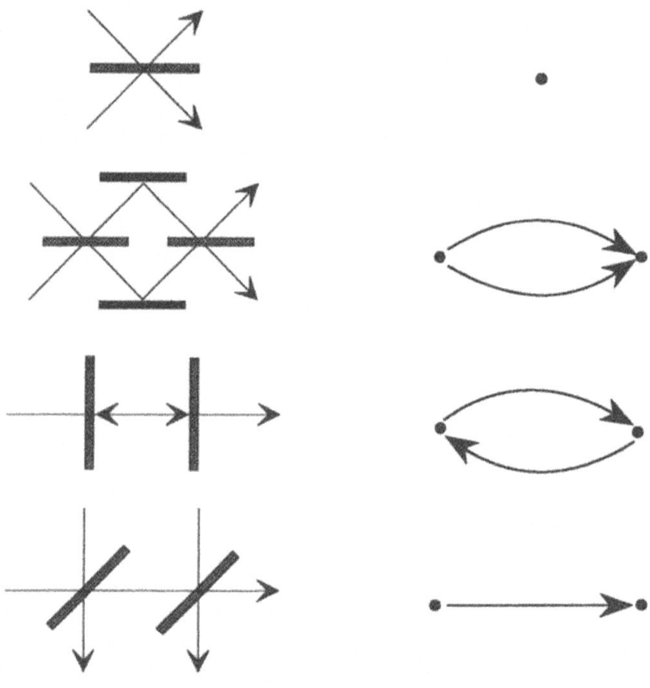

Figure 2: All possible setups with one and two beamsplitters and their corresponding graphs: a single beamsplitter, a Mach-Zehnder interferometer, a Fabry-Pérot interferometer, and a three-way beamsplitter. Each dot in the graph represents a beamsplitter and must be thought supplemented by "dangling" edges (i.e. the external input and output ports) to give a total of two input and two output ports at each beamsplitter.

Quantum phase tracing in the frequency domain assumes all phase shifts are frequency dependent. This gives the system matrix of the experimental setup as a function of the vertices (beamsplitters) and the edges (phase shifts). The frequency-dependent system matrix is then integrated over the input wave packet profile giving the time-dependent output amplitudes.[3,5]

Scattering matrix methods can be used to calculate the system's unitary matrix.[6]

4 Calculating the System Matrix

The system matrix relates the output probability amplitudes with the input:

$$\vec{a}_{\text{out}} = S\vec{a}_{\text{in}}. \tag{1}$$

The system matrix describes the external outputs connected to the detectors as a function of the external inputs which are connected to the sources. It contains two contributions. One describes the scattering at the beamsplitters; the other the propagation between connected beamsplitters. Each beamsplitter may have internal and external ports. The internal ports are those which are connected to another beamsplitter in the system. The number of external ports is equal to the dimension of the system's unitary matrix.

The scattering matrix describing the vertices of the graphs (beamsplitters) can be build from block matrices describing the transfer from internal and external inputs of the beamsplitters to their internal and external outputs:[6]

$$\begin{pmatrix} \vec{a}_{\text{out}}(ext) \\ \vec{a}_{\text{out}}(int) \end{pmatrix} = \begin{bmatrix} S_{ee} & S_{ei} \\ S_{ie} & S_{ii} \end{bmatrix} \begin{pmatrix} \vec{a}_{\text{in}}(ext) \\ \vec{a}_{\text{in}}(int) \end{pmatrix} \tag{2}$$

The submatrix S_{ei}, for example, describes the scattering from the internal inputs of the beamsplitters to the external outputs. The connection matrix Γ describes the propagation between beamsplitters, that is the edges of the graph (with phase shifts):

$$\vec{a}_{\text{out}}(int) = \Gamma\, \vec{a}_{\text{in}}(int). \tag{3}$$

Solving equations (2) and (3) for the external ports gives an expression for the system matrix:

$$\vec{a}_{\text{out}} = \left[S_{ee} + S_{ei}\left(\Gamma - S_{ii}\right)^{-1} S_{ie}\right] \vec{a}_{\text{in}} \tag{4}$$

5 Example: Three-Way Beamsplitter

The experimental setup for a three-way beamsplitter and the corresponding graph are shown in Figure 3.

Each vertex in the graph (simple beamsplitter X with reflectivity r_X and transmittivity t_X) is described by a 2×2 matrix[4,5]:

$$\begin{pmatrix} x_3 \\ x_4 \end{pmatrix} = \begin{bmatrix} t_X & ir_X \\ ir_X & t_X \end{bmatrix} \begin{pmatrix} x_1 \\ x_2 \end{pmatrix} \tag{5}$$

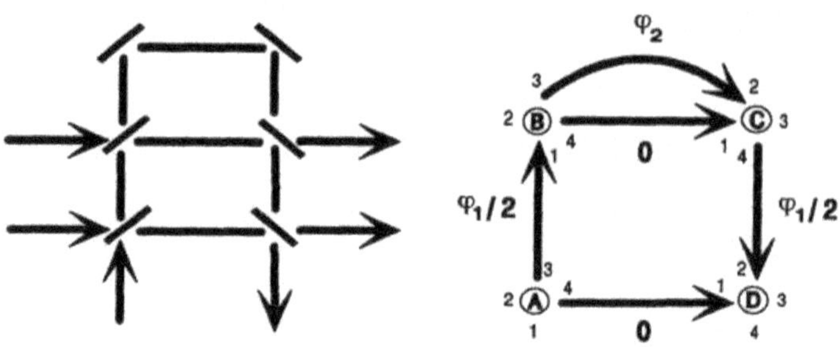

Figure 3: Experimental setup and graph for a three-way beamsplitter. External input ports: A_1, A_2, B_2; Internal input ports: B_1, C_1, C_2, D_1, D_2; External output ports: C_3, D_3, D_4; Internal output ports: A_3, A_4, B_3, B_4, C_4.

The inputs of each beamsplitter are numbered 1 and 2; the outputs are numbered 3 and 4. The scattering matrix for the three-way beamsplitter is given by

$$\begin{pmatrix} C_3 \\ D_3 \\ D_4 \\ \hline A_3 \\ A_4 \\ B_3 \\ B_4 \\ C_4 \end{pmatrix} = \left[\begin{array}{ccc|ccccc} 0 & 0 & 0 & 0 & t_C & ir_C & 0 & 0 \\ 0 & 0 & 0 & 0 & 0 & 0 & t_D & ir_D \\ 0 & 0 & 0 & 0 & 0 & 0 & ir_D & t_D \\ \hline t_A & ir_A & 0 & 0 & 0 & 0 & 0 & 0 \\ ir_A & t_A & 0 & 0 & 0 & 0 & 0 & 0 \\ 0 & 0 & ir_B & t_B & 0 & 0 & 0 & 0 \\ 0 & 0 & t_B & ir_B & 0 & 0 & 0 & 0 \\ 0 & 0 & 0 & 0 & ir_C & t_C & 0 & 0 \end{array}\right] \begin{pmatrix} A_1 \\ A_2 \\ B_2 \\ \hline B_1 \\ C_1 \\ C_2 \\ D_1 \\ D_2 \end{pmatrix} \tag{6}$$

The connection matrix for the three-way beamsplitter is

$$\begin{pmatrix} A_3 \\ A_4 \\ B_3 \\ B_4 \\ C_4 \end{pmatrix} = \begin{bmatrix} e^{-i\varphi_1/2} & 0 & 0 & 0 & 0 \\ 0 & 0 & 0 & 1 & 0 \\ 0 & 0 & e^{-i(\varphi_2+\pi)} & 0 & 0 \\ 0 & 1 & 0 & 0 & 0 \\ 0 & 0 & 0 & 0 & e^{-i\varphi_1/2} \end{bmatrix} \begin{pmatrix} B_1 \\ C_1 \\ C_2 \\ D_1 \\ D_2 \end{pmatrix} \tag{7}$$

A common phase in the connection matrix has been set to zero.

For B and C as 50% beamsplitters, A and D as 66.67% beamsplitters, the multiport has the following system matrix S relating input states with output states:

$$\begin{bmatrix} i\,\dfrac{e^{i\varphi_1/2}(1-e^{i\varphi_2})}{2\sqrt{3}} & \dfrac{e^{i\varphi_1/2}(e^{i\varphi_2}-1)}{\sqrt{6}} & \dfrac{1+e^{i\varphi_2}}{2} \\ i\,\dfrac{2-e^{i\varphi_1}-e^{i(\varphi_1+\varphi_2)}}{3\sqrt{2}} & \dfrac{1+e^{i\varphi_1}+e^{i(\varphi_1+\varphi_2)}}{3} & \dfrac{e^{i\varphi_1/2}(e^{i\varphi_2}-1)}{\sqrt{6}} \\ -\dfrac{4+e^{i\varphi_1}+e^{i(\varphi_1+\varphi_2)}}{6} & i\,\dfrac{2-e^{i\varphi_1}-e^{i(\varphi_1+\varphi_2)}}{3\sqrt{2}} & i\,\dfrac{e^{i\varphi_1/2}(1-e^{i\varphi_2})}{2\sqrt{3}} \end{bmatrix} \tag{8}$$

6 One Photon Description

The photon at input port l of the multiport is described by wave-packet creation operator $\hat{a}_l^\dagger(\omega)$ which creates a photon with a frequency distribution $\epsilon_l(\omega)$:[3,5]

$$\hat{A}_l^\dagger(\alpha) := \int_0^\infty \epsilon_l(\omega)\hat{a}_l^\dagger(\omega)d\omega \tag{9}$$

$$\alpha_l(t) = \frac{1}{\sqrt{2\pi}}\int_0^\infty \epsilon_l(\omega)e^{-i\omega t}d\omega \tag{10}$$

The function $\alpha_l(t)$ gives the probability amplitude for photon detection with a detector placed at input port l before the system.

A coherent superposition of photon states at the input ports l results in a probability amplitude after the system given by

$$\alpha_m(t) = \frac{1}{\sqrt{2\pi}}\sum_l \int_0^\infty S_{ml}(\omega)\,\epsilon_l(\omega)e^{-i\omega t}d\omega$$

with S_{ml} the matrix element relating a specific input port l with a specific output port m.

7 Two-Photon Description

Now we consider the case of two photons propagating through two systems (e.g. see Figure 4). The two photon wave-packet creation operator is:[3]

$$\hat{K}^\dagger(\zeta) := \int_0^\infty\int_0^\infty \zeta(\omega_1,\omega_2)\,\hat{a}_1^\dagger(\omega_1)\hat{a}_2^\dagger(\omega_2)d\omega_1\,d\omega_2 \tag{11}$$

with a normalized joint wave-packet profile.

$$\int_0^\infty\int_0^\infty |\zeta(\omega_1,\omega_2)|^2 d\omega_1 d\omega_2 = 1 \tag{12}$$

If the two photons are directed into separate multiports the coincidence probability during the time T (coincidence window) is the integral over the second order correlation function $G^{(2)}_{1'2'}(t_1,t_2)$ describing the correlation at detectors $1'$ and $2'$ placed after the multiports.

$$P_{1'2'}(T) = \eta^2 \int_{-T/2}^{T/2}\int_{-T/2}^{T/2} G^{(2)}_{1'2'}(t_1,t_2)\,dt_1\,dt_2 \tag{13}$$

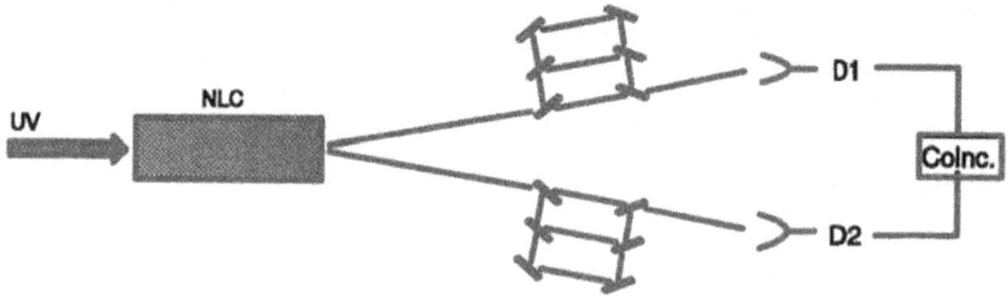

Figure 4: Two photons from a parametric downconversion source are directed into separate multiports and detected in coincidence.

The photodetectors at the output ports have an efficiency η. The second order correlation function is given by:

$$G^{(2)}_{1'2'}(t_1,t_2) = \left| \int_0^\infty \int_0^\infty S_{1'1}(\omega_1) S_{2'2}(\omega_2) \zeta(\omega_1,\omega_2)\, e^{-i\,(\omega_1 t_1+\omega_2 t_2)} d\omega_1\, d\omega_2 \right|^2 \tag{14}$$

with $S_{1'1}(\omega_1)$ and $S_{2'2}(\omega_2)$ the matrix elements relating the respective input and output ports.

8 Two-Photon Coincidence Experiment with Two Three-Way Beamsplitters

We consider two three-way beamsplitters as described above. Signal and idler photons from a parametric downconversion source are fed into input ports A_2 and A'_2. This experiment is a generalized form of that suggested by Franson.[7] In the present case three possible paths can be brought to interference. For simplicity we consider the degenerate case $\omega_s = \omega_i = \omega_p/2$ and two identical three-way beamsplitters (see Figure 4). The parameters that can be varied are the phase shifts φ_1 and φ_2. The UV pump wave is considered ideally monochromatic.[8] We assume the filters have equal bandwidth σ and centers $\omega_0 = \omega_p/2$ before the detectors.

1. If $\varphi_1 \ll \sigma T \ll \omega_0 T$ and $\varphi_2 \ll \sigma T \ll \omega_0 T$, that is the path differences are much less than the filter bandwidths and much less than the detector response times T, the probability for coincidence counts is

$$P^{\Psi}_{12} = \frac{1}{81}\left[3 + 2\cos(\varphi_1) + 2\cos(\varphi_2) + 2\cos(\varphi_1+\varphi_2)\right]^2 \tag{15}$$

   The coincidence probability is the square of the single-photon detection probability and we see single-photon interference fringes with visibility 1.

This is a consequence of the fact that for small enough path differences in the interferometers the time of detection of one photon does not provide information about the path taken by either photon.

2. If $\sigma T \ll \omega_0 T \ll \varphi_1$ and $\sigma T \ll \omega_0 T \ll \varphi_2$, that is the path differences are much greater than the filter bandwidths and much greater than the detector response times T, the probability for coincidence counts then is

$$P_{12}^{\Psi} = \frac{1}{81}\,[3 + 2\cos(2\varphi_1) + 2\cos(2\varphi_2) + 2\cos(2\varphi_1 + 2\varphi_2)] \qquad (16)$$

We observe two-photon interference fringes with visibility 1.

In the second case the coincidence window is so narrow that the second photon's arrival time can be used to determine the path take by the first photon in the other interferometer. Thus there can be no single-photon interference. Detection in coincidence, on the other hand, destroys the two-photon path information, that is the three possible combinations long-long, medium-medium, and short-short are indistinguishable and can interfere. This is a direct result of the entanglement of the two-photon wavefunction.

References

1. F. Harary, Graph Theory, (Addison-Wesley, Reading, Massachusetts, 1969).
2. R. P. Feynman, R. B. Leighton, and M. Sands, *The Feynman Lectures on Physics*, vol. III, (Addison-Wesley, Reading, Massachusetts, 1964).
3. R. A. Campos, B. E. A. Saleh, and M. C. Teich, Phys. Rev. A **42**, 4127 (1990).
4. A. Zeilinger, Am. J. Phys. **89**, 882–883 (1981)
5. H. Fearn and R. Loudon, J. Opt. Soc. Am. B **6**, 917 (1989).
6. J. A. Dobrowolski, *Introduction to Computer Methods for Microwave Circuit Analysis and Design*, (Artech House, Boston, 1991).
7. J. D. Franson, Phys. Rev. Lett. **62**, 2205 (1989).
8. M. Horne, A. Shimony, and A. Zeilinger, "Down-conversion photon pairs: a new chapter in the history of quantum mechanical entanglement," in *Quantum Coherence*, edited by J. S. Anandan, pp. 356–372, (World Scientific, Singapore, 1990).

Acknowledgements

This work was supported by the Fond zur Förderung der Wissenschaftlichen Forschung, Austria, *Schwerpunkt Quantenoptik* project number S6502 and by the U.S. NSF grant PHY 92-13964.

EXTENSIONS OF BELL THEOREM

MAREK ŻUKOWSKI

Institut für Experimentalphysik, Unversität Innsbruck, Technikerstr. 25, A-6020 Innsbruck, Austria

and

Instytut Fizyki Teoretycznej i Astrofizyki Uniwersytet Gdanski, 80-952 Gdansk, Poland

ABSTRACT

A novel way to *derive* the Bell theorem is presented. One does not have to postulate any specific algebraic form that would lead to a Bell inequality. The very shape of the quantum correlation function defines the proper shape of the algebraic form. The algorithm has a simple geometric interpretation. It can be readily applied in new situations. E.g., for the *n*-particle GHZ correlations it limits the maximal visibility of the sinusoidal multi-particle interference pattern, which is still describable by local hidden variable theories, to $2(2/\pi)^n$.

1. Introduction

Quantum interferometric experiments involving entangled states are extremely important for our understanding of the foundations of quantum theory. The Bell theorem tells us that the predictions of the theory, for a certain experiment of this kind, are incompatible with the basic classical notion of local realism (i.e. assumption that a physical system can be described via its objective properties, which do not depend on acts of observation, and that this description is concurrent with Einstein's causality). So far all experimental tests of Bell's inequalities have to be analyzed using some additional assumptions, which are not present in the original reasoning. This is mainly due to the so-called detection loophole. Thus we need alternative experiments, and new ways of interpreting them. New experiments may also increase the possibility of finding practical application for the phenomena (like "quantum cryptography").

The aim of this work is to present a novel way to obtain the Bell theorem. Bell's inequalities will be *derived* with the use of a rather loose algorithm. In the standard reasoning one *postulates* certain algebraic form and shows that when one assumes that the variables constituting this form fulfill are of a local realistic nature, then one obtains certain (Bell) *inequality*. The Bell *theorem* then says that for certain settings of the parameters governing the quantum observables this inequality in violated by a quantum prediction. In the approach presented here the very function which is the quantum prediction defines the proper algebraic form (or, if you like, its coefficients). Thus, the method could be applied in new situations. It is based on simple properties of Euclidean vectors.

2. Remarks on the Bell theorem

The Bell theorem will be only briefly sketched, as the reader may consult many other works. In a Bell-type experiment, one has a source emitting pairs of particles which propagate in two opposite directions; they are in an entangled state $|\Psi(2)\rangle$, *e.g.*

$$|\Psi(2)\rangle = 2^{-1/2}\left(|r_1\rangle_1|r_2\rangle_2 + i|\ell_1\rangle_1|\ell_2\rangle_2\right) \tag{1}$$

with ${}_n\langle r_n|\ell_n\rangle_n = 0$. Each particle enters an apparatus. The operation of the apparatus O_1 (O_2) is controlled by a knob which sets the parameter Φ_1 (Φ_2) (*e.g.*, the phase shift or the orientation of a polarizer). Both apparata measure *dichotomic* observables $O_1(\Phi_1)$ and $O_2(\Phi_2)$ (*i.e.*, with eigenvalues ±1), *e.g.*

$$O_n(\Phi_n) = |+,\Phi_n\rangle_{n\,n}\langle +,\Phi_n| - |-,\Phi_n\rangle_{n\,n}\langle -,\Phi_n|, \tag{2}$$

with

$$|\pm,\Phi_n\rangle_n = 2^{-1/2}\left(\pm|r_n\rangle_n + \exp(i\Phi_n)|\ell_n\rangle_n\right). \tag{3}$$

The correlation function reads:

$$E_{ex}(\Phi_1,\Phi_2) = [n(+,+) + n(-,-) - n(-,+) - n(+,-)]/n \tag{4}$$

where n represents the number of the pairs of particles tested in the experiment (we shall not discuss here the problem of detection errors, typical for real experiments), and *e.g.* $n(+,-)$ represents the number of time correlated results at the spatially separated apparata with the result for $O_1(\Phi_1)$ and $O_2(\Phi_2)$ being respectively $+1$ and -1. As quantum mechanics is a well established theory, we are able to calculate exactly or estimate the function $E_{qm}(\Phi_1,\Phi_2)$, which is a prediction for $E_{ex}(\Phi_1,\Phi_2)$. For the process described by (1-3) the quantum prediction for the (quantum) probability $P(i,j|\Phi_1,\Phi_2)$ reads $[1 + (ij)\sin(\Phi_1+\Phi_2)]/4$, where i and j can be equal to ±1 and represent the eigenvalues of $O_1(\Phi_1)$ and $O_2(\Phi_2)$. The quantum correlation function is given by

$$E_{qm}(\Phi_1,\Phi_2) = \sum_{i\,j} ijP(i,j\,|\,\Phi_1,\Phi_2) = \langle\Psi(2)\,|\,O_1(\Phi_1)O_2(\Phi_2)\,|\,\Psi(2)\rangle = \sin(\Phi_1+\Phi_2). \tag{5}$$

For the *usual* interferometric processes the correlation function is always of the form of the sine or the cosine of the sum or the difference of two local parameters. A suitable redefinition of the parameters can always transform it back to $\sin(\Phi_1+\Phi_2)$.

Let us now look at the above standard experiment from the perspective of a local hidden variable theory[1]. We shall analyze here only the deterministic models (generalizations are possible). The quantum state $|\psi(2)\rangle$ does not determine the result of an individual experiment. One can try to assume a more complete specification of the state of a single pair of particles. Let us denote this specification by λ, and by Λ the space of the states λ, and require that probability measures can be defined on Λ. Each quantum state is now associated with a certain distribution function, ρ, of λ's. by ρ. Of course

$$\int_\Lambda \rho(\lambda)d\lambda = 1. \tag{6}$$

One assumes the existence of functions $O_1(\Phi_1;\lambda)$ and $O_2(\Phi_2;\lambda)$ which determine the results of the measurements of the observables (2). The function O_1 (O_2) depends only on the parameter Φ_1 (Φ_2) but not Φ_2 (Φ_1). This is the *locality* assumption. Thus, the correlation function must have the following structure

$$E_{hv}(\Phi_1,\Phi_2) = \int_\Lambda \rho(\lambda)O_1(\Phi_1;\lambda)O_2(\Phi_2;\lambda)d\lambda. \tag{7}$$

The Bell theorem points out that such a representation of the hidden variable predictions is incompatible with certain quantum mechanical predictions. In the standard approach the main task is to guess (find, derive) the right kind of an algebraic expression that would reveal the above. The most often used form of the Bell inequality has the following algebraic structure

$$|E_{hv}(\Phi_1,\Phi_2) + E_{hv}(\Phi_1',\Phi_2) + E_{hv}(\Phi_1,\Phi_2') - E_{hv}(\Phi_1',\Phi_2')| \leq 2. \tag{8}$$

3. The algorithm

One can try the following simple reasoning. The aim of the Bell theorem is to show the incompatibility between the quantum predictions and assumptions of local hidden variable theories. In simple terms this means that the function $E_{qm}(\theta,\phi)$ representing the quantum prediction is *different* from any function $E_{hv}(\theta,\phi)$, which has the structure of Eq.(7). Now one may follow two paths. Either, one can be interested in comparison of the two functions in several points, *i.e.* for certain specific set of local parameters $\theta=\Phi_1/2$ and $\phi=\Phi_2/2$, or the entire functions (section 5). Let us first discuss the former case. The predictions given by the quantum function at the parameters, say θ_i, with $i = 1,....,k$, and ϕ_j, $j = 1,...,n$ form a Cartesian representation of a (real) vector in a k times l dimensional Euclidean vector space:

$$E_{qm}^{a} = E_{qm}^{ij} = E_{qm}(\theta_i,\phi_j). \tag{9}$$

The single index a of the vector's components is an arbitrary one-to-one function of the pairs of indices i and j. It runs from 1 to kn. We expect the vector $\mathbf{E}_{qm}$ to be different from all possible vectors $\mathbf{E}_{hv}$ which are constructed of possible functions fulfilling the requirements. of the local hidden variable theories:

$$E_{hv}^{a} = E_{hv}^{ij} = E_{hv}(\theta_i,\phi_j) = \int_{\Lambda} \rho(\lambda)A(\theta_i;\lambda)B(\phi_j;\lambda)d\lambda. \tag{10}$$

In this section we denote the observables by A and B. Recall now the condition for two Euclidean vectors to be equal:

$$\mathbf{E}_{hv}\cdot\mathbf{E}_{qm} = \sum_a E_{hv}^{a}E_{qm}^{a} = \|\mathbf{E}_{hv}\|^2 = \|\mathbf{E}_{qm}\|^2 = \mathbf{E}_{hv}\cdot\mathbf{E}_{hv} = \mathbf{E}_{qm}\cdot\mathbf{E}_{qm}, \tag{11}$$

i.e. their scalar product is equal to the square of their length ($\|\mathbf{E}_{hv}\|$ and $\|\mathbf{E}_{qm}\|$). We expect the vector representing quantum results, $\mathbf{E}_{qm}$ to be different from any possible $\mathbf{E}_{hv}$. The trouble is that we do not know the length of the latter one. The only thing that we know about it is its "canonical" structure Eq.(10). As the quantum prediction, $\mathbf{E}_{qm}$, is fully specified We know $\|\mathbf{E}_{qm}\|$. Thus, if one can show that

$$\mathbf{E}_{hv}\cdot\mathbf{E}_{qm} \neq \|\mathbf{E}_{qm}\|^2, \tag{12}$$

the vectors must be different, *i.e.* the quantum predictions are incompatible with (4). This is the general scheme of the reasoning.

The "hidden variable vector" $\mathbf{E}_{hv}$ is given by Eq.(10). Therefore, one can compare with $\mathbf{E}_{qm}$ a new vector $\mathbf{I}(\lambda)$, defined in the following way:

$$I(\lambda)^a = I(\lambda)^{ij} = A(\theta_i;\lambda)B(\phi_j;\lambda). \tag{13}$$

However, if one is able to show that for every λ

$$\mathbf{E}_{qm}\cdot\mathbf{I}(\lambda) \neq \|\mathbf{E}_{qm}\|^2, \tag{14}$$

this still does not imply (12), as the latter one is an average of (14) over all λ's (compare (4)). Nevertheless if one is lucky enough and establishes that for all λ's one has

$$\mathbf{E}_{qm}\cdot\mathbf{I}(\lambda) < \|\mathbf{E}_{qm}\|^2, \tag{15}$$

then with the help of Eq.(6)

$$\mathbf{E}_{hv}\cdot \mathbf{E}_{qm} < \|\mathbf{E}_{qm}\|^2, \tag{16}$$

which is just enough to show the incompatibility between the quantum and local hidden variable descriptions.

It is clear that the algorithm presented above will not always lead to a success. For certain processes one would not get (14). However, the algorithm generates algebraic forms which are suitable candidates for a Bell inequality for a specific quantum process under consideration:

$$\sum_{ij} A(\theta_i;\lambda)B(\phi_j;\lambda)\mathbf{E}_{qm}(\theta_i,\phi_j). \tag{17}$$

for specific value of the hidden variable λ, or

$$\sum_{ij}\mathbf{E}_{hv}(\theta_i,\phi_j)\mathbf{E}_{qm}(\theta_i,\phi_j). \tag{18}$$

for the hidden variable correlation functions. It will be shown here that (8) is an example of the more general (18).

One should now describe briefly some typical situations that may occur in the stage of the algorithm leading to (15). The most characteristic trait of the quantum functions $E_{qm}(\theta,\phi)$, which usually enter the consideration leading to a form of the Bell theorem, is their nonfactorizability. For example, the two photon polarization correlation function is given by

$$E_{qm}(\theta,\phi) = \cos 2(\theta-\phi) = \cos 2\theta \cos 2\phi + \sin 2\theta \sin 2\phi. \tag{19}$$

On the other hand the function $A(\theta;\lambda)B(\phi;\lambda)$ is obviously a factorizable one.

The components of the quantum vector can be re-expressed as

$$E_{qm}(\theta_i,\phi_j) = \cos 2\theta_i \cos 2\phi_j + \sin 2\theta_i \sin 2\phi_j = v(A1)_i v(B1)_j + v(A2)_i v(B2)_j \tag{20}$$

e.i. in the form of a sum of a *diadic* products of vectors $\mathbf{v}(A1)$ with $\mathbf{v}(A2)$ and $\mathbf{v}(B1)$ with $\mathbf{v}(B2)$. The "A" vectors lie in a k–dimensional real Euclidean space, and their components are given by

$$v(A1)_i = \cos 2\theta_i \quad \text{and} \quad v(A2)_i = \sin 2\theta_i \tag{21}$$

whereas the "B" vectors are n–dimensional and given by

$$v(B1)_j = \cos 2\phi_j \quad \text{and} \quad v(B2)_j = \sin 2\phi_j. \tag{22}$$

One can always demand of the vectors $\mathbf{v}(A1)$ and $\mathbf{v}(A2)$ ($\mathbf{v}(B1)$ and $\mathbf{v}(B2)$) to be orthogonal and of the same norm a (b). This can always be done with suitable choice of θ_i and ϕ_j. Having this property, one immediately sees that

$$\|\mathbf{E}_{qm}\|^2 = \mathbf{E}_{qm} \cdot \mathbf{E}_{qm} = \sum_{ij} E_{qm}(\theta_i,\phi_j) E_{qm}(\theta_i,\phi_j)$$

$$= \sum_i \sum_j (v(A1)_i v(B1)_j + v(A2)_i v(B2)_j)(v(A1)_i v(B1)_j + v(A2)_i v(B2)_j) =$$

$$= \sum_{nm} (\mathbf{v}(An)\bullet\mathbf{v}(Am))(\mathbf{v}(Bn)\bullet\mathbf{v}(Bm)) = 2(ab)^2 \tag{23}$$

where $\bullet$ denotes the scalar product in *an* Euclidean space (k or l dimensional). Further, one can create a vector $\mathbf{A}(\lambda)$ which belongs to the k (l) dimensional space, and its components are given by $A_i = A(\theta_i;\lambda)$; for $\mathbf{B}(\lambda)$ one has $B_j = B(\phi_j;\lambda)$). Thus,

$$\mathbf{E}_{qm}\cdot\mathbf{I}(\lambda) = \sum_{ij} E_{qm}(\theta_i,\phi_j)A(\theta_i;\lambda)B(\phi_j;\lambda)$$

$$= \sum_i\sum_j(\mathrm{v}(A1)_i\mathrm{v}(B1)_j + \mathrm{v}(A2)_i\mathrm{v}(B2)_j)A_iB_j = \sum_n(\mathbf{v}(An)\bullet\mathbf{A}(\lambda))(\mathbf{v}(Bn)\bullet\mathbf{B}(\lambda)). \tag{24}$$

The last expression can be reinterpreted as yet another scalar product.

Let us briefly discuss the case of k and/or n being bigger than 2. The two vectors $\mathbf{u}(A1)$ and $\mathbf{u}(A2)$ ($\mathbf{u}(B1)$ and $\mathbf{u}(B2)$) defined by

$$\mathbf{u}(An) = \mathbf{v}(An)/\|\mathbf{v}(An)\| = \mathbf{v}(An)/a \qquad \left(\mathbf{u}(Bn) = \mathbf{v}(Bn)/\|\mathbf{v}(Bn)\| = \mathbf{v}(Bn)/b\right) \tag{25}$$

are orthonormal, and therefore if k>2 (n>2) they span a non-trivial 2 dimensional subspace of the k- (n-) dimensional Euclidean vector space. The $\mathbf{A}(\lambda)$ ($\mathbf{B}(\lambda)$) vector is a sequence of k (n) numbers ±1. Thus, it is a vector of the k (n) dimensional space, which has highly constrained properties. Its projection $\mathbf{A}(\lambda)^{\parallel}$ ($\mathbf{B}(\lambda)^{\parallel}$) into the 2 dimensional subspace SA (SB) spanned by suitably chosen $\mathbf{u}(A1)$ and $\mathbf{u}(A2)$ ($\mathbf{u}(B1)$ and $\mathbf{u}(B2)$) is never be a trivial one, by which we mean that $\|\mathbf{A}(\lambda)^{\parallel}\|<\|\mathbf{A}(\lambda)\|$ (with an analogous relation for $\mathbf{B}(\lambda)^{\parallel}$). Thus,

$$\mathbf{E}_{qm}\cdot\mathbf{I}(\lambda) = ab\sum_n(\mathbf{u}(An)\bullet\mathbf{A}(\lambda))(\mathbf{u}(Bn)\bullet\mathbf{B}(\lambda))$$

$$= ab\|\mathbf{A}(\lambda)^{\parallel}\|\|\mathbf{B}(\lambda)^{\parallel}\|\sum_n(\mathbf{u}(An)\bullet\mathbf{a}(\lambda))(\mathbf{u}(Bn)\bullet\mathbf{b}(\lambda)^{\parallel}) \leq ab\|\mathbf{A}(\lambda)^{\parallel}\|\|\mathbf{B}(\lambda)^{\parallel}\|, \tag{26}$$

where $\mathbf{A}(\lambda)^{\parallel}/\|\mathbf{A}(\lambda)^{\parallel}\| = \mathbf{a}(\lambda)$ ($\mathbf{B}(\lambda)^{\parallel}/\|\mathbf{B}(\lambda)^{\parallel}\| = \mathbf{b}(\lambda)$). The last estimate is via the Cauchy inequality applied to two 2 dimensional unit vectors, Cartesian components of which are given by $\mathbf{u}(An)\bullet\mathbf{a}(\lambda)$ and $\mathbf{u}(Bn)\bullet\mathbf{b}(\lambda)$.

After all these manipulations our Bell inequality is given by

$$\sum_{ij} E_{hv}(\theta_i,\phi_j)E_{qm}(\theta_i,\phi_j) \leq ab\,\mathrm{Max}_{\lambda}\left(\|\mathbf{A}(\lambda)^{\parallel}\|\|\mathbf{B}(\lambda)^{\parallel}\|\right), \tag{27}$$

and if we are lucky the Bell theorem is proven for the given physical process, and given set of the settings of the parameters θ_i, and ϕ_j, provided

$\mathrm{Max}_{\lambda}\left(\|\mathbf{A}(\lambda)^{\parallel}\|\|\mathbf{B}(\lambda)^{\parallel}\|\right) < 2ab$ (compare with Eq.(23)).

4. Example: the standard Bell Inequality

The above algorithm is capable to reproduce the standard form of the Bell theorem, *i.e.* to show the incompatibility of the $E_{qm}(\theta,\phi)$, given by Eq.(20), with the $E_{hv}(\theta,\phi)$, given by Eq.(10), with the use of *two pairs* of settings of the variables θ and ϕ.

Compare two 4-dimensional vectors (10) and (20), with $i,j = 1,2$. One gets

$$\|\mathbf{E}_{qm}\|^2 = \sum_{ij} E_{qm}(\theta_i,\phi_j)E_{qm}(\theta_i,\phi_j) = \sum_{ij} \cos^2 2(\theta_i-\phi_j). \tag{28}$$

and

$$\mathbf{E}_{qm}\cdot\mathbf{I}(\lambda) = \sum_{ij} E_{qm}(\theta_i,\phi_j)A(\theta_i;\lambda)B(\phi_j;\lambda)$$
$$= \sum_i \cos 2\theta_i A(\theta_i;\lambda)\sum_j \cos 2\phi_j B(\phi_j;\lambda) + \sum_i \sin 2\theta_i A(\theta_i;\lambda)\sum_j \sin 2\phi_j B(\phi_j;\lambda). \tag{29}$$

The two vectors given by Eq.(21) are orthogonal, and of unit norm, if $\theta_2 = \theta_1 + \pi/4 + k\pi/2$ (the same condition for ϕ's). Therefore, $\|\mathbf{E}_{qm}\|^2 = 2$ (compare (22)). Let us now choose $k=0$, thus

$$\mathbf{E}_{qm}\cdot\mathbf{I}(\lambda) = \cos 2(\theta_1 - \phi_1)\Big(A(\theta_1;\lambda)B(\phi_1;\lambda) + A(\theta_2;\lambda)B(\phi_2;\lambda)\Big)$$
$$+ \sin 2(\theta_1 - \phi_1)\Big(A(\theta_1;\lambda)B(\phi_2;\lambda) - A(\theta_2;\lambda)B(\phi_1;\lambda)\Big). \tag{30'}$$

Now, as $(\pm 1)^{-1} = \pm 1$, one can show that the last bracket is equal to

$$B(\phi_1;\lambda)B(\phi_2;\lambda)\Big(A(\theta_1;\lambda)B(\phi_1;\lambda) - A(\theta_2;\lambda)B(\phi_2;\lambda)\Big), \tag{30''}$$

and thus if the modulus of the value of the one of the brackets is 2, the other one must give 0. Therefore, $|\mathbf{E}_{qm}\cdot\mathbf{I}(\lambda)|$ reaches its maximum, equal $2^{-1/2}$, for $2(\theta_1 - \phi_1) = \pi/4 + l\pi$. Thus we have the Bell theorem.

Finally one can show that for $2(\theta_1-\phi_1) = \pi/4$ one recovers from (30) the usual algebraic form leading to the standard Bell inequality (8). Indeed, one has

$$\Big(A(\theta_1;\lambda)B(\phi_1;\lambda) + A(\theta_1+\pi/4;\lambda)B(\phi_2+\pi/4;\lambda)\Big)$$

$$+ \left(A(\theta_1;\lambda)B(\phi_2 + \pi/4;\lambda) - A(\theta_1 + \pi/4;\lambda)B(\phi_1;\lambda)\right) \leq 2. \tag{31}$$

5. Example: Bell inequalities for the entire correlation function

Questions: (i) Is it possible to show that the *entire* quantum correlation function which describes the experimental tests of Bell inequalities is incompatible with local hidden variables? (ii) Can we achieve this task without any use of special settings of the parameters governing the measuring apparata, in contrast with the classic formulations of the Bell theorem? The answer to both questions is **positive**. Due to its periodicity, we can analyze the properties of the quantum correlation function within the formalism of the Hilbert space of real square integrable functions defined on the set $[0,2\pi]\times[0,2\pi]$, *i.e.* $\mathcal{L}^2([0,2\pi]\times[0,2\pi])$. *Two real functions* E_{qm} and E_{hv} belonging to a Hilbert space *are equal if and only if their scalar product* (E_{qm},E_{hv}) *is equal to the square of the norms of the two functions* $\|E_{qm}\|^2$ *and* $\|E_{hv}\|^2$. Thus, if one shows that $|(E_{qm},E_{hv})| < \|E_{qm}\|^2$, then $E_{qm} \neq E_{hv}$. The appropriate analysis[2] gives

$$(E_{qm},E_{hv}) = \int_0^{2\pi} d\Phi_1 \int_0^{2\pi} d\Phi_2 E_{hv}(\Phi_1,\Phi_2)\sin(\Phi_1+\Phi_2) \leq 4^2 < \|E_{qm}\|^2 = (E_{qm},E_{qm}) = 2\pi^2 \tag{32}$$

for any E_{hv} of the canonical form Eq.(7). The result can be obtained following a suitable version of the algorithm of Section 3, and with the use of the following simple lemma:

$$\begin{aligned} \operatorname*{Max}_{\alpha} \left(\pi^{-1/2}\int_0^{2\pi} d\Phi\ O(\Phi,\lambda)\sin(\Phi+\alpha)\right) &= \|O(\cdot,\lambda)^{\parallel}\| \leq 4\pi^{-1/2} \\ < \left(\int_0^{2\pi} d\Phi\ |O(\Phi,\lambda)|^2\right)^{1/2} &= \|O(\cdot,\lambda)\| = (2\pi)^{1/2} \end{aligned} \tag{33}$$

where $O(\Phi,\lambda)^{\parallel}$ is a projection of $O(\Phi,\lambda)$ into a 2-dimensional subspace of $\mathcal{L}^2([0,2\pi])$spanned by the basis vectors $\pi^{-1/2}\sin\Phi$ and $\pi^{-1/2}\cos\Phi$. The first inequality (32) is the Bell inequality for the entire range of the correlation function (violated by quantum mechanics by the factor of $\pi^2/8$ – second inequality).

We shall apply now this method to the GHZ correlations[3]. One has to compare the quantum prediction for the N–particle GHZ process[2,3,4],

$$E_{qm} = \langle\Psi(N)|O_1(\phi_1)O_2(\phi_2)...O_N(\phi_N)|\Psi(N)\rangle = \sin(\phi_1+\phi_2+...+\phi_N) \tag{34}$$

where O_n is given by Eq.(2), and

$$|\Psi(N)\rangle = 2^{-1/2}\Big(|r_1\rangle_1|r_2\rangle_2...|r_N\rangle_N + i|1_1\rangle_1|1_2\rangle_2...|1_N\rangle_N\Big), \tag{35}$$

with the canonical representation of a correlation function by local hidden variable theories

$$E_{hv} = \int_\Lambda \rho(\lambda)[O_1(\Phi_1;\lambda)O_2(\Phi_2;\lambda)...O_N(\Phi_N;\lambda)]d\lambda. \tag{36}$$

Following the method leading to (32) one gets[2]

$$|(E_{qm},E_{hv})| \le 4^N < \|E_{qm}\|^2 = 2^{-1}(2\pi)^N. \tag{37}$$

Thus, (for $N\to\infty$) either E_{hv} has the norm which *vanishes* in comparison with the norm of the function predicted by quantum mechanics, or the functions E_{hv} and E_{qm} are *orthogonal* (such *global* conclusions are not accessible via the standard methods). Further, quantum mechanics violates the new inequalities by an amount that grows exponentially with N at hitherto unexpectedly high rate $(\pi/2)^N$. According to the current literature[4] the maximal violation grows at the rate of $2^{N/2}$. Thus *this result restricts the visibility of the sinusoidal multi particle interference pattern which may be still describable by local hidden variable theories to* $2(2/\pi)^N$.

6.Acknowledgements

The author wishes to thank the Department of Natural Sciences of University of Innsbruck, and Professor Anton Zeilinger personally, for enabling him to spend a fruitful year in Innsbruck.

7.References

1. J.F. Clauser and A. Shimony, Rep. Prog. Phys. 41 1881 (1978);
2. M. Zukowski, *Phys. Lett. A* (summer 1993) in print.
3. D.M. Greenberger, M. Horne, A. Shimony, and A. Zeilinger, *Am. J. Phys.* **58** 1131 (1990); N.D. Mermin, *Phys. Rev. Lett.* **65** 3373 (1990).
4. N.D. Mermin, *Phys. Rev. Lett.* **65** 1838 (1990); S.M. Roy and V. Singh, *Phys. Rev. Lett.* A **67** 2761 (1991); S.L. Braunstein, A. Mann and M. Revzen, *Phys. Rev. Lett.* **68** 3259 (1992); M. Ardehali, *Phys. Rev.* A **46** (1992) 5375.

Parallel and Series Configuration of Large-Path-Difference Quantum Interferometers: A Test of Bell's Inequality and a Quantum Eraser in the Time Domain

J. BRENDEL

Physikalisches Institut der Universität Frankfurt am Main
6000 Frankfurt am Main, Germany

Abstract

Two-photon and single-photon interferences in spatially separated, large-path-difference Michelson interferometers are investigated experimentally. When each photon of a pair passes one of the parallel arranged interferometers, non-local interferences are observed, which yield a violation of Bell's inequality. In the case of successively arranged interferometers, the second interferometer works as an eraser of the *welcher Weg* information carried by the single photons leaving the first. Monitoring the transmission time of the photons allows us to restore interferences with high visibilities. Both configurations open up applications in recently suggested quantum-cryptography schemes and show their experimental feasibility.

First we consider the case of parallel arranged interferometers as shown in Fig. 1. Photon pairs are generated by parametric down-conversion and each photon of the pairs is sent to one of the two Michelson interferometers. Due to the energy- and time-correlation of the photons, such devices have been proposed for spin-free tests of Bell's inequality by several authors.[1-3] A first test involving phase and momentum correlations has been performed by Rarity and Tapster.[4] Our experiment is based on energy and time correlations and follows closely a suggestion by Franson.[3] The results have been discussed in detail in a recent publication[5] and are here only summarized briefly.

The total two photon transmission probability (one photon of the pair reaches detector D1 and the other one reaches detector D2) $P_{22}(\delta_1, \delta_2)$ is obtained by summing up the probability amplitudes for the four possible transmission processes:

$$P_{22}(\delta_1, \delta_2) = \langle R^2T^2|(1 + e^{i(\delta_1+\delta_2+\delta_{rs}+\delta_{ri})} + e^{i(\delta_1+\delta_{rs})} + e^{i(\delta_2+\delta_{ri})})|^2\rangle \tag{1}$$

Here δ_1 and δ_2 are the phase differences between the interfering beams and the phases δ_{rs}, δ_{ri} denote random deviations from these phase differences due to the frequency distributions within the signal and idler beam. The processes represented by terms 3 and 4 of the sum constitute a non-interfering constant background since the *large*

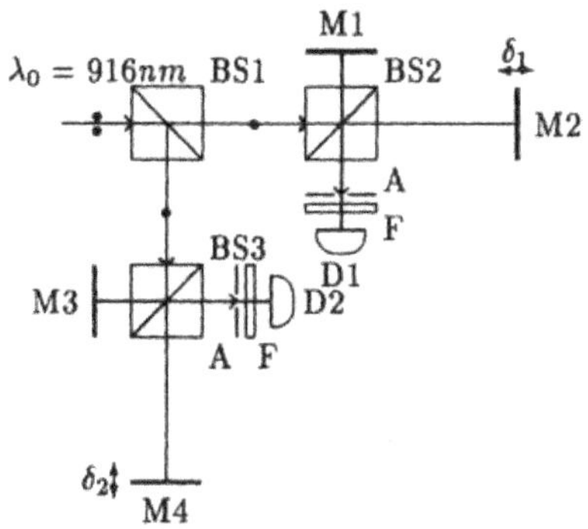

Figure 1: Experimental setup for the parallel configuration.

path differences in the interferometers exceed the coherence length of the light by far. Exclusion of these processes can be achieved experimentally by time-resolved coincidence detection[6] and leads to the transmission probability

$$P_{22}(\delta_1, \delta_2) = 2R^2T^2(1 + V(\delta_1 - \delta_2)\cos(\delta_1 + \delta_2)). \tag{2}$$

The visibility function $V(\delta_1, \delta_2)$ reaches the value of 1 if the phase differences in both interferometers are equal. In practice this optimal value is reduced due to non-perfect alignment and limited time-resolution to a value $0 < V_0 < 1$. The violation of Bell's inequality is usually expressed by a quantity S which is restricted to $S \leq 2$ if the correlations are calculated on the basis of local realistic theories.[7] In contrast, the quantum mechanical calculations yield the sinusoidal correlation function (2). Normalizing this correlation function we find for a reduced visibility V_0: $S = (1 + \sqrt{2}V_0)$. Therefore the inequality $S \leq 2$ is violated if V_0 exceeds a value of $1/\sqrt{2} \approx 0.71$.

The details of our experimental set-up are shown in Fig. 1. Broadband two-photon light with center wavelength $\lambda_0 = 916$ nm is obtained by degenerate parametric down-conversion of pump light from an argon-ion laser in a barium-borate (BBO) crystal. The beam splitter BS1 splits half of the photon pairs and directs correlated single photons to the Michelson interferometers composed of beam splitters BS2 and BS3 and mirrors M1 to M4. The path differences in both interferometers are set to a value of 0.5 m which exceeds the coherence length of the photon pair light by nearly five orders of magnitude, but which is still one order of magnitude smaller than the coherence length of the pump laser light. The phase differences δ_1 and δ_2 can be varied by moving the mirrors M1 and M4 by piezotranslators. The detectors D1 and D2 are photon counting silicon avalanche photodiodes. The output signals of the detectors are fed into a time-to-amplitude converter. By pulse height analysis, the distribution of the time differences between registered photons is obtained. A single-channel analyzer permits setting a variable coincidence-window on this distribution.

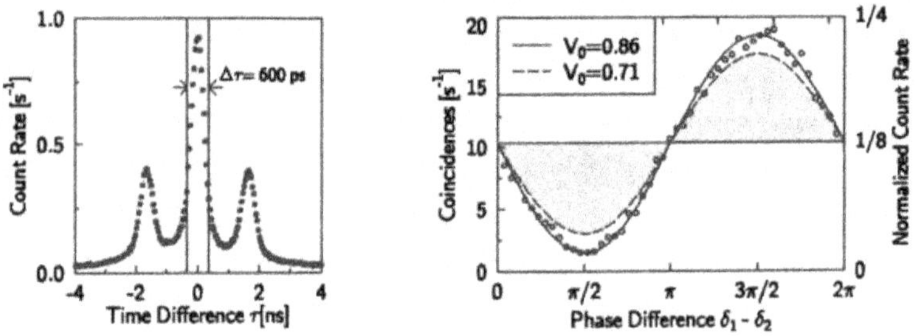

Figure 2: Distribution of the time differences in the detected photon pair signals (left). The vertical lines indicate the applied time-difference window. The figure on the right is a plot of the coincidence count rate showing one period with maximum visibility. The grey region indicates the visibility limit for a violation of Bell's inequality

Fig. 2 shows a plot of the time difference distribution. The central peak corresponds to photon pairs arriving at the detectors simultaneously. In order to cancel the non-interfering events represented by the side-peaks, a coincidence window of 600 ps width is chosen, as indicated by the vertical lines. That way, high visibilities can be achieved. On the right in Fig. 2 a typical interference pattern recorded close to the position of maximum visibility ($\delta_1 \approx \delta_2$)is displayed. The fully drawn sinusoidal line is a fit to the experimental data represented by circles. The value of the fit parameter V_0 is 0.86. From several similar runs we find a mean value of $V_0 = 0.855 \pm 0.016$ which leads to a violation of the inequality by 9 standard deviations.

As recently suggested, the non-local correlations observed in experiments testing Bell's inequality can be applied in quantum cryptography schemes of secure communication.[8] Another kind of quantum cryptography schemes based on single photons as carriers of the information has been discussed by Bennett et al.[9,10] In most experiments proposed and performed until now, the information is transmitted via photon polarization, and single photon states have been replaced by pulsed coherent light beams with mean values of much less than one photon per pulse. Instead of polarization this transmission of information may also be based on energy or phase measurements with interferometers.

In the following we discuss an experimental realization of such a quantum cryptography scheme which consists of two large-path-difference Michelson interferometers in succession as shown in Fig. 3. In contrast to previous experiments, we apply a localized one-photon state[11,12] by use of parametric down-conversion. This state can be realized if one of the two generated photons serves as a trigger for the second photon which passes the interferometers.[13,14]

We first consider the passage of a single photon through the first interferometer. With the same notation as above, we find the transmission probability P_{11} in

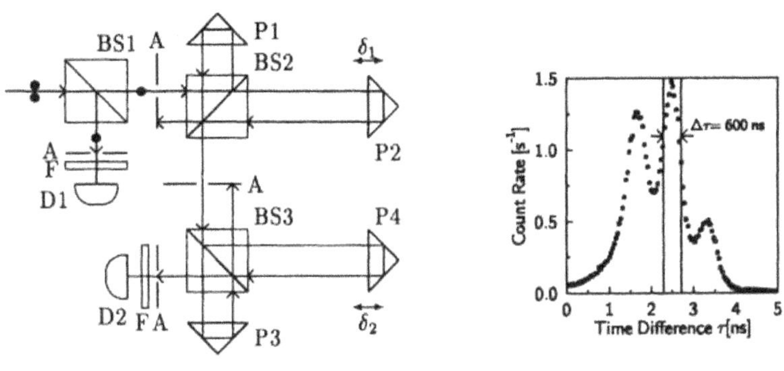

Figure 3: Experimental setup for the series configuration (left) and distribution of the transit time of single photons passing the interferometers (right)

dependence on the phase difference δ_1:

$$P_{11}(\delta_1) = \langle RT|1 + \exp[i(\delta_1 + \delta_r)]|^2\rangle. \tag{3}$$

The average yields the well known single photon transmission function

$$P_{11}(\delta_1) = 2RT[1 + V(\delta_1)\cos(\delta_1)] \tag{4}$$

with the visibility $V(\delta_1)$, which is given by the Fourier transform of the spectral distribution of the light. In our case, the path difference exceeds the coherence length of the light by far, therefore we have $V(\delta_1) = 0$. In other words, the temporal spread of the wave packet is much smaller than the time delay caused by the passage through the long arm in comparison to the short arm. Therefore *welcher Weg* information can be obtained by monitoring the transmission time and no interferences will be found.

When the photons now pass a second Michelson interferometer with phase difference δ_2, the total transmission probability $P_{11}(\delta_1, \delta_2)$ is given by the product of the probability amplitudes for the single interferometers (Eq. 3).[15] This product yields

$$P_{11}(\delta_1, \delta_2) = \langle R^2T^2|1 + \exp[i(\delta_1 + \delta_r)] + \exp[i(\delta_2 + \delta_r)] + \exp[i(\delta_1 + \delta_2 + 2\delta_r)]|^2\rangle. \tag{5}$$

Under the assumption of large path differences and with the substitution $V(\delta_1 - \delta_2) = \langle\cos(\delta_1 - \delta_2)\Delta\omega/\omega_0\rangle$ we find

$$P_{11}(\delta_1, \delta_2) = 2R^2T^2[2 + \cos(\delta_1 - \delta_2)V(\delta_1 - \delta_2)]. \tag{6}$$

For $\delta_1 = \delta_2$ the function $V(\delta_1 - \delta_2)$ reaches its maximum value of 1 and interferences with a visibility of 50% can be observed. The visibility is limited since only the probability amplitudes of the short-long and the long-short transmission processes

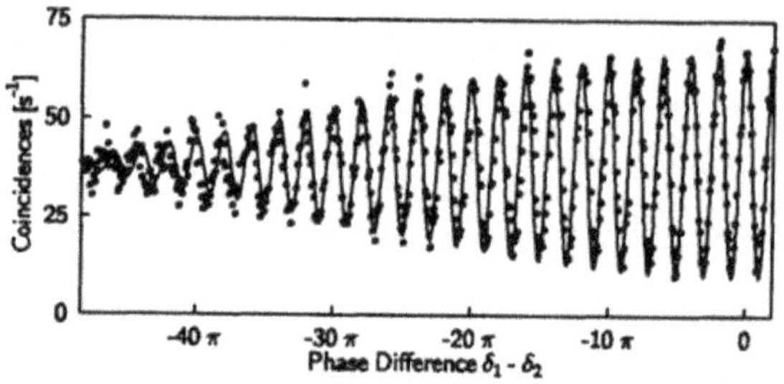

Figure 4: Count rate of detector D2 as a function of the phase difference $\delta_1 - \delta_2$. Only coincidences with detector D1 within the coincidence-window $\Delta\tau$ are recorded.

interfere, whereas the other processes contribute to a non-interfering background. Eliminating this background again by time resolved coincidence detection, we obtain

$$P_{11}(\delta_1, \delta_2) = 2R^2T^2[1 + \cos(\delta_1 - \delta_2)V(\delta_1 - \delta_2)]. \qquad (7)$$

The visibility now can reach a maximum value of 100%.

For the details of our experimental device we refer again to Fig. 3. The photon pair source and the coincidence-electronics is similar to the case of the parallel interferometers. Now one of the photons is directed by beamsplitter BS1 to the detector D1 which provides the trigger signal. The second photon passes the interferometers consisting of the beamsplitters BS2 and BS3 and the retro reflecting prisms P1 to P4. The detector D2 counts the single photons which have traversed both interferometers. The phase differences δ_1 and δ_2 can be varied by piezotranslators which shift the prisms P2 and P4. Both interferometers are set to a path difference of about 25 cm, corresponding to a delay time of about 0.8 ns.

The time difference distribution between photons registered by the monitor detector D1 and the detector D2 is plotted in Fig. 3 (right). The three peaks correspond to the three possible transmission times of the photons. The left peak (short time) corresponds to the short-short process, the center peak (medium time) to the short-long and long-short process and the right peak (long time) to the long-long process. The application of the 600 ps time-window indicated by the vertical lines excludes the non-interfering processes represented by the side peaks from detection.

Fig. 4 is a plot of the coincidence count rate within this window, while the path difference in the second interferometer is scanned over a region of about the coherence length of the light. For large phase differences $\delta_1 - \delta_2$ practically no interference is found whereas fringes with a maximum visibility of 78% are observed at the position of equal path differences in both interferometers.

This recovery of interference by the second interferometer can be interpreted as erasing[16,17] the *welcher Weg* information imprinted on the photons by the first interferometer. Since this information is carried by single photons only, our experimental device can be called a *quantum eraser.*

In the terminology of quantum cryptography we imprint a phase information on the single photons leaving the first interferometer which can be revealed using a second interferometer set to the same path difference. Since the transmission is based on a one-photon state, an eavesdropper cannot detect the photon without being discovered. If the eavesdropper tries to measure the photons energy, he will disturb the correspondence between the data of the sender and the receiver. Due to the high visibility observed in our experiments, such discrepancies can be detected with high certainty. In comparison with classical pulsed light beams, this true single photon scheme may allow higher transmission rates and higher security.

Another interesting question is how this eraser-scheme works if both photons of the pairs are sent through the interferometers. Then we may have an eraser for two-photon *welcher Weg* information and a quantum cryptography channel with a two-photon state. Experimental results concerning this phenomenon will be subject of a future publication.

Financial support by the Deutsche Forschungsgemeinschaft is greatefully acknowledged.

References

1. M. A. Horne and A. Zeilinger, in *Symp. Foundations of Modern Physics*, edited by P. Lahti and P. Mittelstaedt (World Scientific, Singapore, 1985), pp. 435–439.
2. M. Żukowsky and J. Pykacz, *Phys. Lett.* **127** (1988) 1.
3. J. D. Franson, *Phys. Rev. Lett.* **62** (1989) 2205.
4. J. G. Rarity and P. R. Tapster, *Phys. Rev. Lett.* **64** (1990) 2495.
5. J. Brendel, E. Mohler, and W. Martienssen, *Europhys. Lett.* **20** (1992) 575.
6. J. Brendel, E. Mohler, and W. Martienssen, *Phys. Rev. Lett.* **66** (1991) 1142.
7. J. F. Clauser and A. Shimony, *Rep. Prog. Phys.* **41**, (1978) 1881.
8. A. K. Ekert, *Phys. Rev. Lett.* **67** (1991) 661.
9. C. H. Bennett and G. Brassard, *Phys. Rev. Lett.* **68** (1992) 557.
10. C. H. Bennett *et al.*, *J. Cryptology* **5** (1992) 3.
11. C. K. Hong and L. Mandel, *Phys. Rev. Lett.* **56** (1986) 58.
12. E. R. Pike and S. Sarkar, *Europhys. Lett.* **7** (1988) 581.
13. P. Grangier, G. Roger, and A. Aspect, *Europhys. Lett.* **1** (1986) 173.
14. J. Baldzuhn, E. Mohler, and W. Martienssen, *Z. Phys. B* **77** (1989) 347.
15. R. P. Feynman, R. B. Leighton, and M. Sands, *The Feynman Lectures on Physics* (Addison-Wesley, Reading, MA, 1965), Vol. III (Quantum Mechanics).
16. M. O. Scully and K. Drühl, *Phys. Rev. A* **25** (1982) 2208.
17. P. G. Kwiat, A. M. Steinberg, and R. Y. Chiao, *Phys. Rev. A* **45** (1992) 7729.

TWO-PARTICLE-FRINGES DEPENDENT ON THE SUM OF THE COORDINATES

M. HORNE
Stonehill College, North Easton, MA 02357, USA

D. GREENBERGER
City College of New York, New York, NY 10023, USA

A. ZEILINGER
Univ. Innsbruck, A-6020 Innsbruck, Austria

ABSTRACT

A theoretical description is given of a new type of two-particle fringe. With standard conditioned fringes, the joint probability for detecting the two particles depends on the *difference* $x_1 - x_2$ of their positions. The joint probability considered here depends on the *sum* $x_1 + x_2$ of the positions. Two experiments are proposed that would exhibit these fringes. The first manipulates two beams of down-conversion radiation. The second depends on creating transverse standing waves inside the nonlinear medium by manipulating the pump radiation.

Consider two sources, A and B, emitting two particles which are subsequently caught at point x_1 and x_2 on a receiving plane. The quantum description of such two-particle emissions admits two different extremes of entanglement, one of which has not been previously pointed out.

In the first, the well-emphasized extreme, it is known that *the two particles came from different sources* (or alternatively via different routes from a single source), but it is unknown which particle came from which source (or via which route).[1] Then the two-particle amplitude density $a(x_1, x_2)$ at points x_1 and x_2 is the entanglement

$$a(x_1, x_2) \sim a(x_1|A)a(x_2|B) + a(x_1|B)a(x_2|A),$$

where $a(x_1|A)$ is the single particle amplitude to land at x_1 starting from source A (or to land at x_1 via beam A), etc. If the radiation is monochromatic with wave number k, then $a(x_1|A)$ is given by $e^{ikr_{1A}}/r_{1A}$ where r_{1A} is distance from A to point x_1 (or $e^{i\vec{k}_A \cdot \vec{r}_1}$ where $\vec{k}_A$ is the wave vector of beam A and $\vec{r}_1$ the vector position of point x_1). Under Fraunhofer conditions the denominators can be ignored and we obtain

$$a(x_1, x_2) \sim \cos\left[k \sin\Theta\, (x_2 - x_1)\right],$$

where 2Θ is the angular separation of the sources as seen from the receiving plane (or the crossing angle of the beams A and B as they illuminate the plane). This amplitude implies that the coincident count rate in a pair of small detectors will oscillate sinusoidally as the *relative* separation of the detectors is changed[2].

For the new extreme, suppose it is known *that the two particles came from the same source* (or via the same path from a single source), but it is unknown which source (or path) that is. Then the entanglement is

$$a(x_1,x_2) \sim a(x_1|A)a(x_2|A) + a(x_1|B)a(x_2|B),$$

which leads to

$$a(x_1,x_2) \sim \cos\left[k \sin\Theta(x_2 + x_1)\right].$$

This amplitude again implies sinusoidal oscillations of the coincident count rate, but now the fringes have a completely different character: they depend on the *absolute* positions of the two detectors. Because of this particular position dependence, L. Mandel has suggested at this conference that such sum fringes could be simply observed by translating a special single detector: special in that it registers only pairs.

Of course, to observe the fringes they must first be prepared. There are a variety of ways this could be done by using down-conversion radiation. For example, recall that when two symmetric beams of down-conversion radiation are brought via mirrors to a crossing, the difference amplitude is prepared[2]. However, if a beam splitter is placed at that crossing and the outgoing beams are made to cross a second time, the sum amplitude is prepared. Thus one would only need to remove the second beam splitter in the arrangement of Rarity et al[3] and then study that second crossing region either with two detectors[2] or else with a single two-particle detector as suggested here by Mandel. As a second example of preparation, the down-conversion crystal may itself be illuminated by crossing beams of pump radiation. Then the sum amplitude will be present at the first crossing of two-beams of the down-conversion radiation.

This work was supported by the Austrian Science Foundation FWF, project No. S6502 (Schwerpunkt Quantenoptik), and the US National Science Foundation, grant No. PHY92-13964.

References

1. L. Mandel, *Phys. Rev.* **A28** (1983) 929. Our group independently discussed such situations at conferences since 1985, see e.g. M. Horne, A. Zeilinger: A Possible Spin-Less Experimental Test of Bell's Inequality, in *Microphysical Reality and Quantum Formalism*, eds. A. van der Merwe, F. Selleri, G. Tarozzi, (Kluwer, Dordrecht; 1988) 401.
2. R. Ghosh and L. Mandel, *Phys. Rev. Lett.* **59** (1987) 1903.
3. J. G. Rarity, P. R. Tapster, E. Jakeman, T. Larchuk, R. A. Campos, M. C. Teich, B. E. A. Saleh, *Phys. Rev. Lett.* **65** (1990) 1348-1351.

SQUEEZING

Quantum non-demolition measurements in optics and quantum optical repeaters

Jean-François Roch, Jean-Philippe Poizat, and Philippe Grangier

Institut d'Optique, B.P. 147 - F91403 ORSAY Cedex - France

ABSTRACT

The efficiency of an optical quantum non-demolition (QND) measurement can be characterized using three criteria, which describe respectively the quality of the quantum measurement, the non-destruction of the signal, and the conditional variance of the output signal beam, given the output meter beam (quantum-state-preparation criterion). Quantitative limits can be defined with respect to these criteria, delimiting "classical" and "quantum" domains of operation. We then briefly describe the implementation of two experiments which fulfill these criteria, using either three-level atoms inside a doubly-resonant optical cavity, or semiconductors emitters and receivers.

1 Introduction

A quantum measurement usually perturbs the quantity which is measured, adding back-action noise to the system under study. However, quantum non-demolition (QND) measurements can be performed, leaving the observed quantity unperturbed, while adding the back-action noise into another complementary observable. Quantum non- demolition measurements were first proposed to monitor the motion of mechanical oscillators [1] [2] but it was then realized that they were easier to implement in the optical domain [3] [4] [5]. In this case, the measurements are made on a mode of the electromagnetic field considered as a quantum harmonic oscillator, and the coupling responsible for the measurement is created using either χ^2 or χ^3 optical non-linearities [6] [7] [8]. In this paper, we will focus on the QND measurements of "travelling wave" light beams, rather than fields stored in high Q cavities [9]. Moreover, we will assume that the quantum fluctuations of the beams are very small compared to the mean intensities, so that a linear treatment of these quantum fluctuations is possible [10]. A "travelling wave" optical QND scheme usually involves the coupling of two field modes : one "signal" mode, which is to be measured, and one "meter" mode, which is directly detected and yields some informations about the signal mode. A perfect QND measurement device should leave the measured quadrature component of the signal mode unchanged, and turn the detected meter quadrature component into some kind of "macroscopic copy" of the signal. However, such a device does not exist at present time, and it appears necessary to be able to characterize the "non-ideality" of practical QND schemes. We will briefly recall the criteria which have been proposed for this characterization [11] [12] [13], and we will define quantitative limits with respect to these criteria, delimiting "classical" and "quantum" domains of operation. Finally, two experiments, meeting all criteria for QND measurements, will be described in the last sections of the paper.

2 Characterization of a QND device

In QND experiments, the performance of the device has usually been assessed from a measurement of some components of the output covariance W^{out}, which contains the information about the noises and the quantum correlations of the two output beams. This procedure allows one to characterize the "quantum state-preparation" (QSP) properties of the device. A convenient quantity for that purpose is the conditional variance W_{QSP} of the output signal (taken here as X_s^{out}), given the output meter (taken here as Y_m^{out}) [12]. This quantity is equivalent to a squeezing spectrum, and it can be defined from a suitable combination of the various matrix elements of the output covariance W^{out} :

$$W_{QSP} = W^{out}_{X_sX_s} - | W^{out}_{X_sY_m} |^2 / W^{out}_{Y_mY_m} \tag{1}$$

An important advantage of W_{QSP} is that it can be measured directly in an experiment, as a noise level on a spectrum analyser. For that purpose, the signal and meter photocurrents must be recombined, introducing an electronic gain on the meter photocurrent. This gain (or attenuation) must be optimized, in order to obtain the maximum reduction below the shot-noise level of the signal beam. The criterion $W_{QSP} < 1$ means that the device is able to establish sub-shot-noise correlations between the two output beams. It is instructive to give the value of W_{QSP} for a very simple model, corresponding to the following "parametric" coupling between the quadratures components X and Y of the signal and meter modes :

$$\begin{array}{ll} X_s^{out} = X_s^{in} & Y_s^{out} = Y_s^{in} + GX_m^{in} \\ X_m^{out} = X_m^{in} & Y_m^{out} = Y_m^{in} + GX_s^{in} \end{array} \tag{2}$$

This model can be implemented for instance using crossed Kerr effect between the two modes, obtained using two-photon non-linearities [12] [14]. In this case, X and Y are respectively the amplitude and phase fluctuations, defined relatively to the mean values of the fields. For a shot-noise limited meter beam, one then obtains:

$$W_{QSP} = 1/(1 + G^2) \tag{3}$$

A perfect QSP device is obtained for an infinite value of G. Another useful model is a simple beamsplitter ("linear coupler"), in which the input meter beam can be in a squeezed state [7]. Denoting by r, t the amplitude reflection and transmission coefficients of the beamsplitter, the input-output transformation is simply :

$$\begin{array}{ll} X_s^{out} = tX_s^{in} + rX_m^{in} & X_s^{out} = t\,Y_s^{in} + r\,Y_m^{in} \\ X_m^{out} = -\,rX_s^{in} + tX_m^{in} & Y_m^{out} = -\,r\,Y_s^{in} + t\,Y_m^{in} \end{array} \tag{4}$$

In that case, the information on the signal amplitude X_s^{in} is obtained on the meter output amplitude X_m^{out}. If the corresponding quadrature of the input meter beam has a noise power squeezing S^{in}, the QSP properties of this scheme is then given by :

$$W_{QSP} = \frac{S^{in}}{(r^2 + t^2 S^{in})} \tag{5}$$

If the input meter beam is in the vacuum state $S^{in} = 1$, one obtains immediately $W_{QSP} = 1$, for any value of r and t . On the other hand, the inequality $W_{QSP} < 1$ is obtained as soon as the meter beam is squeezed, and perfect QSP is obtained for $S^{in} = 0$. This evidences the close relationship between the conditional variance and a degree of squeezing. However, this criterion does not characterize completely the QND properties of the device, because one must also prove that the output signal and meter beams really yield an information about the input signal beam [12] [13].

The basic idea that we will use is that two types of quantities can actually be measured experimentally. The first one is the output covariance matrix W^{out} of the quadrature components. The second quantity is the gain matrix $\lambda(\omega)$, which represents the transfer function of the system for fluctuations of the input fields, and can be determined using weak classical probe beams. Knowing the appropriate coefficients of $\lambda(\omega)$ and W^{out}, one may write linear "estimators" for the measured quadrature component of the signal. For instance, in the parametric case, defined by Eq.(2), one has:

$$G^{-1}\,Y_m^{out} = X_s^{in} + G^{-1}\,Y_m^{in} \tag{6}$$

$$G^{-1}\,W_{Y_mY_m}^{out}(G^{-1})^* = W_{X_sX_s}^{in} + N_m^{eq} \tag{7}$$

$$N_m^{eq} = |G|^{-2}\,W_{Y_mY_m}^{out} - W_{X_sX_s}^{in} \tag{8}$$

The quantity N_m^{eq} appears as the noise added to input signal variance, and is therefore an "equivalent input noise". Here, N_m^{eq} is due to the shot-noise of the meter beam, which appears as a noise even in the parametric case. For real devices, a second noise source is due to the excess noise added by the non-linear system, and a third noise source appears to be the quadrature of the signal conjugated from the observed one. This "unwanted" quadrature can be coupled in the output signal or meter by the non-ideal device. These three degradations of the measurement efficiency can then be included in N_m^{eq}, equivalent input noise on the meter channel [13]. Similar definition can be also be used for relating the output signal to the input signal, corresponding to N_s^{eq}, equivalent input noise on the signal channel.

Another possible characterization of the same properties is obtained by looking at the faithfulness of the transfer of an input signal-to-quantum-noise ratio (SNR or simply R in the equations), toward the signal and meter output channels [6] [15] [16] [17]. The input SNR is defined as the ratio of the power of a classical modulation at a given frequency by the quantum noise power at the same frequency for a given observable X,

$$R = \frac{\langle X\rangle^2}{\langle \Delta X^2\rangle}, \tag{9}$$

where all quantities are defined in the frequency domain. The signal non-degradation property is therefore evaluated by (the subscript s refers to the signal, and m to the meter)

$$T_s = \frac{R_s^{out}}{R_s^{in}} = \frac{W_s^{in}}{W_s^{in} + N_s^{eq}}, \tag{10}$$

where W_s^{in} is the variance of the input light beam normalized to the shotnoise ($W_s^{in} = 1$ for a coherent state), and N_s^{eq} is the equivalent input noise for the signal. The measurement capability is given by

$$T_m = \frac{R_m^{out}}{R_s^{in}} = \frac{W_s^{in}}{W_s^{in} + N_m^{eq}}, \tag{11}$$

where N_m^{eq} is the equivalent input noise for the meter channel.

Let us finally give the values of N_m^{eq} and N_s^{eq} for our model devices. In the parametric crossed-Kerr effect model, one obtains :

$$N_s^{eq} = 0 \qquad N_m^{eq} = |G|^{-2} \tag{12}$$

In the linear coupler with squeezed meter beam, one has :

$$N_m^{eq} = (t/r)^2 S^{in} \qquad N_s^{eq} = (r/t)^2 S^{in} \tag{13}$$

$$N_s^{eq} N_m^{eq} = (S^{in})^2 \tag{14}$$

If the input meter beam is in the vacuum state $S^{in} = 1$, the product of the equivalent input noises is equal to 1, for any value of r and t .

3 Classical limits for QND measurements

In order to discriminate various modes of operation of a "quantum coupler", it is interesting to know where the "quantum limits" are. As shown above, a useful reference is given by a simple beamsplitter with a meter beam in the vacuum state, for which one has $W_{QSP} = 1$ and $N_s^{eq}.N_m^{eq} = 1$ .We have shown more generally that the domain of operation of a "classical coupler", described as a phase-insensitive device, can be defined by the two following inequalities [15] :

$$W_{QSP} \geq 1 \tag{15}$$

$$N_s^{eq}.N_m^{eq} \geq 1 \tag{16}$$

where 1 is by convention the standard shot-noise level. The inequality (19) means that "conditional squeezing" is forbidden in the classical domain, just like ordinary squeezing. The inequality (20) means that the *product* of the signal and measurement equivalent noises is restricted : in the classical domain, small signal degradation implies poor measurement. We note that the definition of "classical" used in inequality (20) is not the usual one in quantum optics (based on the properties of the P function), but is rather the one connected to phase-sensitive vs phase-insensitive amplifiers in the theory of linear amplifiers [18].

Another form of the same inequality was introduced in Ref. [19], for the particular case of a beamsplitter, using the signal to noise ratios (SNR) on the signal and meter channels :

$$(SNR)_s^{out} + (SNR)_m^{out} \leq (SNR)_s^{in} \tag{17}$$

$$T_s + T_m \leq 1 \tag{18}$$

This alternative form is equivalent to the previous one, under some conditions which are fulfilled in general. A device violating this "classical" limit will be called a "quantum optical tap". On the other hand, obtaining $W_{QSP} < 1$ is "quantum state preparation". A true QND device will be defined as a quantum optical tap yielding also quantum state preparation. Finally, let us emphasize that the two conditions (16) and (17) are independant, and cannot be linked by any "triangular" inequality. This can be shown by realizing that a "twin-beam" generator will yield W_{QSP} arbitrarily small, but $N_s^{eq}.N_m^{eq} > 1$ [18]. On the other hand, it is also easy to check that a phase sensitive amplifier followed by a beamsplitter will yield $N_s^{eq}.N_m^{eq}$ arbitrarily small, but $W_{QSP} > 1$. Therefore, both inequalities must be considered to ensure "quantum operation" of a QND device [15].

4 QND experiments with three-level systems

The basic idea of this experiment is to use the optical Kerr effect to couple the intensity fluctuations of the signal beam to the phase fluctuations of the meter beam ("crossed-phase modulation effect") [8] [20] . The information is then read out using some kind of phase-sensitive (interferometric) detection scheme of the meter beam.

The first attempt to use the optical Kerr effect for QND measurements has been made using non-resonant $\chi^{(3)}$ in optical fibers [3] [21] [22]. Other experiments were done using resonant $\chi^{(3)}$ in solid-state systems [12] [23]. Despite the simplicity of the scheme, it has not been possible to obtain noiseless $\chi^{(3)}$ media: either guided acoustic wave Brillouin scattering (GAWBS) in optical fibers, or spontaneous emission noise in resonant systems, introduces an added noise.

In nearly resonant three-level systems, it is possible to obtain a pure crossed-Kerr coupling, without self-phase modulation, by using two-photon non-linearities [5] [14]. For a proper choice of the detunings between the two laser beams and the two transitions, the two-photon dispersion can be made dominant, and the ideal transformation given above can be recovered in principle [24].

The experimental set-up used for this experiment has been improved several times and is described in Ref. [5] [15] and [16]. The non-linear coupling of the signal and meter fields is obtained in a beam of three-level atoms in a ladder configuration ($3s_{1/2}-3p_{3/2}-3d_{5/2}$ levels of sodium atoms). This atomic system has the advantage that the resonant enhancement of the non-linear effect is much larger than in solid- state non-linear systems, and the phase damping of the transitions much smaller. In the latest version of the experiment [16], the sum of the signal and measurement SNR transfers is $T_s + T_m = 1.35$, which can be seen as a 1.3dB improvement of the overall information transfer of any classical optical tap (a beam splitter for example). A perfect quantum optical tap would have $T_s = T_m = 1$ which would give a perfect 3dB improvement, and could also be regarded as a quantum duplicator. The experimental results concerning the quantum state preparation ability of this device can be characterized by the conditional variance W_{QSP} introduced in section 2. The experimental value is $W_{QSP} = 0.85$, which is also clearly in the quantum domain.

5 Quantum Optical Repeaters using semiconductor emitters and receivers

It has been known for several years that the conversion from light to electrical current and from electrical current to light can be realized below the optical standard quantum limit, using high efficiency semiconductor electro-optical devices. The quantum properties of the inverse transformation are due to the work of Yamamoto and coworkers on semiconductor lasers [25] (see also Ref. [26] [27]). More recently, the appearance of commercial high efficiency Light Emitting Diodes (LED) made possible to use these simple devices for generating non-classical light [28] [29] [30]. A crucial idea in these experiments is that the noise of an electrical current at RF frequencies is usually dominated by thermal (Johnson-Nyquist) noise , whose power can be made much smaller than the shot-noise level corresponding to the same average current intensity [27] [28]. In this regime, the electrical current behaves classically, and can therefore be measured, duplicated, or amplified without the quantum constraints attached to a light field. If one is concerned by the non destructive measurement of the intensity of a light beam, one may imagine to convert the light via a photodiode into an electrical current, to measure this current, eventually to amplify it, and to convert it back to a light field using a semiconductor light emitter [31].

A check of these ideas can be carried out using the experimental set-up schematically depicted in Fig.1. The measurement apparatus itself consists in a large area p-i-n silicon photodiode PD_1 of quantum efficiency $\eta = 0.90 \pm 0.05$, followed by a low noise electronic amplifier, and a light emitting diode LED_2. The current to current conversion rate of the LED - PD system is $\epsilon = 0.16$ at room temperature and $\epsilon = 0.28$ at 100K . The detectors are placed as close as possible to the LEDs to maximize light collection, and introduced in a liquid nitrogen cryostat. The bandwidth of operation is from 150kHz to 350kHz, the lower limit being due to electronic cut-off, and the upper one to the finite response time of the LEDs. From these numbers, it appears that if the LED is series connected with the photodiode, without any amplifier, the transfer coefficients for a shot-noise-limited

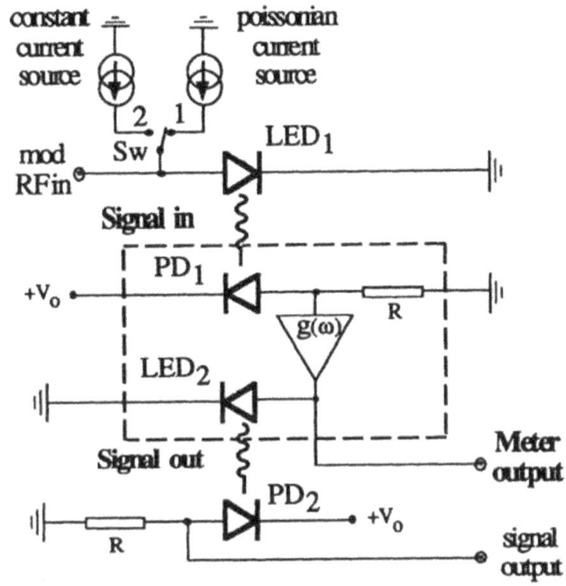

Figure 1: Schematic experimental set-up. The measurement device is enclosed in the dashed box. $LED_{1,2}$ and $PD_{1,2}$ are cooled at 100K. The poissonian current source consists in three parallel connected photodiodes, illuminated by white incandescent lamps. When switch Sw is in position 1, LED_1 is driven by a shotnoise limited current; when in position 2, LED_1 is driven by a constant (filtered) current, and shines therefore squeezed light on PD_1. LED_2 can be also driven by the poissonian current source, in order to get a shotnoise reference on PD_2. Note that, for clarity, all the electrical filtering is not represented in the figure.

input will be respectively $T_s = \eta\epsilon$ and $T_m = \eta$, leading to a rather small effect ($T_s + T_m = 1.15 \pm 0.06$ at 100K) mainly due to the low conversion rate of the LED. In order to overcome this limitation, the current coming from the photodiode is amplified prior to being sent to the LED, which makes the conversion back to light less sensitive to vacuum fluctuations [32], and therefore greatly improves the non-degradation property of the system without changing the others.

For a shot-noise-limited input, and assuming a negligible amplifier noise, the transfer coefficients read

$$T_m = \eta \tag{19}$$

$$T_s = \eta \frac{1}{1 + \frac{1}{g_{\text{eff}}} \frac{(1-\epsilon)}{\epsilon}} . \tag{20}$$

where $g_{\text{eff}} = g^2/G$ with g being the AC gain of the amplifier and G the DC gain. The distinction between AC and DC gain, allowed in our experimental set-up, is an extra degree of freedom given by electronics compared to optics. Note that for

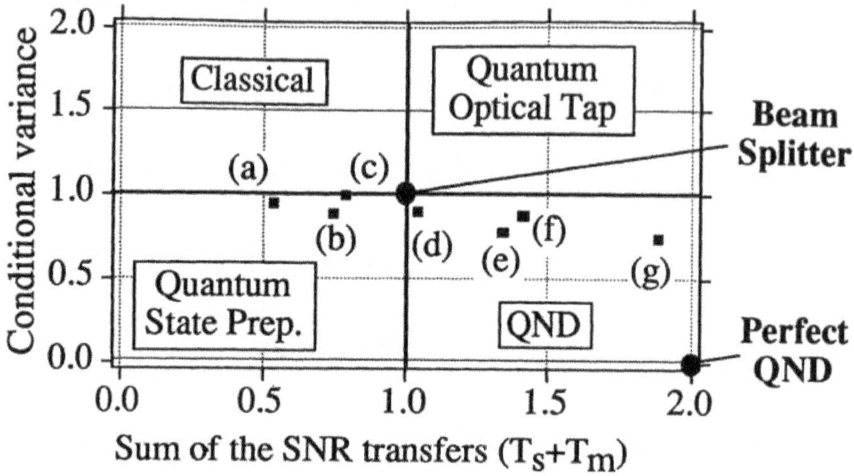

Figure 2: Experimental results plotted in the plane $x = T_s + T_m, y = W_{QSP}$. The point ($x$=1, y=1) is simply a lossless beamsplitter . Points (a), (b), (c), (d), (e) correspond respectively to the experiments described in Ref. [3], [4], [5], [15] and [17]. Points (f) and (g) are the results described in the present paper, which fulfill all criteria for QND measurements.

practical purposes in information transmission, one has to choose $G = g$, which leads to $g_{\text{eff}} = g$. In the high gain limit ($g_{\text{eff}} \to +\infty$), the sum of these two coefficient goes to $T_s + T_m = 2\eta$, and do not depend on LED_2's conversion rate ϵ anymore.On the other hand, the conditional variance is limited by the conversion rate of LED_2, and is given by

$$W_{QSP} = 1 - \epsilon. \tag{21}$$

The experimental results at the high intermediate gain limit ($g_{\text{eff}} = 100$) are $T_s = T_m = \eta = 0.90 \pm 0.05$, i.e. $T_s + T_m = 1.8 \pm 0.1$, and $W_{QSP} = 0.76 \pm 0.02$ (with a theoretical value of 0.72). All these results clearly show that our scheme does operate in the QND regime. The perspective of emergence of very high efficiency single-mode LED [33], or alternatively the technological realization of low threshold laser diodes [34], makes this type of system very promising for a realistic implementation in ultra-low noise optical telecommunication networks.

6 Conclusion

On Fig.2, the results obtained in some optical QND experiments realized to present date are plotted as a function of the values of the conditional variance and of $T_s + T_m$. The best QND result in the usual sense (signal gain close to unity) is the three-level experiment described above (point f), while the LED experiment just described (point g) appears to be the best "quantum repeater" experiment (large signal gain). However, a fair comparison should also take into account also the frequency bandwidth over which the QND effects are achieved. On that respect, pulsed experiments (e.g. point e) have been so far the most successful.

7 Acknowledgements

This work was completed in the framework of the EC Contract ESPRIT III BRA 6934, with support from the Centre National d'Etude des Télécommunications (France Telecom). We acknowledge efficient technical assistance of G. Roger, M. Allain, F. Contet, J. C. Rodier and A. Villing. Institut d'Optique Théorique et Appliquée is URA 14 of the Centre National de la Recherche Scientifique.

References

[1] C.M. Caves, K.S. Thorne, R.W.P. Drever, V.D. Sandberg, and M. Zimmermann, *Rev.Mod.Phys.* **52** (1980) 341 (1980).

[2] V.B. Braginsky, Y.I. Vorontsov and K.S. Thorne, *Science* **209** (1980) 547.

[3] M.D. Levenson, R.M. Shelby, M. Reid and D.F. Walls, *Phys.Rev.Lett.* **57** (1986) 2473.

[4] A.LaPorta, R.E.Slusher and B.Yurke,*Phys.Rev.Lett.* **62**, (1989) 28.

[5] P Grangier, J.F. Roch and G. Roger,*Phys. Rev. Lett.* **66** (1991) 1418.

[6] B. Yurke, *J.Opt.Soc.Am.* **B2** (1985) 732.

[7] J.H. Shapiro, *Optics Letters* **5** (1980) 351.

[8] G.J. Milburn and D.F. Walls, *Phys. Rev.* **A28** (1983) 2065.

[9] M. Brune, S. Haroche, V. Lefèvre, J.M. Raimond and N. Zagury, *Phys. Rev. Lett.* **65** (1990) 976.

[10] J.M. Courty, P.Grangier, L. Hilico and S. Reynaud, *Opt. Commun.* **83** (1991) 251.

[11] N. Imoto and S. Saito, *Physical Review* **A39** (1989) 675.

[12] M.J. Holland, M. Collett, D.F Walls and M.D. Levenson, *Phys. Rev.* **A42** (1990) 2995.

[13] P. Grangier, J.M. Courty and S. Reynaud, *Opt. Commun.* **89** (1992) 99.

[14] P.Grangier, J.F. Roch and S. Reynaud, *Opt. Commun.* **72** (1989) 387.

[15] J. F. Roch, Thèse de l'Université Paris XI, (unpublished, 1992).

[16] J.F. Roch, G. Roger, P. Grangier, J.M. Courty and S. Reynaud, *Appl. Phys.* **B55** (1992) 291 (1992).

[17] A. Levenson, I. Abram, P. Fayolle, Th. Rivera, J.C. Garreau and P. Grangier, *Phys. Rev. Lett.* **70** (1993) 267.

[18] J. F. Roch, Thèse de l'Université Paris XI, (unpublished, 1992).

[19] Y. Yamamoto, *Trans. Inst. Electron. Inf. Commun. Eng.* Sect. E**73** (1990) 1598.

[20] N. Imoto, H.A. Haus and Y. Yamamoto, *Phys. Rev.* A**32** (1985) 2287.

[21] N. Imoto, S. Watkins and Y. Sasaki, *Opt. Commun.* **61** (1987) 159.

[22] H.A. Bachor, M.D. Levenson, D.F. Walls, S.H. Perlmutter and R.M. Shelby, *Phys. Rev.* A**38** (1988) 180.

[23] M.D. Levenson, M.J. Holland, D.F. Walls, P.J. Manson, P.T.H. Fisk, and H.A. Bachor,*Phys. Rev.* A**44** (1991) 2023.

[24] J.P. Poizat, M.J. Collett and D.F. Walls, *Phys. Rev.* **45** (1992) 5171.

[25] S. Machida, Y. Yamamoto, and Y. Itaya, *Phys. Rev. Lett.* **58** (1987) 1000.

[26] M. C. Teich and B. E. A. Saleh, *J. Opt. Soc. Am.* B **2** (1985) 275.

[27] P. R. Tapster, J. G. Rarity, and J. S. Satchell, *Europhys. Lett.* **4** (1987) 293.

[28] P. J. Edwards, *Int. J. Optoelectronics* **6** (1991) 23.

[29] H.-A. Bachor, P. Rottengatter, and C. M. Savage, *Appl. Phys.* B **55** (1992) 258.

[30] E. Goobar, A. Karlsson, G. Björk, and P-J. Rigole, *Phys. Rev. Lett.* **70** (1993) 437.

[31] F. Capasso and M. C. Teich, *Phys. Rev. Lett.* **57** (1986) 1417.

[32] H. P. Yuen, *Phys. Rev. Lett.* **56** (1986) 2176.

[33] E. Yablonovitch, T. J. Gmitter, R. D. Meade, A. M. Rappe, K. D. Brommer, and J. D. Joanopoulos, *Phys. Rev. Lett.* **67** (1991) 3380.

[34] Y. Yamamoto and R.E. Slusher, Physics Today, (June 1993) p.66.

THE ADIABATIC-FEEDBACK MEASUREMENT METHOD AND THE SINGLE-PARTICLE QUANTUM-ERASER

F. DE MARTINI, L. DE DOMINICIS, E. DEL RE and F. TRUC
Dipartimento di Fisica dell'Universita' "La Sapienza", Roma 00185 Italy

ABSTRACT

We report the first demonstration of the *single-photon quantum-eraser* an apparatus by which the manifestation of a quantum-superposition state may be cancelled and further retrieved before state reduction by a detector. The single-photon state is generated by application of a new *"adiabatic feedback"* method reported here for the first time. By this method, which realizes a peculiar *quantum back-reaction* process, highly sub-poissonian distributions with different <n> can be generated in a controlled way. A new kind of photon interferometer is presented that can exhibit the features of *welcher-weg non locality*, *quantum erasure* and *delayed-choice*.

THE ADIABATIC-FEEDBACK MEASUREMENT METHOD AND THE SINGLE-PARTICLE QUANTUM-ERASER

F. DE MARTINI, L. DE DOMINICIS, E. DEL RE and F. TRUC
Dipartimento di Fisica dell'Universita' "La Sapienza", Roma 00185 Italy

1. Introduction

Single-particle interferometry is rooted at the core of modern physics since it provides the most direct demonstration of natural properties as complementarity and quantum nonlocality. According to Feynman it is a phenomenon "..which has in it the heart of quantum mechanics. In reality it contains the *only* mystery.." [1]. It appears then highly significant to base on this unique, most fundamental phenomenon any investigation over the subtle aspects of Bohr's complementarity. The aim of the present work is to provide a demonstration by single-photon, first-order (viz., field's) interferometry of the modern concept of *quantum-erasing*. This process, somewhat inspired by J. A. Wheeler's suggestion of the *delayed-choice* effect, was in fact originally proposed under the form of a gedanken experiment in connection with two-atom emission and, more recently, with atomic micromaser interferometry [2]. The key argument underlying the "erasing" concept in connection with single-particle interferometry may still be expressed by another Feynman's statement, following the quoted one, dealing with the issue of determining which path ("*welcher-weg*": ww) the particle has taken without erasing the fringe pattern on the output screen of the apparatus:"...If an apparatus is capable of determining which hole the electron goes through, it *cannot* be so delicate that it does not disturb the pattern in an essential way..." [1]. In agreement with the Bohr's reply to a celebrated ww-argument by Einstein, Feynman referred there to the uncontrollable phase shift induced on the particle's wavefunction $|\Phi\rangle$ by any scattering process that could provide the ww-information [3]. However it was shown theoretically [2], and we demonstrate in this work, that the interaction of the single-particle with the experimental apparatus, here the interferometer, *can* indeed be so delicate as to allow for the ww-information without impairing the possibility of recovering the interference pattern by acting on $|\Phi\rangle$ at the *detection* stage, *after* the particle "has left" the interferometer. In this way we obtain a "quantum erasure" (QE) of the perturbation induced on $|\Phi>$ by the previous interaction and then a recovering of the superposition state, i.e. a "revival" of the coherence related to the unperturbed dynamics of the particle. As already remarked, though all that does not violate complementarity it is somewhat at variance with some aspects of the common understanding of the measurement process [2,4,5].

2. The Adiabatic-Feedback Measurement Method

A Fock photon $|1>$-state at λ=632.8 nm was created by a particular application of a new *"adiabatic feedback"* method aimed at the generation of approximately pure particle $|n>$-states. The method is described with the help of Fig.1. Assume the "experimental area", i.e., a "box" of linear dimension ~L containing the overall experiment. Assume that a single-mode, monochromatic photon-beam with coherence-time τ_c is injected into a single input port of the "box" and that within this box no photon dissipation or transformation into undetectable radiation takes place. According to the standard quantum-mechanical conception, the final outcome of *any* experiment (as well as the physical meaning of "experiment" itself) implies a state reduction by one, or by a set of detectors by which the "significant" measurement is performed. Within our technique this standard quantum-measurement scheme is completed by providing *every* output port of the experimental *box* with a detector (with quantum-efficiency η). The output signals of *all* detectors (including the *significant* ones that are also used to carry through the measurement for which the experiment is devised) feed a fast computer pre-programmed to allow injection into the "box" of n photons whithin τ_c before activating an electro-optic (EO) shutter which intercepts the input beam for a time about equal to τ_c. We call this a *"total measurement"* procedure. Assume for simplicity, and with no lack of generality, a laser excitation. If $\tau_c >> (L/c+t_f) \equiv T_d$, the time delay of the feedback-loop, being t_f the overall delay due to the electronics, the state of radiation in the "box" is a sharply clipped Poisson distribution with pre-programmed $<n>$, if the average input photon-number $<n> = <m>/\eta$ is not too large. Precisely, for $|1>$-state generation, the probability ratio for injection of n=2 and n=1 is $P_2/P_1 \cong <n>(T_d/\tau_c)$ [6]. The experimental photon-counting distribution presented by the multichannel-analyzer (MCA) histogram of Fig.2 is related to the adiabatic feedback provided by only *one* detector placed right at the input port in the *box* shown in Fig.1. The histogram shows that indeed a very small value of the ratio of the distribution peaks for photocounts m=2 and m=1 can be obtained by the "adiabatic feedback" configuration. In the present experiment this value was P_2/P_1=0.0035, to be compared with the ratio P_2/P_1=0.58 related to the count distribution, also shown in Fig.2, obtained by the *same* experiment with the feedback-loop shut-off. Note that the latter case is well represented by a Poisson distribution, as expected with a laser excitation. The method has been succesfully extended to the generation of sub-poissonian radiation with photocount distribution-clipping at m larger than one [6]. A conceptually interesting, and perhaps important feature of the present method consists of the realization of a peculiar *quantum back-reaction* of the measuring-apparatus on the input quantum-state. Since the feedback process takes place within a short time $\Delta t << \tau_c$, the EO-shutter operation modifies within the *box* the *same* radiation wavepacket that has triggered it. In other words,

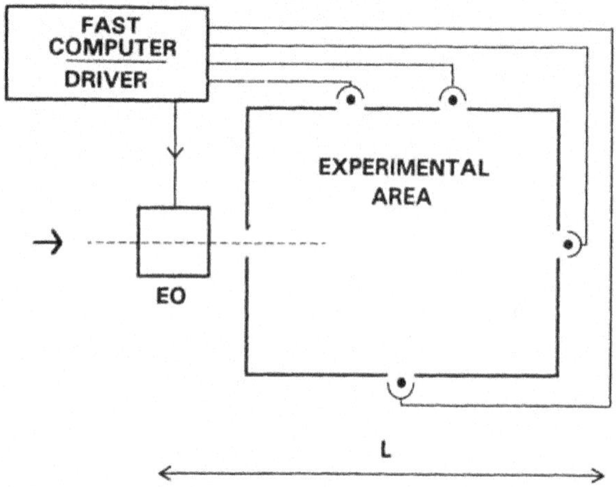

Fig.1: Layout of the "adiabatic-feedback" measurement configuration with four detectors.

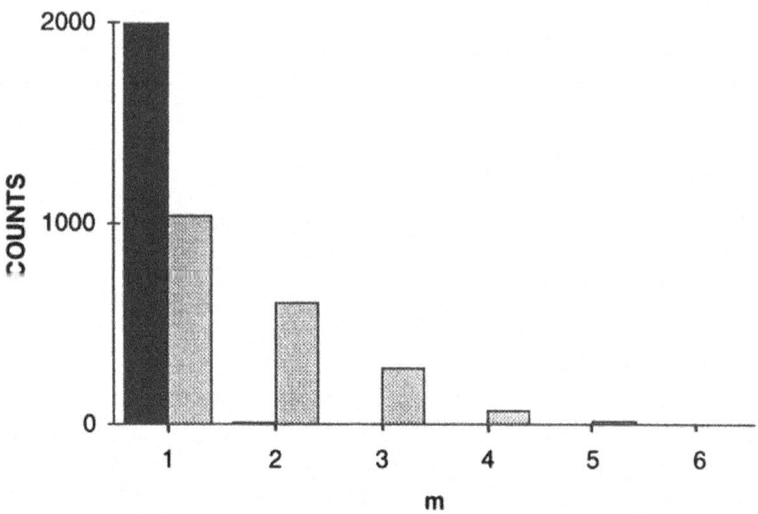

Fig. 2: Multi-Channel-Analyzer histogram showing the experimental photocount distribution P_m for adiabatic-feedback measurement with one detector: P_2/P_1=0.0035 (dark histogram). The disactivation of the feedback-loop leads to the quasi-Poisson distribution shown by the light histogram.

a partial state reduction process back-reacts on the state itself: an interesting quantum-measurement conceptual loop indeed. Note that the "adiabatic feedback" method realizes an actually *usable* n-squeezed radiation-state by making recourse to the nonlinear process which is intrinsic of the time-clipping of the input light beam [5]. Let's return to the generation of the one-photon state |1>. In addition to the MCA result of Fig.2, the latter peculiar quantum condition is demonstrated by the result of a standard Hanbury Brown-Twiss (HBT) test carried out within the *box* and leading to the highly non-classical value of the degree of second-order coherence at zero time-delay: $g^{(2)}(\tau=0)\cong 0$. This result is reported in the inset of Fig.4 where the corresponding result obtained after disactivation of the feedback is also shown [6,8]. Further details of the HBT test are given in Ref. 8, below.

3. Single-Particle Quantum-Eraser

The single-particle quantum-eraser (QE) consists of a modification of the Mach-Zehnder interferometer, Fig 3. The input beam was provided by a highly attenuated, stabilized He-Ne laser (Spectra-Physics 117A) having a coherence-time: $\tau_c > 10^{-7}$sec. [7]. The length of the "adiabatic-feedback" *box* was $L \cong 40$ cm. The light splitting devices were two symmetrical lossless beam-splitters (BS) A and B while the light detection was provided by two equal, cooled avalanche-silicon detectors (ASD) [modified SPCM-200PQ/25] with $\eta \cong 0.7$ and a noise-rate $\cong 7$ sec^{-1} connected to a SR400 coincidence counter. Before being processed, the corresponding output signals feeded independent feedback loops, with $t_F \cong 35$ ns. By standard ensemble-averaging they also provided evidence of two independent, 90^0 out-of-phase fringe patterns with "visibility" $V=g^{(1)}=1$ upon phase ø-changes induced on light beam 1 by an EO ø-modulator. This result corresponds to the celebrated single-particle ww-quantum non locality effect referred to by Feynman in [1], and reproduces the well known result by P. Grangier et al. [7]. In the above experiment both polarizations of the interfering beams 1,2 were set parallel to the vertical axis (**x**-axis). A strong perturbation of the dynamics of the photon in the interferometer was then introduced by inserting in beam 2 a ½π polarization (**p**) rotator, a z-cut quartz slab [2]. By accounting for the entanglement of the |n>-state basis with the **p**-information, the above perturbation changes the field's wave-function which is relevant for the process of photon counting by *either* detector, from $|\Psi\rangle = (\frac{1}{2})^{\frac{1}{2}} \bullet |x\rangle_p \bullet [|0,1\rangle + |1,0\rangle]$ to the entangled: $|\Psi\rangle = (\frac{1}{2})^{\frac{1}{2}} \bullet [|0,1\rangle|x\rangle_p + |1,0\rangle|y\rangle_p]$, where the first (second) place in any representation |h,k> relates to beam 1 (2). Evaluate the output photocurrent of either one detector, say C, in terms of the single-mode operators $\hat{a}_x$, $\hat{a}_y$, $[\hat{a}_i^\dagger, \hat{a}_j]=\delta_{ij}$ where index i (j) refers to the orthogonal **p**-component along x (y) and $\hat{\mathbf{E}}^{(+)} \propto \cdot(\varepsilon_x \hat{a}_x + \varepsilon_y \hat{a}_y)$: $\langle \mathbf{I} \rangle = Tr[\hat{\rho}\, \hat{\mathbf{E}}^{(-)}(z,t) \cdot \hat{\mathbf{E}}^{(+)}(z,t)]$. We may verify that, while the first case $\hat{\rho}=|\Psi_o\rangle\langle\Psi_o|$ leads to a one-particle interference pattern with V=1 as said, in the second case, $\hat{\rho}=|\Psi\rangle\langle\Psi|$, the induced field perturbation cancels any interference effect: V=0. In spite of this perturbation, the interference can nevertheless be retrieved outside the interferometer, and then the effect of the perturbation *"erased"*, by a smart manipulation of the measurement apparatus C. This is done by inserting in front of the detector C, on the output mode **c**, a **p**-analyzer, or Nicol, selecting the direction of the field to be measured $\hat{\mathbf{E}}^{(+)} \propto (\varepsilon_{x'} \hat{a}_{x'})$, $[\hat{a}_{x'}^\dagger, \hat{a}_{x'}]=1$, with **p** parallel to axis $x'=(x \cos\chi + y \sin\chi)$ belonging to the xy plane, being χ the rotation-angle of the Nicol [8]. The ensemble-average of the new measured observable

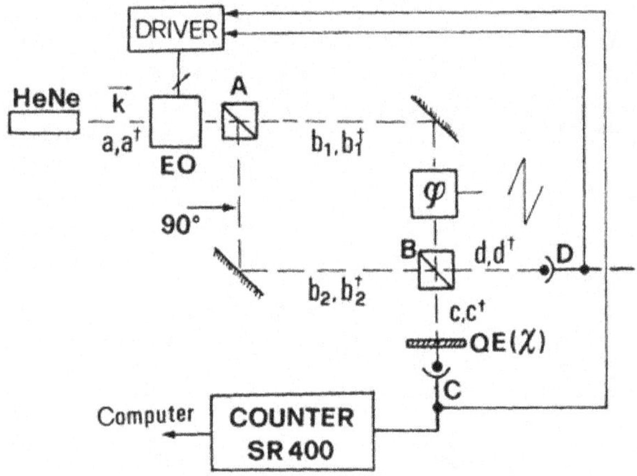

Fig. 3: Layout of the single-particle quantum-eraser (QE) with adiabatic-feedback.

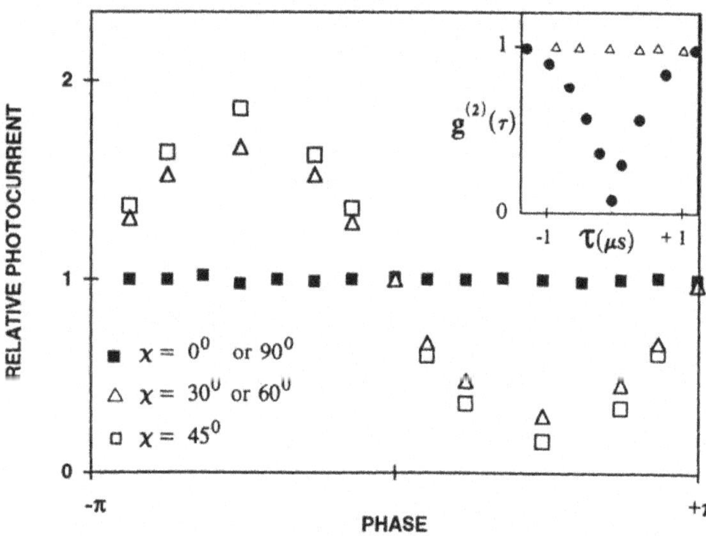

Fig.4: Interference pattern as function of the phase of beam 1 for various values of the QE-analyzer angle χ. INSET: results of a Hanbury Brown-Twiss test showing the sub-poissonian excitation of the quantum-eraser of Fig.3 (solid circles). The corresponding data obtained with the QE-feedback shut off are also shown.

$<\Psi|\hat{\mathbf{I}}_{x'}|\Psi> \equiv <\hat{\mathbf{I}}_{x'}>$ may be evaluated by the projected operator $\hat{a}_{x'}=(\hat{a}_x \cos\chi + \hat{a}_y \sin\chi)$. The fringe contrast is found to depend on the setting of QE(χ) and is expressed by: $V(\chi)=|\sin(2\chi)|$. In agreement with this result, the experimental data reported in Fig.3 show that the interference-fringe visibility is $V=0$ for $\chi=\frac{1}{2}\pi n$, n=0 or 1, and reaches its maximum value V=1 at angles $\chi=\frac{1}{4}\pi(2n+1)$ corresponding, as far as quantum photodetection is concerned, to the maximum single-photon *welcher-weg uncertainty*, i.e. Feynman-path-indistinguishability in the interferometer. Note that the addition of a Pockels-cell to QE(χ), inducing at appropriate times fast **p**-angle changes $\Delta\chi=\pm45°$ or 0° on the output beam before entering the QE(χ) set at a steady angle $\chi=45°$, can add to our *quantum-eraser* the feature of a one-particle *delayed-choice* process[2]. Note that, in analogy with standard interferometry, the behaviour of the device could be easily understood in terms of classical electrodynamics if the excitation is provided by a "classical" source [1,6]. In our case, the quantum conceptual content of the process and the given quantum analysis follow necessarily from the quantum nature of the excitation, as shown by the MCA and HBT results given by Figures 2 and 4. We summarize our results as follows:

(a) a new reliable strategy for the generation of pre-programmed highly sub-poissonian photon-distributions is reported.

(b) a new interferometer has been developed which demonstrates peculiar features of the single-particle in a *perturbed* quantum superposition-state.

From this viewpoint, starting from the original standard single particle Mach-Zehnder interferometer condition, in our QE experiment two successive partial $|\Psi>$-reduction processes determined by similar collision events have mutually reversed (viz., *erased*) their effects. All that may lead to an important general conclusion according to which an apparatus may indeed disturb a quantum system by cancelling the information (e.g., the interference) which is complementary to the one associated to any (in principle) observable, e.g., the photon's ww-localization. However, even if that cancellation is complete but the actual measurement is not carried through, e.g., the ww-information is not realized as a "phenomenon" in Bohr's sense [3], the overall interaction process *needs not* to be irreversible. In fact that cancellation may not destroy the superposition state of the system: it can be erased and the original coherence retrieved as this one has been stored in the system's $|\Psi>$. In this sense the present demonstration may somewhat be interpreted as a manifestation of the concept of *coherence-revival*, e.g. arising in the more complex domain of the atom-field interactions in a micromaser and of spin-echoes [9]. Furthermore, the crucial role plaid by the QE analyzer in front of the detector and then the decisive role of the overall measurement apparatus in assessing the *elements of reality* of the investigated observable, on one side is a striking demonstration of the Feynman's theory of probability-amplitudes characterizing quantum-paths [1]. On another side it leads to recognize the gnoseological wisdom of a well known statement by J.A.Wheeler: "...No elementary phenomenon is a phenomenon until is a registered phenomenon... A phenomenon is not yet a phenomenon until it has been brought to a close by an irreversible act of amplification...." [10]. Thanks are due to R.Tommasini for contribution to the early part of the work and to R.Y.Chiao and M.L. Dalla Chiara for stimulating discussions•

4. References

1. R. P. Feynman, R. Leighton and M. Sands, *The Feynman Lectures in Physics*, (Addison Wesley, Reading, 1965), Ch.1.; R. P. Feynman and A.Hibbs, *Quantum Mechanics and Path Integrals*, (Wiley, New York, 1963).
2. M. O. Scully and K. Druhl, *Phys.Rev* A 25 2208 (1982). M.O.Scully, B. Englert and H. Walther, *Nature* 351, 111 (1991). J. A. Wheeler, in *Mathematical Foundations of Quantum Theory*, A.Marlow, ed. (Academic, New York, 1990). Note that the quartz-plate inducing a ½π **p**-flip in our QE is not rigorously a $\lambda/2$ plate, and then no modification of the vector-character of the interferometer modes is introduced. Since there the **p** flip is due to the *rotatory-power* of quartz, an analogy can be established between the quartz-plate and a magnetic-coil spin-flipper in a neutron interferometer. Similarly,the **p**-analyzer in front of QE detector may be thought of as a kind of Stern Gerlach apparatus in agreement with the Bohm's optical version of the EPR process: D. Bohm, ref. 4 below, Ch.22.
3. N. Bohr, in *Albert Einstein Scientist Philosopher*, P. A. Schlipp, ed. (Cambridge University Press, Cambridge, 1982).
4. D. Bohm, *Quantum Theory*, (Dover, New York, 1979).
5. The feedback process adopted within our method is reminiscent of the dead-time-deletion effect proposed by J. Walker and E. Jakeman, in *Proc.Soc.Photo-Opt. Instr. Eng.*, 492 (1985); J. Shapiro, G. Saplakoglu, S. Ho, P. Kumar, B. Saleh and M. C. Teich, *J. Opt. Soc. Am. B 4*, 1604 (1987). However our new measurement strategy, by which the back-reaction is provided by the *full set of detectors* operating *adiabatically* across the apparatus, is indeed the key solution that allows the *actual use* of the generated n-squeezed radiation within any experiment. As shown theoretically by J. Shapiro et al. above, the n-squeezed radiation generated by the previously reported, standard feedback schemes could not be estracted from the feedback-loop without *losing* its sub-poissonian character. Then it *could not* be adopted as the input source of quantum-light excitation for any real experiment.
6. R. Loudon, *The Quantum Theory of Light*, (Clarendon, Oxford, 1983), Ch.5. For related recent neutron work cf. H. Rauch, Phys. Lett. A 24, 173 (1993)
7. P. Grangier, G. Roger and A. Aspect, *Europhys.Lett.* 1, 173 (1986).
8. The idle escape-beam of the output Nicol was intercepted by a third ASD detector feeding an independent feedback loop acting on the photon-gate. This loop is not reported in Fig.3 where the complex overall output **p**-analyzer is indicated by QE(χ). The HBT test, Fig.4, was carried out by removing from the interferometer the beam-splitter A and QE(χ) and by evaluation of $g^{(2)}(\tau) = t' N_{cd}(\tau) / [\Delta t N_c N_d]$, being N_c, $N_d \cong N_c$, N_{cd} the number of singles and coincidences detected, with a resolution Δt = 8ns, within the effective overall measurement time t' related to each datum of Fig.4. Each datum there was obtained by averaging over 10^4 samples.
9. E. Hahn, *Phys. Rev.* 80, 580 (1950), G. Rempe, H. Walther and N. Klein, *Phys. Rev. Lett.* 58, 353 (1987). The analogy stated in the text implies that in all these processes the disappearance of the manifestation of a quantum-superposition lies on the deterministic evolution of $|\Psi\rangle$ rather than on an irreversible dissipation process. For

time, since it is related to the deterministic spin-dephasing process implied by the *inhomogeneous* linewidth of the resonance and *not* to the irreversible dephasing, viz., associated with the *homogeneous* linewidth. In this sense the interpretation of *spin-echo* as an example of violation of the statistical *arrow-of-time* can be mislea-ding: cf. P.C.Davies, *The Physics of Time Asymmetry*, (Surrey Un., London 1974).

10. J. A. Wheeler, "Law without law" in *Quantum Theory and Measurement*, J. A. Wheeler and W. H. Zurek ed. (Princeton Un., 1983). The results of our experiments, while supporting Wheeler's view, disprove the quantum-logical realistic conception about one-particle interference expressed by Hilary Putnam in "Is Logic Empirical?" *Boston Studies in the Philosophy of Science* 5, 199 (1969). cf.: P. Gibbins, *Particles and Paradoxes* (Cambridge University Press, 1987).

PHASE-SENSITIVE BACKACTION MEASUREMENTS IN PENNING TRAPS: SQUEEZING THE THERMAL NOISE

Irene Marzoli[1]
Dipartimento di Matematica e Fisica
Università di Camerino; 62032 Camerino Italy

Paolo Tombesi
Dipartimento di Matematica e Fisica
Università di Camerino; 62032 Camerino Italy
and
Dipartimento di Fisica
Università di Roma "La Sapienza"; 00185 Roma Italy
e-mail tombesi@roma1.infn.it

Abstract

We show that, by using a feedback, the thermal noise in the axial oscillation, due to the back-action of the current of the external electronic device, can be squeezed up to 50%.

1 Introduction

Penning traps[1] for electrons, protons and ions have been extensively used for high precision measurement of fundamental constant and laws of Nature like, for instance, the g factor of the electron and the CPT invariance[2]. With such a trap a single quantum object, like an electron for instance, can be confined in a very small volume practically at rest and continously observed[3]. A Penning trap consists of a combination of a constant magnetic field and a quadrupole electric field in which a charged particle can be confined[4].The motion of the trapped particle is a combination of three independent harmonic motions; the axial, cyclotron and magnetron motions[4].If we do not consider other sources of errors due to non ideal traps, which on the other hand can be controlled by tuning the compensation electrodes of the trap, the ultimate limit in the precision measurements of the various frequencies of the motions seems to be imposed by the thermal noise, due to the backaction of the measurement device, in the axial motion of the electron[4]. The coupling of the axial motion to other motions is obtained using a spatial modulation of the magnetic field, produced by a ferromagnetic nickel wire wound around the ring electrode, through a so called magnetic bottle. The magnetic bottle produces a

[1]Present address: Joint Institute for Laboratory Astrophysics, Boulder CO 80309-0440, USA

static quadratic field that is superimposed on the homogeneous magnetic field in such a way to produce a total magnetic field along the symmetry axis equal to

$$B_z = B_0 + B_2 z^2 \tag{1}$$

and consequently the observed cyclotron angular frequency depends on the axial coordinate as

$$\omega_c(z) = \omega_c(1 + \epsilon z^2) \tag{2}$$

where, in the current experiments, $\epsilon = B_2/B_0 \simeq 0.003\ cm^{-2}$. This coupling, on the other hand, is also responsible for the back-action of the readout circuit on the axial motion of the electron and limits the minimum shift in frequency which can be measured. Indeed, the coupling of the quantum system with the external world is provided by the axial oscillation of the electron[4] which induces time varying electric charges on the two end-cap electrodes of the trap forming a capacitor C. Therefore, a current passes through an effective resistor R of the external electronic device and the voltage across R can be detected. The resistor is in equilibrium at temperature $T \simeq 4K$. The Johnson noise in the resistor gives the backaction in the oscillatory axial motion.The equipartition of the energy gives

$$\frac{m\omega_z^2}{2} < z^2 >= \frac{k_B T}{2} \tag{3}$$

and, from (2,3), we obtain a linewidth parameter

$$\Delta\omega_c = \omega_c \epsilon < z^2 >= \omega_c \epsilon \frac{k_B T}{m\omega_z^2} \tag{4}$$

where $\omega_z/2\pi$ is the axial frequency, n its quantum number and k_B the Boltzmann constant. Thereby, because we are far away from the quantum limit, the ultimate precision measurement of these frequencies is connected with $< z^2 >= k_B T/m\omega_z^2 = n\hbar/m\omega_z$. The experimental value of $\omega_z/2\pi$ is $64MHz$ then, n is quite big[4]; i.e. $n \cong 10^3$ .We will not consider the magnetron motion which, although important for the cooling mechanism of the whole system[4], is very slow with respect to the other motions.In our analysis we are interested only in the axial motion and in its influence in the motions coupled to it. This is also called the back-action effect and, in the limit of a quantum limited readout this back-action enforces the Heisenberg principle on the electron axial degree of freedom, leading to the standard quantum limit (SQL) for the monitoring of the axial motion

$$\Delta z^2_{SQL} = \frac{\hbar}{2m\omega_z} \tag{5}$$

In order to overcome the quantum limit of sensitivity for detecting small displacements in macroscopic harmonic oscillators quantum non-demolition (QND) measurement strategies have been proposed[5]. In the classical limit (the axial coordinate, in the current scheme, is far from the quantum limit and a classical framework is adequate) the analogy to the quantum noise is the classical, back-action noise due to the amplifier and a QND strategy allows to evade this back-action on a quadrature phase. In the classical limit a QND strategy is more commonly named Back-Action Evasion (BAE) strategy. Recently, after the successful production of the squeezed light[6], the intriguing possibility of squeezing the thermal noise was experimentally demonstrated[7], and the classical amplitude squeezing for precision measurement was also emphasized[8].

2 The Model

In this note we present a model for the amplitude squeezing of the axial motion which requires a slight modification of the existing apparatus. We will show that a negative feedback is suitable to control over the thermal fluctuations in the current loop of the device. Our approach is related to the stochastic cooling considered by Beverini*et al.*[9]. Our aim is to go beyond the thermal limit.

As we mentioned above, the frequencies of the trapped electron, once the magnetic bottle has been introduced, depends on the mean square displacement of the axial motion. The last motion can be considered, to a great extent, harmonic[4]. We will show how this mean square could be reduced or "cooled" below the thermal value, therefore permitting an improvement of the measurement of these frequencies. Our analysis concerns the axial motion; however, the same argument could be applied to the cyclotron motion when the detection scheme is obtained by splitting the ring electrode into two or more segments[4]. As we mentioned above, the best way of measuring the properties of the various motions of the electron is to measure the current induced by the axial motion of the electron along the z-axis[4]. Indeed, the electric charges induced by the oscillatory motion on the end-cap generate a current which can be measured.

The axial motion plus the read-out are described by the following Hamiltonian

$$H = \frac{p_z^2}{2m_0} + \frac{m\omega_z^2}{2}z^2 + \frac{1}{2C}(az+Q)^2 + \frac{\phi^2}{2L} + \frac{1}{2}\int_o^\infty d\Omega\{[p(\Omega) + \kappa(\Omega)Q]^2 + \Omega^2 q^2(\Omega)\} \quad (6)$$

where we have considered a thermal bath with a continuous distribution of modes linearly coupled to the electronic circuit; ϕ is the electric flux in the inductance L, Q is the electric charge on the capacitor C which is the capacity of the trap; az represents the induced charge due to the axial motion of the electron with $a = \alpha e/2z_0$ where $2z_0$ is the distance between the two end-caps and α is a constant of order of unity which takes into account the curvature of the capacitor surfaces[4].

By making a Markov approximation in the equation of motion for the variables of the thermal bath, we can write the following equations[10]

$$\begin{cases} \dot{z} &= \frac{p_z}{m} \\ \dot{p}_z &= -m\omega_z^2 z - \frac{a}{C}Q \\ \dot{Q} &= \frac{\phi}{L} \\ \dot{\phi} &= -\frac{Q}{C} - \frac{az}{C} - \gamma\dot{Q} + \xi(t) \end{cases} \quad (7)$$

where γ represents the rate at which the axial motion dissipates its energy due to the coupling with the rest of Universe represented by the read-out apparatus. Of course, in this case one has to sustain the axial oscillation with an oscillating external potential $V(t)$ tuned at the axial frequency of the electron.

In the experimental situation is always

$$\hbar\omega_z << k_B T \tag{8}$$

with k_B the Boltzman's constant. Then, it is possible to show [10] that the statistics of the noise term $\xi(t)$ is that of a white noise with expectations

$$\begin{aligned} \langle\xi(t)\rangle &= 0 \\ \langle\xi(t)\xi(t')\rangle &= 2\gamma k_B T\,\delta(t-t'). \end{aligned} \tag{9}$$

By introducing the Fourier transforms defined by

$$f(t) = \frac{1}{\sqrt{2\pi}} \int_{-\infty}^{+\infty} d\omega \tilde{f}(\omega) e^{i\omega t} \tag{10}$$

we can write the linear system:

$$\left\{ \begin{array}{lllll} i\omega\tilde{z}(\omega) & -\dfrac{\tilde{p}(\omega)}{m} & & & = 0 \\ m\omega_z^2\tilde{z}(\omega) & +i\omega\tilde{p}(\omega) & +\dfrac{a\tilde{Q}(\omega)}{C} & & = 0 \\ & & i\omega\tilde{Q}(\omega) & -\dfrac{\tilde{\phi}(\omega)}{L} & = 0 \\ \dfrac{a\tilde{z}(\omega)}{C} & & +\left(\dfrac{1}{C}+i\omega\gamma\right)\tilde{Q}(\omega) & +i\omega\tilde{\phi}(\omega) & = \tilde{\xi}(\omega) - \tilde{V}(\omega) \end{array} \right. \tag{11}$$

The determinant of the homogeneous system is

$$\Delta = (\omega_z^2 - \omega^2)(\omega_e^2 - \omega^2 + i\gamma_e\omega) - \frac{a^2}{m}\omega_e^2 \tag{12}$$

with

$$\omega_e = \sqrt{\frac{1}{LC}} \qquad \gamma_e = \frac{\gamma}{L} \tag{13}$$

i.e., the characteristic frequency and the bandwidth of the electronic circuit respectively.

The solution of the algebraic system is easily obtained and we get:

$$\tilde{z}(\omega) = -\frac{a\omega_e^2}{m\Delta}\left(\tilde{\xi}(\omega) - \tilde{V}(\omega)\right) \tag{14}$$

$$\tilde{p}(\omega) = -\frac{ia\omega_e^2}{\Delta}\left(\tilde{\xi}(\omega) - \tilde{V}(\omega)\right) \tag{15}$$

$$\tilde{Q}(\omega) = \frac{\omega_z^2 - \omega^2}{L\Delta}\left(\tilde{\xi}(\omega) - \tilde{V}(\omega)\right) \tag{16}$$

$$\tilde{\phi}(\omega) = \frac{\omega_z^2 - \omega^2}{\Delta} \left(\tilde{\xi}(\omega) - \tilde{V}(\omega)\right). \tag{17}$$

We see that at $\omega = \omega_z$ both $\tilde{Q}$ and $\tilde{\phi}$ are zero and the current which dissipates energy in the resistor is only due to the induced charge on the end-caps. When the external electronic circuit is tuned with the axial frequency, the induced current produces Ohmic losses in the resistor which, in its turn, feeds the thermal noise back to the axial motion.

In this case the Hamiltonian of the system coupled to the external world reduces to

$$H = \frac{p_z^2}{2m} + \frac{1}{2} m\omega_z^2 z^2 + \frac{1}{2}\int_o^\infty d\Omega\{[p(\Omega) + \kappa(\Omega)Q]^2 + \Omega^2 q^2(\Omega)\}. \tag{18}$$

The axial oscillator can be considered electrically equivalent to an LC circuit[11] which dissipates its energy in the resistor R in presence of a stochastic force due to the Johnson noise. Here is again $Q = \alpha e z/(2z_o)$. The current that dissipates in the resistor is $I = dQ/dt = \alpha e\dot{z}/(2z_o)$.

3 Feedback

We now modify the previous scheme by introducing a feedback in one end-cap electrode. The induced current is then pumped back at twice the frequency of the axial motion and dephased by an angle φ. Between the two end-cap electrodes (the capacitor C of the trap) there will be an extra field due to the feedback. The potential energy acting on the electron at the position z from the centrum of the trap, as a consequence of the feedback, is

$$U(t) = -eE(t)(z_o - z); \quad E(t) = \frac{\alpha}{2z_o} IRG\cos(2\omega_z t + \varphi) \tag{19}$$

where φ is the phase controlled by the experimenter and G the gain or the "efficiency" of the feedback. The new Hamiltonian thus becomes;

$$\begin{aligned} H = \frac{p_z^2}{2m} + \Big(\frac{m\omega_z^2}{2} + \frac{k^2(t)}{2m}\Big)z^2 - \frac{k(t)}{2m}\,[p_z(z_o - z) + (z_o - z)p_z] + \\ + \frac{1}{2}\int_o^\infty d\Omega\Big\{[p(\Omega) + \frac{\alpha e}{2z_o}\kappa(\Omega)z]^2 + \Omega^2 q^2(\Omega)\Big\} \end{aligned} \tag{20}$$

where the conjugate momentum is $p_z = m\dot{z} + (z_o - z)k(t)$ with $k(t) = k_o\cos(2\omega_z t + \varphi)$ and $k_o = (\alpha e/2z_o)^2 RG$. We now neglect, for sake of simplicity, the negligibly small corrections to the axial frequency due to the feedback and the thermal bath; then, we get the following equations of motion of the axial observables.

$$\begin{aligned} \dot{z} &= \frac{p_z}{m} + \frac{k(t)}{m}(z_o - z) \\ \dot{p}_z &= -m\omega_z^2 z + \frac{k(t)}{m} p_z - \gamma\dot{z} + \frac{k^2(t)}{m} z_o - \xi(t) \end{aligned} \tag{21}$$

These equations are valid for $t > t_o$ where t_o represents the initial time of the interaction of the system with the meter. They are easily obtained, as shown by Gardiner [10], in the Markovian limit when one considers

$$\gamma = \left(\frac{\alpha e}{2z_o}\right)^2 \frac{\kappa^2(\Omega)}{(\pi/2)} = \left(\frac{\alpha e}{2z_o}\right)^2 R \tag{22}$$

and $\gamma_z = \gamma/m$ is the rate at which the energy is dissipated into the resistor R; $\xi(t)$ is a stochastic force depending only on the state of the thermal bath at time t_o, with expectation values given in Eq.(9).

The Johnson noise $\epsilon(t)$ is connected with the stochastic force $\xi(t)$ by the relation $\epsilon(t) = 2z_o\xi(t)/(\alpha e)$.

It is convenient to introduce the classical quadratures of the harmonic motion, defined by:

$$Z_1 = z\,cos\omega_z t - \frac{p_z}{m\omega_z} sin\omega_z t; \quad Z_2 = z sin\omega_z t + \frac{p_z}{m\omega_z} cos\omega_z t\ . \tag{23}$$

Within the rotating wave approximation (i.e. neglecting all terms rotating faster than ω_z) the equations of motion of the two quadratures become

$$\frac{d}{dt}\begin{pmatrix} Z_1 \\ Z_2 \end{pmatrix} = \begin{pmatrix} a - \frac{\gamma_z}{2} & b + c \\ b - c & -a - \frac{\gamma_z}{2} \end{pmatrix} \begin{pmatrix} Z_1 \\ Z_2 \end{pmatrix} + \begin{pmatrix} sin\omega_z t & cos(\omega_z t + \varphi) \\ -cos\omega_z t & -sin(\omega_z t + \varphi) \end{pmatrix} \begin{pmatrix} \frac{\xi(t)}{m\omega_z} \\ \frac{k_Q z_Q}{2m} \end{pmatrix} \tag{24}$$

where in the nonhomogeneous terms we have neglected oscillating terms with negligibly small amplitudes with respect to those considered,

$$a = G\frac{\gamma_z}{2}\left(\frac{\gamma_z}{2\omega_z} sin\varphi - cos\varphi\right); \; b = G\frac{\gamma_z}{2}\left(\frac{\gamma_z}{2\omega_z} cos\varphi + sin\varphi\right); \; c = G^2\frac{\gamma_z^2}{4\omega_z^2}. \tag{25}$$

By means of a unitary transformation the homogeneous part of Eqs.(24) can be diagonalized with the eigenvalues given by

$$\lambda_{1,2} = -\frac{\gamma_z}{2}(1 \pm G) \tag{26}$$

where $0 < G < 1$ and we have neglected terms of the order of $(\gamma_z/\omega_z)^2$. For $G = 0$ we get the known result. The value $G = 1$ gives the marginal stability of the equations. In such a case as time goes on the amplitude of the oscillation grows linearly with the time and the electron disappears on one end-cap electrode. The mean square value of the displacement is given by

$$\begin{aligned} < z^2(t) > =& < Z_1^2(t) > cos^2\omega_z t + < Z_2^2(t) > sin^2\omega_z t \\ &+ \frac{sin2\omega_z t}{2}\left(< z_1(t)z_2(t) > + < z_2(t)z_1(t) >\right) \end{aligned} \tag{27}$$

We are actually interested in the long time limit of the various expectation values in Eq.(27). After some lenghty but straightforward algebra we get, for $i = 1,2$;

$$\lim_{t\to\infty} < Z_i^2(t) > = \frac{k_B T}{m\omega_z^2}\,\frac{1 + (-1)^i G cos\varphi}{1 - G^2} \tag{28}$$

$$\lim_{t\to\infty} < Z_i(t)\, Z_j(t) >= \lim_{t\to\infty} < Z_j(t)\, Z_i(t) >= \frac{k_B T}{m\omega_z^2} \frac{G sin\varphi}{1-G^2} \tag{29}$$

where we neglected oscillating terms with a very small amplitude. Finally we obtain

$$\lim_{t\to\infty} < z^2(t) >= \frac{k_B T}{m\omega_z^2} \frac{1 - G cos(2\omega_z t + \varphi)}{1-G^2} \tag{30}$$

and

$$\lim_{t\to\infty} < p_z^2(t) >= m k_B T \frac{1 + G cos(2\omega_z t + \varphi)}{1-G^2} \tag{31}$$

We can conclude that, with a feedback tuned at twice the angular frequency of the axial motion, by stroboscopic sampling at times

$$t_k = \frac{2k\pi - \varphi}{2\omega_z} \tag{32}$$

the linewidth of the cyclotron frequency in Eq.(4) can be reduced up to a factor $(1+G)^{-1}$ with $G \simeq 1$. Thus the maximum reduction will be limited at 50%.

Acknowledgement

This research has been supported by the European Economic Community by the Uman Capital and Mobility program.

REFERENCES

1. F.M. Penning, *Physica* **3**, 873 (1936);
2. R.S. Van Dyck jr., P.B. Schwinberg and H.G. Dehemelt, *Atomic Physics 9* Eds. R.S. Van Dyck jr. and E.N. Forstron (World Scientific, Singapore 1984);
3. H.G. Dehmelt, *Am. J. Phys* **58**, 17 (1990);
4. L.S. Brown and G. Gabrielse, *Rev. Mod. Phys.* **58**, 233 (1986) and references therein;
5. V.B. Braginsky and F.Ya. Khalili, *Quantum Measurements* (Cambridge University Press, 1992);
6. R. Loudon and P. Knight, *J. Mod. Opt.* **34**, 709 (1987) and references therein;
7. D. Rugar and G. Grutter, *Phys. Rev. Lett.* **67**, 699 (1991);
8. F. DiFilippo, V. Natajan, K.R. Boyce and D.E. Pritchard, *Phys. Rev. Lett.* **68**,2859 (1992);
9. N. Beverini, V. Lagomarsino, G. Manuzio, F. Scuri, G. Testera and G. Torelli, *Phys. Rev.* **A38**, 107 (1988);
10. G.W.Gardiner: *Quantum Noise,* (Springer, 1992).
11. D.J. Wineland and H.G. Dehmelt, *J. Appl. Phys.* **46**, 919 (1975);

GENERATION OF NON-CLASSICAL LIGHT BY THE OPTICAL ACTIVE MICROCAVITY.

M. MARROCCO and F. DE MARTINI
Dipartimento di Fisica, Università di Roma "La Sapienza", P.le Aldo Moro 2
Roma 00185, Italy

ABSTRACT

We report the experimental demonstration that the active optical microcavity under a short-pulse excitation can be adopted as the source of non-classical light. In the present experiment the excitation was provided by a highly attenuated pulsed laser beam interacting with a very small number of fluorescent molecules trapped in a plane microcavity with optical "finesse" $f \approx 1.6\ 10^3$. The photon-distribution of the emitted photon-beam is investigated by a standard Hanbury Brown-Twiss method by which the nonclassical behaviour of the *coincidence-parameter* α, or of the degree of second-order coherence, could be assessed for our experimental conditions. A theory of the *coincidence-parameter* showing its relation with $g^{(2)}(0)$ is also given.

1. Introduction.

The generation of non-classical, antibunched light is a most important topic of modern quantum optics as it is intended to provide the basic tools which are necessary for the experimental demonstration of the fundamental quantum mechanical processes involving the interferometry of photons. For this purpose many sophisticated techniques have been adopted in the past, most of which are essentially based on the *dead-time-deletion* effect arising in atomic fluorescence when the external excitation sets a single, or a small number of atoms into an appropriate dynamical Rabi regime [1,2,3]. A recent work reports the generation of antibunched light from a single dye molecule trapped in a solid host [4]. These techniques are generally of difficult application because of the complex and expensive trapping procedure of a well defined number of emitting atoms e.g., in a solid matrix or in a radio-frequency trap. They are generally also very inefficient because the detection of the emitted photons is strongly affected by the total unpredictability of the emission direction, viz., of the photon momentum $\hbar k$. In the present work we present a new method in which most of these problems are solved by trapping a small number of emitting molecule in an optical micro-cavity, i.e. a Fabry-Perot plane cavity with relevant dimension, or mirror-separation, equal to $d = m\,\lambda/2$, where λ is the wavelength of the emitted radiation (i.e., the resonant frequency of the atom) and m is a small odd number larger than zero [5]. The peculiar, Casimir-type, *extreme* vacuum-confinement realized by this device leads to an increasing inhibition of the number of emission modes of the confined radiation field as the microcavity size d is reduced. When the size of the cavity is set its minimum value $d = \lambda/2$, $m = 1$ the differential fluorescence cross-section is highly peaked in the direction of the optical axis of the cavity, orthogonal to the mirror plane, and then virtually all emitted photons are funneled in that direction. Furthermore, the number of photons emitted in a fixed amount of time dt is determined by a corresponding number of dye molecules excited by a pulsed excitation beam within the (microscopic) volume $dV = \sigma\lambda/2$ corresponding to the focalization of that beam in the microcavity plane on a focal area $\sigma \approx \lambda^2$. We demonstrate by the present work that, by a previous calibration of the intensity of the excitation pulses and of the dye concentration, the system can provide a convenient source of an antibunched light emitted over a well defined single radiation mode, with large efficiency. In the following text we provide a description of the apparatus and the presentation of some relevant experimental results.

2. The microcavity and the coherent excitation system.

The experimental layout of the process shown in Fig. 1 includes the Hanbury Brown-Twiss (HBT) detection configuration adopted to test the quantum properties of the emitted light. Optical pulses with duration dt=4 nsec emitted with a frequency of 20 pulse-per-second were generated by a highly stabilized Nd-Yag laser (Continuum Surelite) operating in second harmonic generation at a wave-length λ_p=532 nm. The pulsed beam, highly attenuated by neutral-density filters, excited the active medium contained in a microcavity with m=1 and "finesse" $f \cong 1.6\ 10^3$. The beam was then focused on the cavity active plane by a short focal length lens (3.5 cm) in order to excite within the cavity a small pumping volume $dV \cong 10^{-14}\ cm^3$ of active molecular dye (sulforhodamine-B in ethilene glycol) filling the active volume with a calibrated low concentration, ρ. A previous accurate optimization, with no further need for adjustements, of the position of the pump focal region over the cavity mirror plane was necessary at the beginning of the experiment. The fluorescence time *in vacuum* of the dye molecules at the fluorescence wavelength λ=633 nm was 10^{-8}sec. while the coherence-time of the emitted photons in the cavity was estimated to be $\tau_c = 3 \cdot 10^{-12}$ sec. The pump-intensity and the dye-concentration parameters were such that in our experiment the average number of fluorescence photons generated by any pump laser pulse was of the order of one. Two cooled phototubes (RCA 31034A-02), PM1 and PM2, having quantum-efficiencies $\xi_1 \approx \xi_2 \approx 0,13$, provided the light detection across a symmetric, lossless beam-splitter operating in the HBT configuration. By a laser-triggered, computer interfaced gated photon-counter (Stanford Research SR400) the number of counting coincidences and anti-coincidences could be registered and averaged over sets of large numbers of pumping pulses (typically $10^4 \div 10^5$) in order to determine the value of the degree of second-order coherence $g^{(2)}(\tau)$ of the emitted light [6] or the one of the corresponding *coincidence-parameter* α [7]. The emission rate of the anti-bunched photons with our system was approximately 0.5 sec^{-1}.

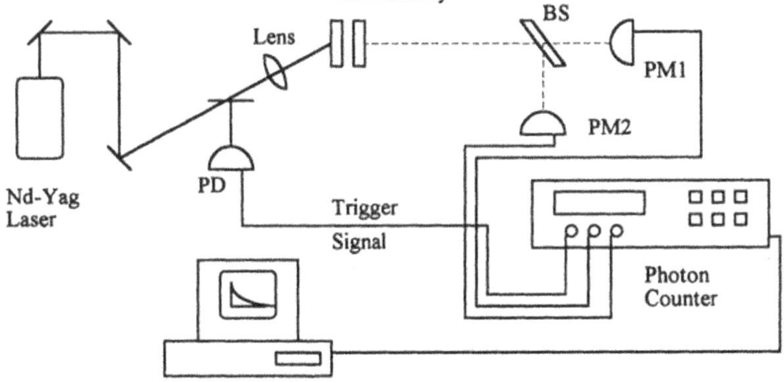

Fig. 1. Experimental set-up.

3. The coincidence-parameter.

The results of the experiment just described will be conveniently analyzed in terms of the coincidence-parameter α introduced by P.Grangier et al. [7]. Before doing that we give here a detailed theory of the process of photon-detection taking place within a Hanbury Brown-Twiss experimental test-configuration, the one adopted here to assess the quantum properties of the fluorescence radiation emitted by the excited molecules in our work.

Let's consider in a photon-counting experiment N gates during which the photon counting is performed. A *yes-gate* is the one corresponding to the detection of one or more photons by either one photomultiplier, PM1 or PM2. Suppose that among N gates, N_1 and N_2 are yes-gates corresponding to PM1 and PM2, respectively. Let N_c be the number of coincidences recorded during the whole measure and let's define the following *coincidence-parameter* α:

$$\alpha = \frac{N_c\ N}{N_1\ N_2} = \frac{p_c}{p_1\ p_2} = \frac{\langle p_c(t,T)\rangle}{\langle p_1(t,T)\rangle\ \langle p_2(t,T)\rangle} \tag{1}$$

where α is also written in terms of the corresponding detection probabilities and the angle brackets denote the average over the time t, the time at which the gate opens. The probability of one correlated event during the gate T is given by:

$$p_c(t,T) = \sum_{m1=1}^{\infty} P_{m1}^{PM1}(t,T) \sum_{m2=1}^{\infty} P_{m2}^{PM2}(t,T) \tag{2}$$

where $P_{m1}^{PM1}(t,T)$ and $P_{m2}^{PM2}(t,T)$ are the photon-count distributions for each detector, while the probability to detect something at each photomultiplier is:

$$p_1(t,T) = \sum_{m1=1}^{\infty} P_{m1}^{PM1}(t,T) \qquad p_2(t,T) = \sum_{m2=1}^{\infty} P_{m2}^{PM2}(t,T) \tag{3}$$

By inserting these results in the definition of α we obtain, for single-mode excitation [6]

$$\alpha = \frac{\mathrm{tr}\left\{\hat{\rho}\hat{N}\left[1-\exp\left(-\xi_1\hat{a}_1^{+}\hat{a}_1\right)\right]\left[1-\exp\left(-\xi_2\hat{a}_2^{+}\hat{a}_2\right)\right]\right\}}{\mathrm{tr}\left\{\hat{\rho}\hat{N}\left[1-\exp\left(-\xi_1\hat{a}_1^{+}\hat{a}_1\right)\right]\right\}\ \mathrm{tr}\left\{\hat{\rho}\hat{N}\left[1-\exp\left(-\xi_2\hat{a}_2^{+}\hat{a}_2\right)\right]\right\}} \tag{4}$$

where $\hat{\rho}$ is the density-operator which contains the statistical behaviour of the source, $\hat{N}$ is the normal-ordering operator and ξ_1, ξ_2 are the quantum efficiencies. By use of the n-state representation of the density operator we obtain,

$$\alpha = \frac{N_c\ N}{N_1\ N_2} = \frac{\sum_n P_n\left\{1+(1-\xi_1 R-\xi_2 T)^n-(1-\xi_1 R)^n-(1-\xi_2 T)^n\right\}}{\left\{\sum_n P_n\left[1-(1-\xi_1 R)^n\right]\right\}\left\{\sum_n P_n\left[1-(1-\xi_2 T)^n\right]\right\}} \tag{5}$$

where P_n is the probability distribution of the source, R and T are the optical coefficients of the beam splitter. This expression can be used to obtain the parameter α for a single-mode excitation and for some relevant photon light distributions (Fig.2):

Anti-bunched light: $P_n = \delta_{n,\bar{n}}$

$$\alpha = \frac{1+(1-\xi_1 R-\xi_2 T)^{\bar{n}}-(1-\xi_1 R)^{\bar{n}}-(1-\xi_2 T)^{\bar{n}}}{\left\{1-(1-\xi_1 R)^{\bar{n}}\right\}\left\{1-(1-\xi_2 T)^{\bar{n}}\right\}} \tag{6}$$

Poissonian light: $P_n = \frac{\bar{n}^n}{n!}e^{-\bar{n}}$

$$\alpha = 1 \tag{7}$$

Chaotic light: $P_n = \frac{\bar{n}^n}{(1+\bar{n})^{1+n}}$

$$\alpha = \frac{2+\xi_1 R\bar{n}+\xi_2 T\bar{n}}{1+\xi_1 R\bar{n}+\xi_2 T\bar{n}} \tag{8}$$

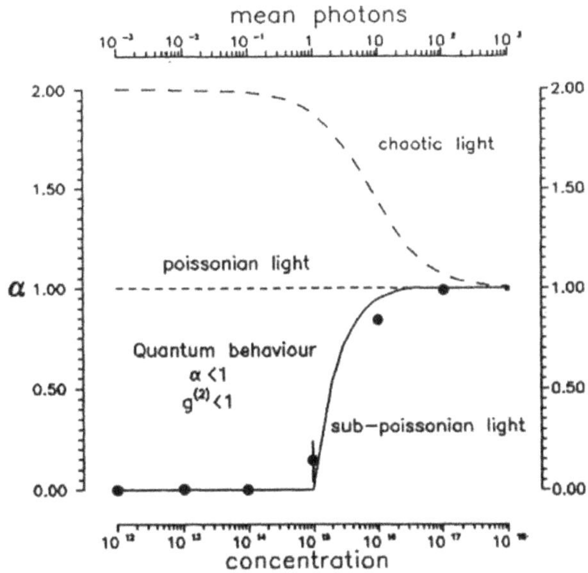

Fig. 2. The coincidence-parameter α as function of the mean-number of emitted photons per pulse and of the concentration, ρ. Some corresponding experimental data are also shown.

Figure 2 shows the behaviour of α as function of the mean number of emitted photons as a result of a test experiment involving a simmetrical beam splitter, ξ=0.1 and for different kinds of single-mode excitation of the HBT apparatus. The experimental points there correspond to different values of the dye concentration, ρ and to an excitation provided by laser-pulses with duration dt=4 nsec and associated energy η = 30 nJ. Note in Fig.2 that the quantum regime appears to be well separated from the classical one. Accordingly, our results show the good performance of the microcavity as a novel source of highly non-classical light. In this respect the results of the present experiment compare very favorably with the corresponding results reported earlier in [7]. It is not difficult to show, with the help of the above theory, the strict relation existing between α and the correlation function of second order $g^{(2)}(0)$, i.e., the degree of second-order coherence [6]. By a first order expansion of α, a relation of this quantity with the degree of second-order coherence at zero time-delay is easily obtained:

$$\alpha = g^{(2)}(0)\; A(\bar{n}) + B(\bar{n}) \tag{9}$$

A and B are suitable functions of the average number photons associated with the single-mode field.

4. Results.

The experimental results reported in Figure 2 exhibit a clear quantum behaviour taking place at low volume-concentrations ρ of the active medium and disappearing at values of ρ larger than $\approx 10^{-16}$ for a pump pulse energy $\eta \cong 30$ nJ. Further results corresponding to a ten times smaller value of η are shown in Fig. 3 to fit well the corresponding curve also determined by the given theory. In all experiments the gate-aperture was set at T=5 nsec. It appears that, at the chosen values of η and ρ there exists a saturation effect that regulates the photon emission from the cavity. This is precisely the nonlinear process that, according to the present understanding of the light-squeezing process, gives rise to the reported nonclassical effect of the output photon distribution (Teich, Saleh and Perina: Ref. 1).

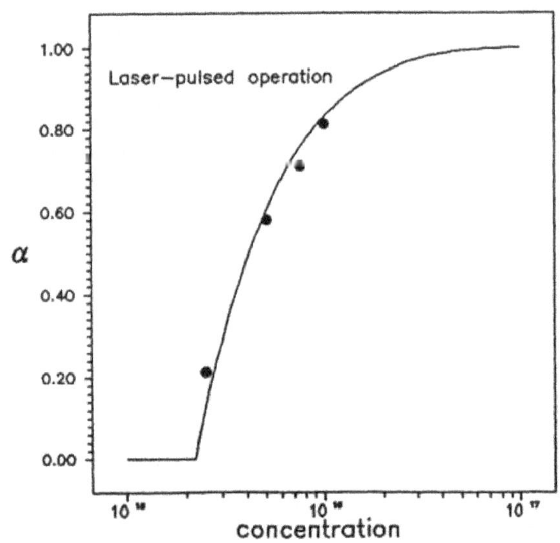

Fig. 3. The coincidence parameter α for a 4 nsec pulse excitation η=3 nJ.

In conclusion, we have reported the experimental demonstration that the peculiar vacuum-confinement properties of the optical microcavity under coherent pulsed excitation are apt to provide an efficient mechanism for the generation of light possessing a highly non-classical, anti-bunched character. We have identified the molecular-saturation effect in cavity-confinement conditions as the nonlinear dynamical process that is responsible for this kind of n-squeezing of light. In this connection we believe that the peculiar QED field-enhancement properties of the microcavity can play a further positive role, for instance by providing a further efficient control mechanism over the coherent Rabi dynamics of the interacting atoms. These new theoretical aspects, as well as a detailed theory accounting for the crucial interplay taking place in our experiment between the very small number of trapped atoms and the excitation photons in a short-pulse regime, will be investigated in a forthcoming comprehensive paper.

5. References.

1. M. Teich, B. Saleh and J. Perina, *Journal of the Optical Society of America* **B 1** (1984) 366.
2. H. J. Kimble, M. Dagenais and L. Mandel, *Phys. Rev. Lett.*, **39** (1977) 691.
3. F. Dietrich and H. Walther, *Phys. Rev. Lett.*, **58** (1987) 203.
4. Th. Basché, W. E. Moerner, M. Orrit and H. Talon, *Phys. Rev. Lett.* **69** (1992) 1516.
5. F. De Martini, G. Innocenti, G. Jacobovitz and P. Mataloni, *Phys.Rev.Lett.*, **59** (1987) 2955; F. Cairo, F. De Martini and D. Murra, *Phys.Rev.Lett.*,**70** (1993) 1413; F. De Martini, R. Loudon, M. Marrocco, P. Mataloni and D. Murra, *JOSA* **B 10** (1993) 360.
6. R. Loudon, *The Quantum Theory of Light* (Clarendon, Oxford, 1983) Ch.5.
7. P. Grangier, G. Roger and A. Aspect, *Europhys.Lett.* **1** (1986) 173.

BASIC QUANTUM MECHANICS

MODIFYING QUANTUM MECHANICS: MOTIVATIONS AND IMPLICATIONS.

GIANCARLO GHIRARDI
Department of Theoretical Physics, University of Trieste and International Centre for Theoretical Physics, Trieste, Italy.

ABSTRACT

We briefly review a recent proposal of modifiyng quantum dynamics in such a way to overcome the conceptual difficulties met by the formalism in connection with macroscopic systems. We discuss its implications and the possiblity of testing the model against the standard quantum theory.

1. Introduction

We discuss a recently proposed modification of quantum theory[(1)]. The reasons for paying attention to such an attempt derive from our concern, shared by many scientists, about the measurement problem or, more generally, about the so called objectification problem. Even though we are fully aware of the unprecedented successes of quantum mechanics and of the fact that[(2)] it "represents an important, in a certain sense even final, advance in physical knowledge", we believe that there is a fundamental gap in our physical understanding between the "small scale quantum level" (including the behaviour of microscopic entities at the atomic, and/or subatomic level) and the "larger classical level" of the world as we perceive it.

The model we will discuss exhibits non-linear and stochastic dynamical features. Before coming to its detailed description we would like to point out the precise relations of the model to standard quantum mechanics in its orthodox formulation including the postulate of wave packet reduction and to comment on the inclusion of stochasticity at the fundamental level. Such a fact has been sometimes considered as highly unpalatable, due to the irreversibility which goes with it. It seems to us that, in general, the supporters of the orthodox position and more specifically those who on this basis do not consider as acceptable to entertain the idea of stochastic modifications of the basic quantum evolution equation take a quite peculiar position with respect to the probabilistic nature of the theory. Quantum mechanics, when the completeness assumption is made, implies the acceptance of nonepistemic probabilities, i.e. the recognition that in nature there are fundamental stochastic elements which do not admit an ignorance interpretation. We have always found a little bit puzzling the insistence of our teachers, with a sort of complacency, on the fact that quantum indeterminism and the random nature of physical processes emerge only in connection with the act of measurement, whereas the quantum evolution is strictly deterministic. Obviously this is a conceptually relevant point of the standard formulation and as such it certainly deserves a detailed illustration, but what seemed inappropriate was the fact that this feature was presented as a merit of the formalism. Such an attitude was a common one during the early times of quantum mechanics. This can be easily understood if one takes into account that strong emphasis was then given to the disturbance interpretation (traced back to the finiteness of the quantum) of the measurement process. But after such an interpretation has been shown to be untenable it seems quite problematic to accept the fundamentally nonepistemic nature of quantum probabilities but to confine it to particular physical processes and, moreover, to processes whose identification is basically ambiguous. This distinction between the deterministic and the stochastic aspects of the formalism indicates its fundamentally dualistic nature, and as such it appeared to us a serious drawback much before we get acquainted with the puzzling aspects of the measurement process and, more generally, of the difficulties that the theory meets with the description of macroscopic events.

From this perspective it becomes quite natural to entertain the idea of an all-pervading nonepistemic probabilistic nature of physical processes which has to be taken into account at the very fundamental level of microscopic phenomena by modifying the evolution equation. In view of the extreme success of the theory, it appears as a sort of a miracle that one can (as we will show below) actually follow such a line and, in particular, to make the nonepistemic probabilistic nature of measurement outcomes a consequence of the basic nonepistemic stochastic nature of the dynamics characterising all physical processes, without clashing with any (presently) known fact about quantum phenomena.

Said differently, we will exhibit a model theory which is (almost) identical with quantum mechanics plus the wave packet reduction postulate, but which does not require such an independent (and incompatible with the linear nature of quantum evolution) postulate: reduction is a physical process deriving from the unique dynamical principle governing all phenomena.

In our opinion, even the convinced supporters of the standard theory should show some interest in the fact that it can be made internally consistent and can be based on a unique dynamical principle. We do not want to be misunderstood; we are not claiming that the model we will present has to be taken as a serious candidate for replacing quantum mechanics (actually, as we will discuss below, a lot of hard work is still necessary even to simply approach such a stage), we claim that it is of some conceptual relevance to have an explicit proof that at least one consistent model theory exhibiting the above features can be worked out.

Obviously, since the model represents an actual modification and not simply a reinterpretation of quantum theory, it has testable (in principle) consequences contradicting quantum predictions. We consider interesting to present the model to this audience having a deep and detailed knowledge of the most advanced and refined experimental techniques to test quantum theory. It would be nice if some suggestion about possible "experimenta crucis" would be stimulated by this presentation.

2. A Simple Model of Spontaneous Reduction.

Several interesting attempts of modifying quantum evolution in order to induce, at the appropriate level, the objectification of macroscopic properties had been made[(3)], but they did not lead, up to recent times, to a real breakthrough, due to some crucial problems which had been left unsolved. The first one was that of the choice of the preferred basis: which specific properties of individual systems should one require to be dynamically and spontaneously actualised? Secondly, how can the dynamical modifications have a practically negligible effect for all microscopic systems (a necessary requirement due to the already stressed extremely highly accurate predictions of quantum theory for such systems) and simultaneously to be able to induce a very rapid suppression of macroscopically distinguishable superpositions?

The solution came[1] from the identification of the appropriate preferred basis, i.e., the one associated to positions. Subsequently, many interesting and important contributions[4-6] to this line of research have appeared. For our purposes we can limit our considerations to expounding the original model, usually referred to as QMSL, an acronym for Quantum Mechanics with Spontaneous Localizations.

QMSL is based on the following assumption: the statevector of any physical system, besides evolving according to the standard Schrödinger equation, is subjected, at randomly distributed times with an appropriate mean frequency, to spontaneous localisation processes affecting the elementary constituents of the system itself. Such processes are assumed to correspond to some fundamental mechanism of nature.

For definiteness one has to make precise the features of the mechanism, i.e. to specify when the spontaneous localizations take place, how they affect the wave function and where do they take place. The answer to these questions are very simple: first, the localizations occur at random times obeying a Poisson distribution. The mean frequency for their occurrence is denoted by λ, so that λdt is the probability of a process occurring in the time interval dt. Secondly, suppose we are dealing with a system of N particles, and consider the configuration space representation of the state vector $\Psi(r_1,r_2,\ldots,r_N)$. If the i-th particle suffers a localisation around the point x, the above wave function changes according to:

$$\Psi(r_1,r_2,\ldots,r_N) \Rightarrow \Psi_x(r_1,r_2,\ldots,r_N) = \frac{\Phi_x}{\|\Phi_x\|}$$

where

$$\Phi_x(r_1,r_2,\ldots,r_N) = \left[\frac{\alpha}{\pi}\right]^{\frac{1}{4}} e^{-\frac{\alpha}{2}(r_i - x)^2}\,\Psi(r_1,r_2,\ldots,r_N)$$

Finally, the probability density of the localisation taking place at x is given by $\|\Phi_x\|^2$.

To make clear that the spontaneous processes, when they occur, tend to suppress linear superpositions of far away states, let us consider the case of one particle in a state which is a superposition of a wave function $\Psi_A(x)$ localized around A and of one $\Psi_B(x)$, localized around B, with $|A\text{-}B| >> 1/\sqrt{\alpha}$. As a consequence of the previous assumptions one can easily convince himself that localizations can occur (with appreciable probability) only either around A or around B and that, according to the position which is selected, the resulting wave function describes a particle which is well localized around it.

The theory has some extremely interesting features: provided the parameter α is appropriately chosen, in the case of an almost rigid body, the spontaneous localizations affect essentially only the centre of mass and, moreover, the frequency of the localizations is amplified with the number N of constituents. Some comments: the expression "almost rigid body" denotes a body, like an insulating solid, for which the spreads of the positions of its lattice points around their equilibrium positions are much smaller than $1/\sqrt{\alpha}$. The amplification of the mechanism for what concerns the dynamics of the centre of mass motion holds true in all cases. However, if the body is not almost rigid, the internal motion can be appreciably affected. Without entering into technical details one can easily grasp the reason of the amplification mechanism. Consider a macroscopic body (such as the pointer of an apparatus) which is in a superposition of two differently located states:

$$|\text{Pointer Here}\rangle + |\text{Pointer There}\rangle.$$

We remark that the first of the above states corresponds to the situation in which all particles of the pointer are "around here", while the second to the one in which all are "around there". Multiplication of the configuration space wave function by a Gaussian in any one of the coordinates which is centred "here" ("there") and normalization of the state, leads to the suppression of the $|\text{Pointer there}\rangle$ ($|\text{Pointer here}\rangle$) term of the superposition.

Now we are ready to make the choice for the parameters α and λ and to discuss the ensuing implications. Assume:

$$\lambda \cong 10^{-16}\,\text{sec}^{-1}; \frac{1}{\sqrt{\alpha}} \cong 10^{-5}\,\text{cm}$$

The choice of λ implies that a microscopic system suffers a localisation, on the average, every 10^8 years! In the case of a macroscopic system, however, the amplification mechanism has remarkable effects; since in such a case N is of the order of Avogadro's number a localisation takes place on the average every 10^{-7} sec!

From the point of view which concerns us here we can state that the model induces the spontaneous emergence of the property of being localized for a macroscopic object. It has to be noted that the spatial localisation mechanism also forbids, indirectly, the persistence of other types of superpositions of macroscopically distinguishable states. The simplest example is that of a superposition of two states of a macroscopic object corresponding to its being essentially at the same place but possessing different momenta. The localisation mechanism cannot be immediately effective but, as soon as the different velocities lead to different positions, one of the two terms of the superposition is suppressed.

Concerning the problem mentioned in Section 1, i.e. that within standard quantum mechanics the borderline between micro- and macrophenomena is fundamentally ambiguous, the model yields a very precise answer: the suppression of the superpositions (i.e. the violations of linearity) is governed by the number of particles which are displaced by a distance appreciably larger than the characteristic localisation distance.

3. Improving the Model and Discussing its Testability.

Quantum Mechanics with Spontaneous Localisations has still some limitations which require to be cured before it can be taken as a consistent theoretical framework. The first one derives from the fact that it does not respect the symmetry properties of the wavefunction in the case of identical constituents. Its generalisation satisfying such a requirement has been exhibited and discussed in various papers(4). We mention that the final formulation of the model, which has been referred to as CSL (Continuous Spontaneous Localisation), has a remarkable mathematical elegance and is based on the introduction of a continuous stochastic equation which makes the statevector to perform a sort of Brownian motion in the Hilbert space. For what concerns us here we simply recall that the CSL formalism is quite simple and can be summarized as follows: suppose one is dealing with a system containing many identical constituents and with a superposition of two states Ψ_1 and Ψ_2. The decoherence process for the two states in the superposition is then governed by the following mechanism. Consider the whole three-dimensional configuration space as divided into cells of volume of about 10^{-15} cm^3. Call $n_1^{(i)}$ and $n_2^{(i)}$ the occupation numbers of the generic cell for the two considered states (in the model the index i is replaced by a continuous variable). The decoupling takes then place continuously in time and its rate is determined by an exponential involving the expression $\sum_i (n_1^{(i)} - n_2^{(i)})^2$, the sum being extended to all cells. The reader can easily understand that the new mechanism respects the symmetry properties of the wavefunction and at the same time has implications which are strictly similar to those of QMSL.

Another important problem is that of finding generalizations of CSL which respect the requirements of relativistic invariance. J. S. Bell has repeatedly stressed the great relevance of doing this. In one of his last papers(7), after having considered the only (in his opinion) acceptable alternatives to the unsatisfactory standard formulation of quantum mechanics, i.e. the Pilot Wave and CSL, he wrote: "*the big question, in my opinion, is which, if either, of these two precise pictures can be redeveloped in a Lorentz invariant way*". Such a problem has been considered in refs.(5). These attempts have led to some interesting results, in particular to a quite general formulation of the concept of stochastic invariance and they have stimulated new hints and ideas about quantum nonlocailty(8). In spite of this it has not been possible, up to now, to build a completely consistent relativistic model of dynamical reduction.

Before concluding we address the question raised at the end of Sect.1: is it possible to test dynamical reduction models against standard quantum mechanics? As already stated, in principle, such a distinction is obviously possible. However, due to the specific manner in which the standard theory has been modified, i.e. in such a way that linearity holds to an extremely high degree of accuracy for microsystems, from a practical point of view it turns out to be quite difficult, if not impossible, with the present experimental techniques, to devise crucial tests for the model. We would like to recall a joke which P. Pearle has made several times and which has been misinterpreted by J. Horgan leading him to state(9), that A. Zeilinger had proved experimentally that CSL is wrong. The joke is based on the following fact: the decoherence which is induced between the two terms in a superposition of states describing far away localizations of a physical system leads, in the long run, to the suppression of interference effects. So, such a decoherence can lead to discrepancies with quantum predictions about neutron interferometry. But this happens after 10^8 years, so that, as stated by Pearle, the experiment would simply require an improvement of 10^{15} in the accuracy of neutron interferometry experiments!

To come back to our question we list now some of the effects that dynamical reduction mechanisms of the type considered above can have for microsystems and macrosystems.

1. Using the simple language of QMSL one can easily understand that if an atom suffers a spontaneous localisation process its wave function is modified and as a consequence a certain probability of excitation and/or dissociation of a bound state arises. In the case of an hydrogen atom the explicit computation leads to predict few excitations per mole per second.

2. The irreversible nature of the dynamics leads to energy nonconservation. Another way of understanding this is to remark that localizations introduce high momentum components and, as a consequence, an increase of the kinetic energy. One can easily evaluate the corresponding effect on the average increase of the temperature of the universe. The explicit computation shows that the model implies a total temperature increase from the beginning of the universe of about 10^{-3} ^{0}K, a value to be compared with the background radiation of 2.7 ^{0}K. A detailed discussion of the two above effects has been given in ref.(10).

3. Effects of QMSL on scattering processes have been discussed in detail in ref. (11). Once more the effects of dynamical reduction turn out to be largely screened by many other effects such as the interactions with the environment, the background radioactivity, quantum gravity effects and so on.

4. Recently, E. Squires and P. Pearle (private communication) have remarked that if one assumes spontaneous localizations to take place for quarks, the analogous of the dissociation of atoms would be a change of the quark state and in particular this would give rise to the possibility of excitation of the proton state, leading to an apparently testable proton decay probability. Subsequently they seem to have reached the conclusion that this effect too is very difficult to detect; we cannot be more precise on this point.

5. In the recent literature there have been various proposals of testing QMSL or CSL in processes involving macroscopic quantum coherence. In refs.(12) explicit evaluations of the effect of the modified dynamics for superconductivity have been given. The model actually implies an increase of the resistivity of a superconductor, but its amount turns out to be various orders of magnitude out of experimental testability.

The conclusion is that, for the moment, it seems out of question to look at effects of dynamical reduction which could discriminate between dynamical reduction models and quantum mechanics. This being the situation, to consider seriously or to reject the models of dynamical reduction is, to a large extent, a matter of taste. The real interest of the program stems, in our opinion, from its constituting an explicit example of a theory which reconciliates the two contradictory dynamical principles of standard quantum mechanics, i.e. the linear and deterministic Schrödinger's evolution and the the nonlinear and stochastic reduction of the wave packet. Finally we would like to mention that, as it should be obvious and as it has been repeatedly stressed(6), taking the dynamical reduction attitude allows the adoption of a macrorealistic position about natural phenomena, a relevant conclusion for people interested in the epistemological consequences of our scientific theories.

References.

1. G.C. Ghirardi, A. Rimini and T. Weber, in *Quantum probability and Applications*, L. Accardi, W. vonWaldenfels eds., Springer, Berlin, 1985; Phys. Rev., **D 34**, 470 (1986); *ibid.*, **D 36**, 3287 (1987); J.S. Bell in: *Schrödinger-Centenary celebration of a polymath*, C. W. Kilmister ed., Cambridge University Press, Cambridge, 1987; G.C. Ghirardi, A. Rimini and T. Weber, Found. Phys., **18**, 1 (1988).

2. A. Einstein in: M.Born, *The Born-Einstein Letters*, Walker and Co., New York, 1971.

3. F. Karolyhazy, Nuovo Cim., **A 42**, 390 (1966); P. Pearle, Phys. Rev., **D 13**, 857 (1976); Int. Journ. Theor. Phys., **48**, 489 (1979); Found. Phys., **12**, 49 (1982); Phys. Rev., **D 29**, 235 (1984); N. Gisin, Phys. Rev. Lett., **52**, 1657 (1984), *ibid.*, **53**, 1776 (1984).

4. L. Diosi, Phys. Lett., **A 129**, 419 (1988); P. Pearle, Phys. Rev., **A 39**, 2277 (1989); N. Gisin, Helv. Phys. Acta, **62**, 363 (1989);G.C. Ghirardi, P. Pearle and A. Rimini, Phys. Rev., **A 42**, 78 (1990); G.C. Ghirardi, R. Grassi and A. Rimini, Phys. Rev., **A 42**, 1057 (1990).
5. P. Pearle, in *Sixty-two Years of Uncertainty*, A. Miller ed., Plenum, New York, 1990; G.C. Ghirardi, R. Grassi and P. Pearle, Found. Phys., **20**, 1271 (1990); in: *Symposium on the Foundations of Modern Physics 1990*, P. Lahti and P. Mittelstaedt eds., World Scientific, Singapore, 1990; P. Pearle in: Quantum Chaos-Quantum measurement, P. Cvitanovic, I. Percival and A. Wirzba eds., Kluwer Academic Publishers, Dordrecht, 1992.
6. J.S. Bell in: *Themes in Contemporary Physics II, Essays in Honour of Julian Scwinger's 70th birthday*, S. Deser, and R.J. Finkelstein eds., World Scientific, 1989; D.Z. Albert and L. Vaidman, Phys. Lett., A 139, 1 (1989); A. Shimony in:*PSA 1990*, A. Fine, M. Forbes and L. Wessel eds., East Leasing, Michigan, 1990; F. Aicardi, A. Borsellino, G.C. Ghirardi and R. Grassi, Found. Phys. Lett., **4**, 109 (1991); G.C. Ghirardi, in: *Bell's Theorem and the Foundations of Modern Physics*, A. van der Merwe, F. Selleri and G. Tarozzi eds., World Scientific, Singapore, 1992.
7. J.S. Bell, in: *Sixty-two Years of Uncertainty*, A. Miller ed., Plenum, New York, 1990.
8. G.C. Ghirardi, R. Grassi, J. Butterfield and G.N. Fleming, Found. of Phys., **23**, 341 (1993).
9. J. Horgan, *Quantum Philosophy*, in the July 1992 issue of Scientific American.
10. G.C. Ghirardi and A. Rimini, in: *Sixty-two Years of Uncertainty*, A. Miller ed., Plenum, New York, 1990.
11. M.E. Tegmark: *Apparent Wave Function Collapse Caused by Scattering*, preprint, March 1992.
12. A.I.M. Rae, J. Phys., **A23**, 57 (1990); in: *Bell's Theorem and the Foundations of Modern Physics*, A. van der Merwe, F. Selleri and G. Tarozzi eds., World Scientific, Singapore, 1992; M.R. Gallis and G.N. Fleming, Phys. Rev., **A42**, 38 (1990).

EVIDENCE FOR THE "LAW OF THE EXCLUDED MUDDLE"

DANIEL M. GREENBERGER
City College of the City University of New York
New York, New York 10031

ABSTRACT

In a recent attempt to prove a Bell Theorem without inequalities for two spinless particles, use had to be made of an additional assumption, the "Law of the Excluded Muddle", (also called the EPNT assumption). We present some arguments here to show the overwhelming plausibility of the assumption.

Introduction

The "Law of the Excluded Muddle", also called the EPNT assumption, for "Emptiness of the Paths not Taken" (an acronym we shall use from now on, for convenience), was introduced in an attempt to prove a Bell Theorem Without Inequalities for two particles[1]. This assumption was needed as an addition to the postulates of "reality" and "locality", taken from the famous Einstein-Podolsky-Rosen attempt[2] (hereafter called EPR) to prove the incompleteness of quantum theory.

Bell showed[3] that if one accepted the EPR assumptions literally, then it follows that there is a maximum possible correlation between two spin-1/2 particles in a singlet state. He then showed that the correlations produced by quantum theory are greater than this maximum. Thus, if one accepts that the results of quantum theory are correct, it follows that the EPR assumptions cannot be correct. Of course, historically, the conflict between the EPR results and quantum theory was taken as a challenge to see whether one could decide the issue experimentally, and by now the results overwhelmingly show that quantum theory is correct.[4]

Bell's Theorem takes the form of an inequality that any theory obeying the postulates of EPR must satisfy, but which is violated by the results of quantum theory. Specifically, in the special case where the two particles are perfectly correlated, so that one can derive what EPR call an "element of reality" for the system, namely when the measurement directions for the spins of the two particles differ by 0° or 180°, quantum theory and the EPR postulates lead to consistent results. However in the more general case, where such elements do not exist, namely for arbitrary angles, quantum theory violates Bell's inequality.

It has recently been shown that for a system of three spinless particles one may derive a stronger result, the GHZ theorem[5]. In this case, even when the particles are perfectly correlated, and an element of reality exists, it is impossible to explain these correlations consistently, using the EPR postulates. No inequalities are needed.

In trying to extend this theorem to two spinless particles, as mentioned above an extra assumption, the EPNT assumption had to be made. According to this assumption:

> If the path taken by a particle is an element of reality then there is no entity associated with this particle that in any way samples alternate paths (i.e., the paths not taken are truly *empty*) (EPNT).

Considerable effort was expended in Ref. (1) to show that the assumption is not inconsistent with the introduction of pilot waves for one-particle interference experiments because in this case there is no element of reality limiting the particle to one of the possible paths. From a quantum mechanical point of view, the interference is caused by the fact that there is an amplitude for the particle to take either route from the source to the detector. For people who would like to assign a path to the particle, one usually introduces some entity, like a pilot wave, to sample the other path. This is necessary because when the paths recombine the particle may be affected by any phase shifters, etc, that were present in the path it did not take. The pilot wave is introduced to provide this non-local knowledge for the particle.

However in two-particle experiments, one may make the two possible paths for one of the particles interfere with each other as before, but one has an alternative here. One may instead catch the companion particle. If the two particles were in an entangled state, then by catching the companion one instantly knows which path the original particle took. In such cases, one will get no interference, as only one path is now possible for the original particle. Thus in this case there is no need to introduce a pilot wave, as there is no interference to explain. The EPNT assumption plays the role of Occam's Razor in this case, and says that whenever the path is explicitly known as an element of reality, there are no such entities as pilot waves exploring alternative paths.

By denying that there can be any such entities exploring alternative paths, the EPNT assumption automatically rules out the possibility of any experiment that might depend in any way on phase shifters, etc, that reside in these alternate paths. This feature makes it possible to exploit entangled two-particle states to show that the EPR assumptions then become so restrictive that one cannot consistently use them, a result that is also true in the three-particle case (where one does not even need the EPNT assumption). But it would seem impossible to prove such an assumption as EPNT. How can one prove that a pilot wave sampling alternative paths does not exist, even if its existence has no consequences?

So we are not attempting such a proof. The best one can hope for is to show that there are cases, even complicated cases, where one can see the possible results of a pilot wave, because an element of reality for the path does not exist. But one can change the experiment by catching the companion particle entangled with the original particle, so as to create such a path element of reality, and if we do then all information that could be explained by assuming the pilot wave also disappears. This kind of experiment reinforces the uselessness of the pilot wave in cases where a path element of reality exists. It does not constitute a proof, but it enhances the plausibility of the EPNT assumption, which we claim is consistent with all known real or gedanken experiments (that are themselves consistent with quantum theory).

In this paper we shall first give an argument showing that if one believes that a particle always takes one path or the other, even when one cannot experimentally tell which one, then one must introduce some such concept as a pilot wave. We will then show that if one can create an element of reality confirming which path the particle took, by measuring a different particle and never touching the particle in question, then nonetheless the necessity for introducing the pilot wave disappears. We shall then examine a more complicated situation and arrive at the same result.

The Necessity of Pilot Waves for Single-Particle Experiments

Consider the following experiment. A single particle passes through a Mach-Zender interferometer consisting of two beam-splitters, each of which transmits 50% of the beam intensity, as in Fig. (1). A phase shifter of phase-shift α is placed in beam a and an absorber that transmits an intensity ρ^2 is placed into beam b.

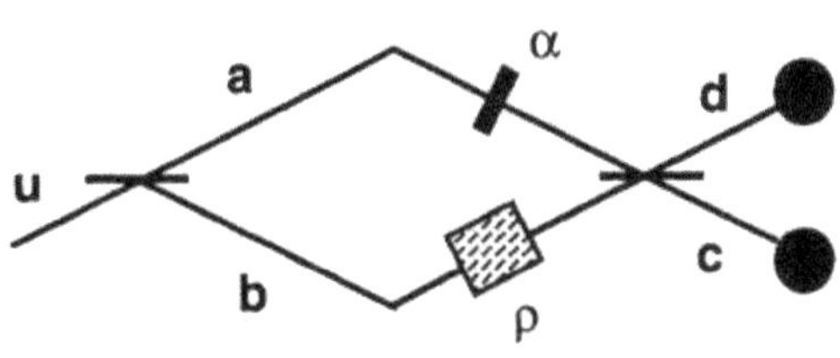

Fig. (1). A One-Particle Interferometer.
The beam is split and recombined by two beam-splitters. There is also a phase-shifter of phase α in beam a and an aborber that transmits intensity ρ^2 in beam b.

In what follows we shall assume (without loss of generality) that the phase shifter produces a 90^0 phase shift in the reflected beam, leaving the phase of the transmitted beam unaffected. Then

$$u \rightarrow \tfrac{1}{\sqrt{2}}(e^{i\alpha}|a\rangle + i\rho|b\rangle) \rightarrow \tfrac{1}{2}[(e^{i\alpha} - \rho)|c\rangle + i(e^{i\alpha} + \rho)|d\rangle],$$

and the probability of a detection at a detector placed in c or d is

$$P_c = \tfrac{1}{4}(1+\rho^2 - 2\rho\cos\alpha),$$
$$P_d = \tfrac{1}{4}(1+\rho^2 + 2\rho\cos\alpha),$$
$$V_{c,d} = \frac{2\rho}{1+\rho^2},$$

where V is the fringe visibility for either beam c or d, (obtained by varying α).

In this experiment, one does not know which beam, a or b, the particle takes, and so there is no element of reality associated with the path of the particle. If one would like nonetheless to ascribe a path to the particle, one can associate a pilot wave to the particle that senses the path not taken, so that if the particle takes beam b, it is still aware of the existence of the phase shift α in beam a.

We can convert the above into a "proof" that there *must* be a pilot wave present in this situation. Note that if the absorber is absent ($\rho = 1$), the visibility of the fringes is 100%. Note also that if the absorber completely absorbs the beam ($\rho = 0$), the visibility of the fringes is 0. In this case one knows that if the particle reaches one of the detectors, it must have taken beam a, and it has a 50-50 chance of ending up in either beam c or d, independently of α. For an intermediate case where the transmission coefficient is ρ^2, a detector placed in beam a after the phase shifter will detect a transmission rate 1/2, while one placed in beam b, after the absorber, will detect a transmission rate $\rho^2/2$, (assuming unity transmission rate in the incident beam u). The total transmission rate has thus been reduced to $(1+\rho^2)/2$.

This means that only $\rho^2/2$ of the particles could have been in beam b. These $\rho^2/2$ particles can contribute to a 100% fringe visibility, since for any particular particle, one does not know which beam it was in. The rest of the particles can only have been in beam a, and so they contribute 0 to the fringe visibility. Thus the maximum possible fringe visibility can be

$$V_{c,d\,Max} = \frac{\frac{\rho^2}{2}\times 1 + \frac{1}{2}\times 0}{\frac{1}{2}+\frac{\rho^2}{2}} = \frac{\rho^2}{1+\rho^2}.$$

This is to be compared with the actual fringe visibility $V_{c,d}$ given above. If ρ^2=.01, then ρ=.1, and so while the maximum visibility, based on a particle argument is only 1%, the actual visibility is 20%! Thus there must be some kind of communication to the beam splitter, telling the particles which way to divide, that goes beyond what can be carried by the density of particles present. Thus some kind of pilot wave, or its equivalent, must exist that can make its presence felt even when there are no particles to carry it.

Failure of the Argument When An Element of Reality Is Present

We shall now show that when an element of reality is present that labels which beam the particle is in, the above argument fails. Consider the two-particle interferometer in Fig. (2), equivalent to the single particle interferometer above.

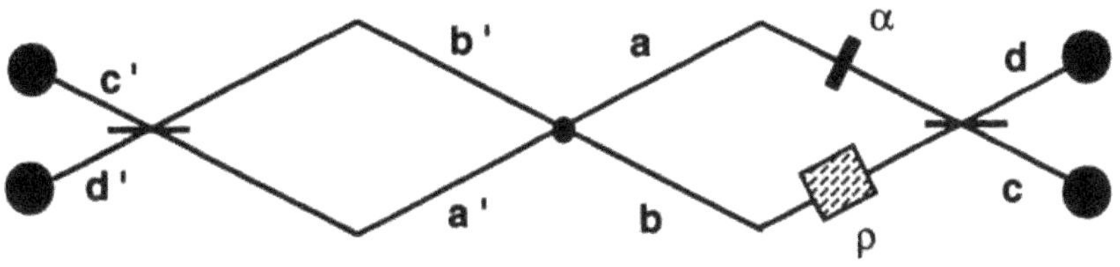

Fig. (2). A Two-Particle Interferometer.
Two particles are emitted from the central source in an entangled state. They are either in the beams a-a', or in the beams b-b'.

In this case the two particles are produced in an entangled state

$$\psi = \tfrac{1}{\sqrt{2}}(|a\rangle|a'\rangle + |b\rangle|b'\rangle).$$

Once again there is a phase shifter of phase α in beam a and an absorber that transmits intensity ρ^2 in beam b. The wave function becomes

$$\psi \to \tfrac{1}{\sqrt{2}}(e^{i\alpha}|a\rangle|a'\rangle + \rho|b\rangle|b'\rangle)$$

$$\to \tfrac{1}{2}[e^{i\alpha}(|c\rangle + i|d\rangle)|a'\rangle + \rho(|d\rangle + i|c\rangle)|b'\rangle]$$

$$\to \tfrac{1}{2\sqrt{2}}[(e^{i\alpha} - \rho)(|c\rangle|c'\rangle - |d\rangle|d'\rangle) + i(e^{i\alpha} + \rho)(|c\rangle|d'\rangle + |d\rangle|c'\rangle)].$$

This leads to the following probabilities for the simultaneous coincidence count rates for counting one photon on each side of the source:

$$P_{cc'} = P_{dd'} = \tfrac{1}{8}(1 + \rho^2 - 2\rho\cos\alpha),$$

$$P_{cd'} = P_{dc'} = \tfrac{1}{8}(1 + \rho^2 + 2\rho\cos\alpha).$$

If one records a count in c or c' as (+1), and one in d or d' as (-1), then a coincidence count will be recorded as the product: thus a count c-c' or d-d' will be recorded as (+1) and a c-d' or d-c' count will be recorded as (-1). The analog to the counts P_c and P_d from the previous one-particle experiment will be $P_+ = P_{cc'} + P_{dd'}$, and $P_- = P_{cd'} + P_{dc'}$. These give the identical results as before,

$$P_+ = P_{cc'} + P_{dd'} = \tfrac{1}{4}(1 + \rho^2 - 2\rho\cos\alpha),$$

$$P_- = P_{cd'} + P_{dc'} = \tfrac{1}{4}(1 + \rho^2 + 2\rho\cos\alpha).$$

Also note that fringes dependent on α only show up in coincidence counts. Any single counter in one of the beams c, c', d, d' will not show any such fringes, for example, $P_c = P_{cc'} + P_{cd'} \neq f(\alpha)$.

The difference between this experiment and the one-particle one is that we may catch one of the particles, such as a' or b' with a detector. If we do so, we immediately know which beam the other particle is in. For example, if we detect that one particle is in beam a', then we immediately know that its companion is in beam a. But if we do this, the probability for a count in c or d will reduce to

$$P_{a'c} = P_{a'd} = 1/4,$$

$$P_{b'c} = P_{b'd} = \rho^2/4.$$

Thus by catching the companion particle, we eliminate all interference between the beams. But the particle on the right cannot know whether we will catch the particle on the left or let it pass through the system. The only difference between the two cases for the particle on the right, is that in one case there is a hidden variable label attached to the particle, defining which beam, a or b, it is in. The existence of this label eliminates any interference, and therefore any knowledge the particle needs about the other beam, and so any need to introduce a pilot wave. But this is exactly what the EPNT assumption entails.

There is a loophole in this proof in that it might be possible for the particle to have encoded in it a set of instructions that tell it to behave such that for single particle counts, there are no correlations, but for coincidence counts there are the appropriate correlations. But this possibility runs afoul of Bell's Theorem[3], which says that it is possible to build such local correlations into the system, but not in a manner that agrees with quantum mechanics. Furthermore, we could have set the system up as a three-particle interferometer. But this attempt to produce multi-particle correlations would run afoul of the GHZ theorem[5], which says that it is impossible to assign any set of consistent correlations. Thus we see that this system behaves completely consistently with the EPNT assumption.

Lucien Hardy[6] has produced a more involved, but perhaps more convincing proof of the ability of a pilot wave to produce an effect when there are no particles present in the beam. But his proof is subject to the exact same two-particle argument as ours, and the conclusion is the same, namely that when an element of reality exists labelling the path of the particle, then the effects of the pilot wave disappear.

A More Complicated Example

We will briefly go through a more complicated example, shown in Fig. (3). Here beams d and e do not interact when they cross (think of them as being in different planes). There are five beam-splitters on the right hand side, and a phase shifter of phase α in beam d and one of phase β in beam e.

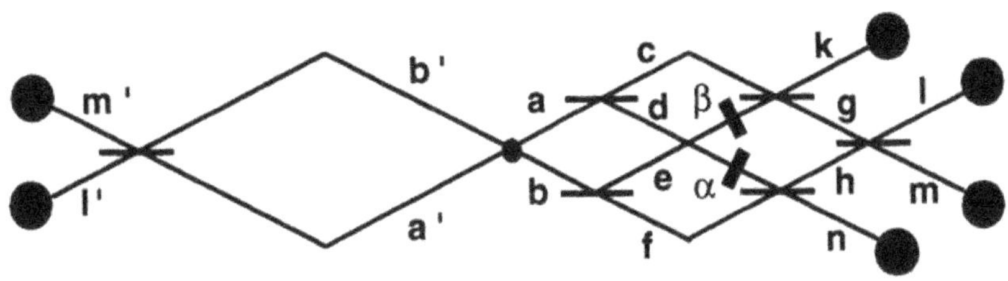

Fig. (3). A More Complicated Two-Particle Interferometer.
Here there are two phase shifters, α and β, and there are five beam-splitters on the right. Note that the beams d and e do not cross or interact.

In this example, the probability of coincidence counts in the various beams is given by

$$P_{mm'} = P_{ll'} = \tfrac{1}{16}(1+\cos(\alpha-\beta)),$$
$$P_{ml'} = \tfrac{1}{16}(3+2\sin\alpha - 2\sin\beta - \cos(\alpha-\beta)),$$
$$P_{lm'} = \tfrac{1}{16}(3-2\sin\alpha + 2\sin\beta - \cos(\alpha-\beta)),$$
$$P_{kl'} = \tfrac{1}{8}(1+\sin\beta), \quad P_{km'} = \tfrac{1}{8}(1-\sin\beta),$$
$$P_{nl'} = \tfrac{1}{8}(1-\sin\alpha), \quad P_{nm'} = \tfrac{1}{8}(1+\sin\alpha).$$

If one examines these results, one can see quite a lot. For example, if one looks at beam k (which can contribute to $P_{kl'}$ and $P_{km'}$), one sees that to get interference one needs contributions from both beams a and b, via the paths a-c-k and b-e-k. Only the phase shift β is in either path, and so the result must be independent of α. On the other hand if one looks at the beam l, one sees that there are one-particle interference contributions from both beams a and b, as well as two-particle contributions from both together. The one-particle contributions from a alone, come from paths a-d-h-l and a-c-g-l. These depend only on α. There are also one-particle contributions from beam b that depend only on β. The two-particle contribution depends on $(\alpha-\beta)$ in a non-separable way. Note by the way that the total count rate at l' or m' is independent of either angle,

$$P_{l'} = P_{kl'} + P_{ll'} + P_{ml'} + P_{nl'} = \tfrac{1}{2}, \quad P_{m'} = \tfrac{1}{2}.$$

However if one catches the companion particle a' or b', then one knows with certainty which beam the right side particle is in and there is an element of reality associated with it. The counting rates then simplify significantly.

$$P_{ka'} = P_{kb'} = P_{na'} = P_{nb'} = \tfrac{1}{8},$$
$$P_{la'} = \tfrac{1}{8}(1-\sin\alpha), \quad P_{ma'} = \tfrac{1}{8}(1+\sin\alpha),$$
$$P_{lb'} = \tfrac{1}{8}(1+\sin\beta), \quad P_{mb'} = \tfrac{1}{8}(1-\sin\beta).$$

Note here that the beam k now shows no interference at all, because there can be only one contribution, from a or b, depending on which particle was caught on the other side. The question that arises then is how the pilot wave "knows" not to contribute in this case, since the right hand side of the experiment cannot be aware of what choices have been made on the left side by the experimenter. The answer given by the EPNT assumption is that there can be no pilot wave here, since there is now an element of reality associated with the path.

The situation is similar for the beams that contribute to l and m. If one catches the particle a' on the left side, there is still a one-particle interference pattern that can contribute to l on the right, via interference between the paths a-c-g-l, and a-d-h-l. These only intercept the phase α, and so there can be no contribution from β. So again the question arises, "How does the pilot wave 'know' not to contribute in this case, which is identical on the right hand side to the two-particle case, except that now there is an element of reality connected with the path?" Again the EPNT assumption gives the same simple answer. All these multi-particle cases are extremely mystifying from the point of view of pilot waves, but extremely simple from the EPNT point of view.

Finally we reiterate our earlier comment that we know of no real or gedanken experiment (agreeing with the results of quantum theory) whose results are inconsistent with the EPNT simplifying assumption. And yet these experiments are very unnatural to explain from the point of view of a pilot wave, whose existence is quite superfluous in experiments where the path can be labelled by an element of reality. The reason we have belabored this point is because a GHZ theorem follows for a two-particle spinless system as mentioned before, if one accepts the EPNT assumption.

Acknowledgment

This work was performed in collaboration with my colleagues Herb Bernstein, Mike Horne, and Anton Zeilinger. It was supported in part by grant #PHY 92-13964 from the National Science Foundation.

REFERENCES

1. H. Bernstein, D. M. Greenberger, M. A. Horne, and A. Zeilinger, Phys. Rev. **A 47**, (1993), 78 .
2. A. Einstein, B. Podolsky, and N. Rosen, Phys. Rev. **47**, (1935), 777-780. We shall refer to this paper as EPR. It is reprinted in J. A. Wheeler and W. H. Zurek, *Quantum measurement Theory* (Princeton U. P., Princeton, 1983).
3 J. S. Bell, Physics **1**, (1964) 195-200. All of Bell's basic writings on the subject have been reprinted in J. S. Bell, *Speakable and Unspeakable in Quantum Mechanics* (Cambridge U. P., Cambridge, England, 1987).

4. The first such experiment was performed by J. Clauser and S. Freedman, Phys. Rev. Lett. **28**, (1972) 938-942. The most definitive of these experiments was that of A. Aspect, J. Dalibard, and G. Roger, Phys. Rev. Lett. **47**, (1982) 1804-1807, because the two particles involved could not be connected by a time-like signal. A critical analysis of all these experiments can be found in F. Selleri, *Quantum Paradoxes and Physical Reality* (Kluwer Academic, Dordrecht, 1990).
5. D. M. Greenberger, M. A. Horne, and A. Zeilinger, *Bell's Theorem, Quantum Theory, and Conceptions of the Universe* (M. Katafos, ed., Kluwer Academic, Dordrecht, 1989). See also D. M. Greenberger, M. A. Horne, A. Shimony, and A. Zeilinger, Amer. J. Physics **58**, (1990) 1131.
6. L. Hardy, Phys. Lett. **A161**, (1992) 326.

COMPATIBLE STATISTICAL INTERPRETATION OF TWO-PARTICLE INTERFERENCE

Mirjana Božić

Institute of Physics, Beograd, P.O.Box 57, Yugoslavia

ABSTRACT

By consistently combining the superposition principle, the probabilistic interpretation of $|\psi(\mathbf{r},t)|^2$ and the classical laws of probability theory, probabilities associated with elementary events in the two-slit experiment, neutron, Mach-Zehnder, and multiple-slit interferometers, were determined recently. New probabilities, called de Broglian probabilities, taken together with probabilities from the standard statistical interpretation, form new statistical interpretation - *compatible statistical interpretation* of quantum mechanics. In this interpretation particle trajectories and wave phenomena (interference) are **coexistent**.

In the present paper we associate with an arbitrary one-particle state, which is a superposition of states, the set of compatible elementary events and we determine their probabilities. Also, we generalize the compatible statistical interpretation to many-particle states and apply de Broglian probabilities to interpret in a new way two-particle interference experiments.

1. De Broglian probabilities associated with one-particle states

The recently proposed statistical interpretation of quantum mechanics,[1,2,3] denoted as *compatible statistical interpretation*, contains in addition to the well known probabilities of the standard statistical interpretation new kind of probabilities denoted as de Broglian probabilities. These are probabilities of classical kind, associated with events which are expected to happen objectively,[4,5] although we might be unable to precisely follow their evolution.[6] For an arbitrary one-particle state $\psi(\mathbf{r},t)$, which is a superposition of eigenfunctions $\varphi_m(\mathbf{r})$ of an observable $\hat{\mathbf{Q}}$

$$\psi(\mathbf{r},t) = \sum_m c_m(t)\varphi_m(\mathbf{r}) \tag{1}$$

de Broglian probabilities are determined[1,2,3] from the following relations:

$$|\,\psi(\mathbf{r},t)\,|^2 = \sum_m P_m(\mathbf{r},t) \tag{2}$$

$$\frac{P_m(\mathbf{r},t)}{P_n(\mathbf{r},t)} = \frac{| c_m(t) |^2 | \varphi_m(\mathbf{r}) |^2}{| c_n(t) |^2 | \varphi_n(\mathbf{r}) |^2} \tag{3}$$

The general solution of this set of equations reads:[3]

$$P_m(\mathbf{r},t) = | c_m(t) |^2 | \varphi_m(\mathbf{r}) |^2 \left[1 + \frac{\sum_{nk}' c_n^*(t) c_k(t) \varphi_n^*(\mathbf{r}) \varphi_k(\mathbf{r})}{\sum_n | c_n(t) |^2 | \varphi_n(\mathbf{r}) |^2}\right] \tag{4}$$

In the physical picture on which this scheme is based, $\psi(\mathbf{r},t)$ is a real, objectively existing wave in space and time. Probability density $P_m(\mathbf{r},t)$ gives the distribution of particle coordinates for those particles (immersed in the wave $\psi(\mathbf{r},t)$), whose eigenvalue of $\hat{\mathbf{Q}}$ is equal to m.

For example, if $\hat{\mathbf{Q}}$ is the momentum operator $\hat{\mathbf{p}}$, then the "wave packet"

$$\psi(\mathbf{r},t) = \sum_{\mathbf{k}} c_{\mathbf{k}}(t) e^{i\mathbf{k}\cdot\mathbf{r}} = \sum_{\mathbf{k}} c_{\mathbf{k}} \cdot e^{-i\omega t} \cdot e^{i\mathbf{k}\cdot\mathbf{r}} \tag{5}$$

is conceived as an objectively existing wave which accompanies any particle in this state. Different particles may have different momenta $\mathbf{p} = \hbar\mathbf{k}/m$ in this state (the distribution of momenta is $| c_{\mathbf{k}} |^2$ - the same as the distribution of measured momenta in the standard statistical interpretation). Space distribution of particles with momentum $\mathbf{p}$ (wave vector $\mathbf{k}$) is given by $P_{\mathbf{k}}(\mathbf{r},t)$.

Probabilities $P_m(\mathbf{r},t)$ associated with one-particle states in the two-slit experiment,[7] neutron,[8] Mach-Zehnder[9] and multiple - slit interferometers could be very usefull in analyzing and claryfying[1,2,3] the conclusions drawn in the discussion on wave - particle duality and complementarity.

2. De Broglian probabilities associated with two-particle wave functions

By applying similar reasoning to two-particle states we could understand in a new way certain features of entangled states of two (many) noninteracting quantum particles - quantons:

$$\psi(\mathbf{r}_1,\mathbf{r}_2,t) = [\phi_\alpha(\mathbf{r}_1,t)\phi_\beta(\mathbf{r}_2,t) + f\phi_\beta(\mathbf{r}_1,t)\phi_\alpha(\mathbf{r}_2,t)]/\sqrt{2} \tag{6}$$

Here $f = \pm 1$. The value of f depends on the spin function and particle statistics (Fermi or Bose). The intriguing feature of those states is that the probability density to find quantons at $\mathbf{r}_1$ and $\mathbf{r}_2$ is not equal to the half sum of products of probability densities of independent quantons; $| \phi_\alpha(\mathbf{r}_1,t) |^2 \cdot | \phi_\beta(\mathbf{r}_2,t) |^2$ and $| \phi_\beta(\mathbf{r}_1,t) |^2 \cdot | \phi_\alpha(\mathbf{r}_2,t) |^2$; but it is equal to:

$$| \psi(\mathbf{r}_1,\mathbf{r}_2,t) |^2 = \frac{1}{2} | \phi_\alpha(\mathbf{r}_1,t) |^2 \cdot | \phi_\beta(\mathbf{r}_2,t) |^2 + \frac{1}{2} | \phi_\beta(\mathbf{r}_1,t) |^2 \cdot | \phi_\alpha(\mathbf{r}_2,t) |^2$$
$$f\, Re\phi_\alpha^*(\mathbf{r}_1,t) \cdot \phi_\beta^*(\mathbf{r}_2,t) \cdot \phi_\beta(\mathbf{r}_1,t) \cdot \phi_\alpha(\mathbf{r}_2,t) \tag{7}$$

As a consequence, it might happen, as it has been confirmed by experiments ,[10,11,12] that the probability to find two particles in a particular pair of points $\mathbf{r}_1$ and $\mathbf{r}_2$ is equal to zero.

The agreement of the measurement of joint positions of photons with those predictions is understood, inside the standard interpretation, as follows:[10] "The vanishing of $P_2(\mathbf{r},\mathbf{r}')$ {$=$ const. $\times \int_{t_0-T/2}^{t_0+T/2} \int dt\,dt' \cdot P_2(\mathbf{r},t,\mathbf{r}',t')$ where T is the time for which the measurement proceeds and $P_2(\mathbf{r},t,\mathbf{r}',t')$ is joint probability of detecting one photon at $\mathbf{r}$ at time t and another at $\mathbf{r}'$ at time t' - *note of the author*} for two photons at widely separated points $\mathbf{r},\mathbf{r}'$ confronts us with another example of quantum mechanical nonlocality, for the outcome of a photoelectric measurement at $\mathbf{r}$ appears to be influenced by where we have chosen to place the $\mathbf{r}'$ detector. At certain positions $\mathbf{r}'$ we can never hope to get a count at $\mathbf{r}$ when there is a count at $\mathbf{r}'$, whereas at other positions $\mathbf{r}'$ it is possible. This conclusion is related to the fact that in quantum mechanics we cannot associate an objective physical reality with the two photons that is independent of the measurement we choose to make. The phenomena involves the same violation of local realism that was recently tested in the experiments of Aspect and his collaborators (*Phys. Rev. Lett.* **47** (1981) 460; **49** (1983) 91), and was first discussed by Einstein, Podolsky, and Rosen (*Phys. Rev.* **47** (1935) 777)".

In the above explanation are denied the objective reality of $\psi(\mathbf{r},t)$ and the objectively existing mutual dependence of particle positions. As a consequence it was necessary to implicitly assume that the mutual dependence of particle positions is created by the measurement process.

Such an explanation is hardly acceptable and we propose an explanation which is based on the elaboration of de Broglie wave mechanics[13] in which $\psi(\mathbf{r}_1,\mathbf{r}_2,t)$ describes objectively existing mixture of waves associated with possible states of individual quantons. This mixing does not influence quantum numbers α and β of individual quantons. Those numbers remain appropriate characteristics of quantons in their joint state. But this wave mixing essentially influences the probability distribution of positions of two particles:

$P_{\alpha\beta}(\mathbf{r}_1,\mathbf{r}_2,t)$ – probability density that a quanton with quantum number α is in $\mathbf{r}_1$ and a quanton with quantum number β in $\mathbf{r}_2$.

$P_{\beta\alpha}(\mathbf{r}_1,\mathbf{r}_2,t)$ – probability density that a quanton with quantum number β is in $\mathbf{r}_1$ and a quanton with quantum number α in $\mathbf{r}_2$.

The sum of the above two probability densities should be equal to $|\psi(\mathbf{r}_1,\mathbf{r}_2,t)|^2$

$$|\psi(\mathbf{r}_1,\mathbf{r}_2,t)|^2 = P_{\alpha\beta}(\mathbf{r}_1,\mathbf{r}_2,t) + P_{\beta\alpha}(\mathbf{r}_1,\mathbf{r}_2,t) \tag{8}$$

The generalization of the hypothesis (3) in this case reads :

$$\frac{P_{\alpha\beta}(\mathbf{r}_1,\mathbf{r}_2,t)}{P_{\beta\alpha}(\mathbf{r}_1,\mathbf{r}_2,t)} = \frac{0.5\,|\phi_\alpha(\mathbf{r}_1,t)|^2 \cdot |\phi_\beta(\mathbf{r}_2,t)|^2}{0.5\,|\phi_\beta(\mathbf{r}_1,t)|^2 \cdot |\phi_\alpha(\mathbf{r}_2,t)|^2} \tag{9}$$

Now, we easily get:

$$P_{\alpha\beta}(\mathbf{r}_1,\mathbf{r}_2,t) = \frac{1}{2} \mid \phi_\alpha(\mathbf{r}_1,t) \mid^2 \cdot \mid \phi_\beta(\mathbf{r}_2,t) \mid^2 \times$$
$$\left\{1 + 2f \frac{Re\, \phi_\alpha^*(\mathbf{r}_1,t) \cdot \phi_\beta^*(\mathbf{r}_2,t) \cdot \phi_\beta(\mathbf{r}_1,t) \cdot \phi_\alpha(\mathbf{r}_2,t)}{\mid \phi_\alpha(\mathbf{r}_1,t) \mid^2 \cdot \mid \phi_\beta(\mathbf{r}_2,t) \mid^2 + \mid \phi_\beta(\mathbf{r}_1,t) \mid^2 \cdot \mid \phi_\alpha(\mathbf{r}_2,t) \mid^2}\right\}$$
$$P_{\beta\alpha}(\mathbf{r}_1,\mathbf{r}_2,t) = \frac{1}{2} \mid \phi_\beta(\mathbf{r}_1,t) \mid^2 \cdot \mid \phi_\alpha(\mathbf{r}_2,t) \mid^2 \times$$
$$\left\{1 + 2f \frac{Re\, \phi_\alpha^*(\mathbf{r}_1,t) \cdot \phi_\beta^*(\mathbf{r}_2,t) \cdot \phi_\beta(\mathbf{r}_1,t) \cdot \phi_\alpha(\mathbf{r}_2,t)}{\mid \phi_\alpha(\mathbf{r}_1,t) \mid^2 \cdot \mid \phi_\beta(\mathbf{r}_2,t) \mid^2 + \mid \phi_\beta(\mathbf{r}_1,t) \mid^2 \cdot \mid \phi_\alpha(\mathbf{r}_2,t) \mid^2}\right\} \tag{10}$$

The experiments in which one measures joint probabilities of particle positions are performed with many pairs of quantons without identifying quantum numbers α and β of each particular quanton. It is why one is not able to measure $P_{\alpha\beta}$ and $P_{\beta\alpha}$ separately, but only their sum. But, this does not mean that those probabilities are of no theoretical interest. On the contrary, the importance of $P_{\alpha\beta}$ and $P_{\beta\alpha}$ lies in the fact that they synthesize in a new way particle and wave aspects of quantons. The coexistence of those aspects is directly seen from numerous interference experiments in which interference patterns were obtained[7,8,9,10,14] through an accumulation process of the individual quanton (quanton pair) arrivals to detectors.

References

1. M. Božić and Z. Marić, *Phys. Lett.* **A 158** (1991) 33.
2. M. Božić, Z. Marić and J. P. Vigier, *Found. Phys.* **22** (1992) 1325.
3. M. Božić and Z. Marić, to be published in *Courants, amers, écueils en microphysique*, ed. George Lochak.
4. F. Selleri and G. Tarozzi, *Il Nuovo Cimento* **43 B** (1978) 31.
5. G. Tarozzi, in *The Wave-Particle Dualism*, eds. S. Diner, D. Fargue, G. Lochak and F. Selleri (Reidel-Dordrecht, 1984).
6. N. C. Petroni, *Phys. Lett.* **A 160** (1991) 107.
7. A. Tonomura, J. Endo, T. Matsuda and T. Kawasaki, *Am. J. Phys.* **57** (1988) 177.
8. J. Summhammer, H. Rauch and D. Tuppinger, *Physica* **B 151** (1988) 103.
9. P. Grangier, G. Roger and A. Aspect, *Europhys. Lett.* **1** (1986) 173.
10. R. Ghosh, C. K. Hong, Z. Y. Ou and L. Mandel, *Phys. Rev.* **A 34** (1986) 3962.
11. R. L. Pflegor and L. Mandel, *Phys. Rev.* **159** (1967) 1084.
12. R. Ghosh and L. Mandel, *Phys. Rev. Lett.* **59** (1987) 1903.
13. L. de Broglie, *Etude critique des bases de l' interprétation actuelle de la mécanique ondulatoire* (Gauthier-Villars, Paris, 1963).
14. Y. Mizobuchi and Y. Ohtaké, *Phys. Lett.* **A 168** (1992) 1.

THE PHOTONIC TUNNELING TIME AND THE SUPERLUMINAL PROPAGATION OF WAVE PACKETS

Raymond Y. Chiao, Aephraim M. Steinberg, and Paul G. Kwiat
Department of Physics, University of California, Berkeley, CA 94720, U. S. A.

ABSTRACT

We have measured the time delay for a photon to tunnel across a barrier. The peak of the essentially undistorted (but greatly attenuated) photon wavepacket appears on the far side of a 1.1-μm thick barrier 1.47 ± 0.21 fs *earlier* than if it had travelled at the vacuum speed of light c. Although the apparent tunneling velocity $(1.7 \pm 0.2)\,c$ is superluminal, Einstein causality is not violated. The measured tunneling time is consistent with with the group delay (or "phase time"), but disagrees with the "semiclassical time." The superluminality of the tunnel effect has led us to consider the propagation of wave packets in the transparent spectral region near dc of an *inverted* two-level atomic medium. This propagation can also be superluminal, with Sommerfeld and Brillouin's phase, group, energy, and signal velocities all exceeding the vacuum speed of light c, again without violating Einstein causality. This propagation may be accompanied by negligible gain or distortion.

1. Introduction

In this lecture, one of us (R. Y. C.) shall review our recent work at Berkeley on the measurement of the photon's tunneling time, which shows that the tunneling delay time can be superluminal, and then present some new ideas on how to extend this discovery to other unusual situations where electromagnetic wave packets can also propagate superluminally. In particular, we shall consider the possibility of superluminal propagation of wave packets in transparent media with inverted atomic populations.

2. Measurement of the photonic tunneling time

Tunneling is one of the best-known consequences of quantum mechanics, and is treated in every quantum text. The *time* it takes a particle to tunnel through a barrier, on the other hand, remains controversial to the present day. The question at once appears paradoxical because the momentum in the barrier region is imaginary. The earliest answer, the "phase time" (the group delay as calculated by the method of stationary phase), can in certain limits be anomalously small, implying barrier traversal at a speed greater than that of light in vacuum[1-3]. This apparent violation of Einstein causality does not arise from the use of the nonrelativistic Schrödinger equation, since it also arises in solutions of Maxwell's equations, which are fully relativistic. It has generally been assumed that such velocities cannot be physical[4, 5], but in the case of tunneling, no resolution has been universally accepted. This time has been questioned both mathematically and interpretationally. Alternate definitions of the tunneling time have also been proposed[6]. Some experiments have already been done[7-9], but more are needed in order to clarify the meanings and ranges of validity of the different tunneling times.

We have recently completed an experiment[10, 11] that offers a direct measurement of the time delay in tunneling (which may not *necessarily* be identified with the "duration" of the process). We employed the process of spontaneous parametric down-conversion, in which pairs of photons are emitted essentially simultaneously (to within tens of femtoseconds). After one particle traverses a tunnel barrier, its time of arrival can be compared with that of its twin (which encounters no barrier), thus offering a clear operational definition of the tunneling time. The magnitude of this time is so small as to be inaccessible to electronic measurement, but a two-photon interference effect[12] can be used to study the overlap of the two photons' wavepackets when they are brought together at a beam splitter, with sub-femtosecond resolution. We have previously used this effect to confirm that single photons in glass travel at the group velocity, and demonstrated that it benefits from a peculiar cancellation of the effects of group-velocity dispersion[13]. Since this technique relies on coincidence detection, the interpretation of the experiment is simple: Each coincidence detection corresponds to a single tunneling event. The particle aspect of tunneling can thus be clearly observed.

In our current apparatus, the tunnel barrier is a multilayer dielectric mirror, which possesses a one-dimensional photonic bandgap[14] that extends approximately from 600 nm to 800 nm; the transmission reaches a minimum of 1% at 692 nm. A particle travelling at c would require 3.6 fs to cross the 1.1 μm-thick barrier. However, over most of the bandgap, the calculated group delay is less than this, and remains relatively constant near 1.7 fs. It is interesting to note that one of the alternate tunneling times, the semiclassical time of Büttiker and Landauer, also dips below 3.6 fs, but only over a narrower range of frequencies; at the center of the bandgap, this time actually vanishes[15]. This time is obtained by dividing the thickness of the barrier by the velocity ($\hbar|\kappa|/m$ for a massive particle) corresponding to the *magnitude* of the imaginary momentum of the particle in the barrier region, neglecting the effects of coherent reflections between the extremities of the barrier. While in itself, this last approximation is reasonable for barriers much wider than the incident wavepacket, the identification of $\hbar|\kappa|/m$ with an actual velocity is not obvious. Indeed, while the semiclassical time grows linearly with barrier width, for very wide barriers the group delay saturates at a finite limit.

Our apparatus is shown in Fig. 1. A KDP crystal with a $\chi^{(2)}$ nonlinearity is pumped by an Argon laser operating single-mode at 351 nm. Conjugate pairs of photons are emitted simultaneously in the process of spontaneous parametric down-conversion. We select out nearly degenerate pairs centered at 702 nm, with an rms bandwidth of approximately 6 nm. One photon of each pair travels through air, while the conjugate photon impinges on our sample. This consists of an étalon substrate of fused silica, which is coated over half of one face with the 1.1 μm coating described above, and uncoated on the other half of that face. This sample is periodically placed in each of two positions: In one position, the photon must tunnel through the 1.1 μm coating in order to be transmitted, while in the other position, it travels through 1.1 μm of air. In each position, it traverses the same thickness of substrate. The two conjugate photons are brought back together by means of mirrors, so that they impinge simultaneously on the surface of a 50/50 beam splitter. A coincidence count is recorded when detectors placed at the two output ports of the beam splitter register counts within 500 ps of one another.

If the two photons' wavepackets are made to overlap in time at the beam splitter, a destructive interference effect leads to a theoretical null in the coincidence detection rate. Thus as the path-length difference is changed by translating a "trombone" prism (see Fig. 1), the coincidence rate exhibits a dip with an rms width of approximately 20 fs, which is the correlation time of the two photons. The rate reaches a minimum when the two wavepackets overlap perfectly at the beam splitter. For this reason, if an extra delay is inserted into one arm of this interferometer (e.g., by sliding the 1.1 μm coating into the beam), the prism will need to be translated in order to compensate for this delay and restore the coincidence minimum.

In our experiment, we found that inserting the barrier into the beam caused the center of the dip to be shifted to a position in which the prism was located *farther* from the barrier. This determines the sign of the effect: The external delay had to be *lengthened*, implying that the mean delay time experienced by the photon inside the barrier was *less than* the delay time for propagating through the same distance in air (see Fig. 1). We performed 12 one-hour runs and one four-hour run, alternating the direction in which we translated the prism in case the direction had a systematic effect on the result (none was observed). In each run, the data with and without the barrier were fitted to separate Gaussians, and the relative shift Δt between their centers was calculated. By averaging the results of all these runs, we found that $\Delta t = -1.47 \pm 0.21$ fs, including estimated systematic errors. The measured delay time is *smaller* by 7 standard deviations than the time it takes to traverse the barrier width at *c*. (Taking into account the width of the wavepackets, this implies that about 52% of the time, a photon which tunnelled would arrive sooner than one which travelled at *c*.) It is consistent with the group delay prediction, $\Delta t = -1.9$ fs, but differs greatly from the semiclassical time, $\Delta t = -3.0$ fs.

The normalized data from the 4-hr run are presented in Fig. 2. Each data point with the barrier in the beam has been averaged for 70 seconds, and each point with no barrier for 7 seconds. Since the transmission probability was nearly constant over the bandwidth of our photons, the *form* and *width* of the dip were not appreciably changed by the insertion of the barrier, aside from the overall reduction of the count rate. For this reason, there is physical significance to following the peak of the photon wavepackets. Furthermore, unlike in the electronic case, the shift cannot be understood as arising from the higher speed of the preferentially transmitted energy components before they reach the barrier. The transmission is a weak function of energy near the middle of the bandgap, and photons of all energies travel at the same velocity, except in a few dispersive optical elements. Even the small dispersive effects of those elements, however, have been shown to cancel out in this type of experiment[13, 16].

To study the energy-dependence of the delay time, we "tuned" the barrier by rotating the sample about the vertical axis, thus shifting the center of the bandgap. For the p-polarized photons we employed, this also reduces the bandgap width. A 45° rotation placed our photons close to the band edge, where the semiclassical time diverges, and the transmission probability rises to about 10%. Our results at 23° and at 45° again supported the group delay prediction.

The analogy between quantum mechanical tunneling and evanescent wave phenomena in electromagnetism is well known[8, 10, 17, 18], but there is an important difference between classical wave propagation and single-particle tunneling. In classical optics, the existence of group velocities greater than c, and even negative ones under certain conditions, is known and has been observed experimentally[19]. This phenomenon is understood as a "pulse reshaping" process, in which a medium preferentially attenuates the later parts of an incident pulse, in such a way that the output peak appears shifted towards earlier times. Einstein causality is not violated in this process; the output peak arises from the forward *tails* of the input pulse in a strictly causal manner, and no abrupt disturbance in the input pulse would propagate any faster than c.[4]

In the case of a single-particle wavepacket in standard quantum mechanics, there is no meaning to the question of which "part" of a minimum-uncertainty wavepacket gives rise to a given detection event. If the peak of such a wavepacket is shifted forward in time, this implies that the mean delay between the single-particle emission event and the corresponding detection event may be smaller than d/c, *whenever the particle is transmitted*. The source of this anomaly is that the incident particles are transmitted with low probability. It has been shown by Aharonov and Vaidman[20] that when a "weak measurement" (one with a sufficiently large uncertainty as to leave the state on which the measurement is performed essentially unperturbed) is made on a subensemble defined both by state preparation and by a post-selection of low probability, mean values can be obtained which would be strictly forbidden for any complete ensemble. This has been observed experimentally[21]. We believe that a similar effect is occurring here.

In tunneling, the anomalously large value of the effective velocity of the peak is related to the fact that in the opaque limit, the group delay becomes independent of d. That is, the tunneling process seems to be described not by a characteristic velocity, but rather by a characteristic time. We interpret this to mean that tunneling is a transient effect: For early times, the charging fields inside the bandgap material have not yet built up to their steady-state values, and thus do not suppress transmission as strongly as they do at later times. The relevant time-scale can be estimated from the time-energy uncertainty relation as the reciprocal of the frequency difference between the incident light and the nearest frequency at which transmission is allowed.

Within the conventional interpretation of quantum mechanics, it has so far proven impossible to go beyond a mere calculation and understand *why* the group delay should be so small. It may be that the Bohm-de Broglie pilot-wave interpretation[22] will prove useful in this connection. Insofar as it gives individual particles deterministic trajectories, thus making it possible even before the wave packet encounters the barrier to speak of which subset of the ensemble will be transmitted, this picture is not unrelated to the weak measurement idea. Leavens has applied this model to the tunneling problem[23] since the durations of these definite trajectories can be easily calculated. Bohm trajectories have the property that they may never intersect one another, and in one dimension, it follows trivially from this that *all* transmitted particles must originate earlier in time than *all* reflected particles. At the present time, this is the closest there is to an explanation of *why* only the early part of an incident wavepacket traverses a tunnel barrier.

Thus our measurements indicate that the peak of the essentially undistorted (but greatly attenuated) single-photon wavepacket appears on the far side of a tunnel barrier *earlier* than it would were it to propagate at the speed of light. There is no genuine violation of Einstein causality, as explained above. This tunneling time does not appear to be a strong function of the angle of incidence, and the data indicate that in our experiment, the group delay (or "phase time") gives a better description of the physically observable *delay* than does the "semiclassical" interaction time. In future experiments, we plan to study the dependence of this delay time on barrier thickness by using the phenomenon of frustrated total internal reflection. We have recently derived a set of correspondences between the delay times for that two-dimensional process and for tunneling through a one-dimensional square barrier[24].

3. Superluminal propagation of wave packets in inverted atomic media

The superluminal tunneling of photon wave packets leads us to ask if there are other situations where one might see similar superluminal effects. One such situation was discussed a long time ago by Sommerfeld and Brillouin[4], who treated the problem of propagation of light in dispersive media. They pointed out that the group velocity in regions of anomalous dispersion near an absorption line can exceed the vacuum speed of light, and at certain frequencies can become infinite, or even negative. Garrett and McCumber[25] showed that Gaussian wave packets propagating in these regions can in fact propagate at these abnormal group velocities. Within certain realistic approximations, they showed that an incident Gaussian wave packet will be reshaped by the absorption process (in which later parts of the wave packet are attenuated to a greater extent than the earlier parts), in just such a way as to produce a smaller, but undistorted, Gaussian wave packet at the exit face of the medium. Thus the transmitted wave packet has essentially the same shape and width as that of the incident wave packet. The peak of the wave packet appears to have moved at an abnormal group velocity inside the medium. As mentioned earlier, Chu and Wong[19] verified that this unusual behavior actually occurred for weak picosecond laser pulses propagating near the center of the bound A-exciton line of a GaP:N sample.

In their treatment of the propagation problem, Sommerfeld and Brillouin enumerated five different kinds of wave velocities:

(1) the phase velocity, at which the carrier wave moves.
(2) the group velocity, at which the peak of a wave packet moves.
(3) the energy velocity, at which energy inside a wave packet moves.
(4) the signal velocity, at which the half-maximum amplitude moves.
(5) the front velocity, at which the first appearance of a discontinuity moves.

Other workers have added to this list[26]. All five wave velocities differ from each other in the region of anomalous dispersion near the center of an absortion line, where the first two, the phase and group velocities, can become superluminal, but where the last three, the energy, signal, and front velocities, are either subluminal or equal to the vacuum speed of light. The question naturally arises: At which of these velocities does the photon propagate? One might guess that since the classical electromagnetic wave represents in some approximation the wave function of the photon, the answer is: at all of them. However, how can the photon then propagate at the superluminal wave velocities? Phrased more precisely, how can the peak of a single-photon wave packet propagate at the group velocity, and hence superluminally? Since there is at present no universally accepted theory of electromagnetic field quantization in dispersive media[26], there seems to be no universally accepted answer to this question. However, one might be able to answer a simpler question at the classical level which also naturally arises: Are there classical situations where the last three wave velocities are all superluminal? Special relativity rules out the possibility that the front velocity exceed c, but it does not rule out the possibility that any of the other velocities be superluminal. Here we present an interesting example where *all* of the wave velocities, except for the front velocity, should be superluminal.

Consider the propagation of wave packets in atomic media with *inverted* populations. It will be demonstrated here that the phase, group, energy, and signal velocities for classical wave packets propagating in the transparent spectral region in such a medium near dc, are all superluminal[27]. (Note that Sommerfeld and Brillouin's "signal velocity" is rather arbitrarily defined as the velocity of the point at which the amplitude of the signal equals *half* of its maximum amplitude. It is their "front velocity" which actually limits the propagation of any real signal. As will been shown below, their signal velocity is equal to the group velocity in our example, and is therefore also superluminal.) Quotation marks will henceforth be used on "superluminal" to emphasize that there is no genuine violation of causality, as Sommerfeld and Brillouin's *front velocity* in this medium never exceeds the speed of light in the vacuum. However, in contrast to Sommerfeld and Brillouin, who considered a discontinuous (step-modulated) signal, which possesses infinitely high frequency components, let us consider here the propagation of a smooth, limited-bandwidth wave packet, whose carrier frequency is near zero frequency. Think of a Gaussian wave packet, for example. The resulting "superluminal" propagation of this wave packet can again be thought of as a pulse-reshaping phenomenon (in which the earlier parts of the wave packet are *amplified* relative to the later parts) without any violation of causality, since there is no information contained in the peak of any smooth wave packet which is not already contained in its forward tail. However, in contrast to the previous cases of anomalous dispersion and of tunneling, here there is no appreciable change in the amplitude of the wave packet after its propagation through the inverted medium, since this medium is essentially transparent when it is excited sufficiently far off resonance. We hope hereby to help correct the common misconception that the group velocity, when it is "superluminal," is somehow unphysical or unuseful[5, 28, 29]. We hope hereby also to help correct some common statements such as "*nothing* can travel faster than light" which, when unqualified, are erroneous.

Consider a gas of *inverted* two-level atoms in a cell of length L. Let us treat the electromagnetic field classically. Let a limited-bandwidth wave packet, whose carrier frequency ω_c is far below the resonance frequency ω_0 of the two-level atoms, be incident on this medium. Let the amplitude of this wave packet be sufficiently small so that only the *linear* response of the medium need be considered. Then it is helpful to use a generalized Lorentz model, in which the oscillator strength f is replaced by $-f$ to characterize the inverted two-level system, since this model is a good approximation to the density-matrix equations of motion for the weakly perturbed two-level system[30]. However, it is important to note that these results do not depend in any crucial way on the Lorentz model; rather, the crucial assumptions here are (1) that the system responds *linearly* to the classical electromagnetic field, and (2) that this response is *causal*, so that the Kramers-Kronig relations are valid.

From the generalized Lorentz model, the refractive index of this inverted medium is

$$n(\omega) = \sqrt{1 - \frac{\omega_p^2}{\omega_0^2 - \omega^2 - i\gamma\omega}}\,, \tag{1}$$

where γ is a (small) phenomenological linewidth, and ω_p is the effective plasma frequency[31],

$$\omega_p = \sqrt{\frac{4\pi N e^2 |f|}{m}}\,, \tag{2}$$

where N is the atomic number density, e is the electron charge, and m is the electron mass. In typical situations, the inequalitites $\gamma << \omega_p << \omega_0$ are obeyed. Note that the minus sign in front of the second term under the square root in Eq. (1) arises from population inversion: It differs from its usual positive sign for a two-level atomic system in its ground state. As a result of this sign change, the index of refraction near zero frequency is less than unity:

$$n(0) = \sqrt{1 - \frac{\omega_p^2}{\omega_0^2}} < 1\,. \tag{3}$$

Figure 3 shows a plot of the real part of the refractive index calculated from Eq. (1), versus the frequency ω of an applied monochromatic plane wave.

The fact that $n(0) < 1$ implies that the phase velocity

$$v_p(0) = \frac{c}{n(0)} > c \tag{4}$$

is greater than the vacuum speed of light c. Near zero frequency, the group velocity

$$v_g(0) = \frac{c}{\left[n(\omega) + \omega \frac{dn}{d\omega}\right]_{\omega\to 0}} = \frac{c}{n(0)} = v_p(0) > c \tag{5}$$

is equal to the phase velocity near zero frequency, and is therefore also "superluminal," i.e., it also exceeds the vacuum speed of light c. Furthermore, in contrast to the *absorptive* media considered by Sommerfeld and Brillouin, since an *inverted* medium can temporarily give up part of its energy to the forward tail of a signal, here the energy velocity can also be "superluminal,"

$$v_E(0) = \frac{S}{u} = \frac{c}{\sqrt{\varepsilon(0)}} = \frac{c}{n(0)} = v_p(0) > c, \tag{6}$$

where S is the time-averaged Poynting vector, u is the time-averaged energy density, and $\varepsilon(0)$ is the dc dielectric constant. That these three velocities are all equal to each other is a consequence of the essentially *transparent, dispersionless* nature of the inverted atomic medium near zero frequency, just as in the case of a normal (uninverted) medium near dc. Note that the slopes of the $n(\omega)$ curves vanish for both these cases near dc in Fig. 3; hence all frequency components of the wave packet travel all the same speed near zero frequency for both the uninverted, subluminal case and for the inverted, "superluminal" case. Hence Sommerfeld and Brillouin's signal velocity is identical to the phase, group, and energy velocities, and is "superluminal" for the inverted case. Of course, special relativity prohibits energy from truly propagating faster than c; the energy velocity as defined by Sommerfeld and Brillouin considers only the energy stored in the field and not the energy used to maintain the atomic inversion. As the field "borrows" energy from the atoms in virtual emission processes, the field energy considered alone can propagate superluminally without violation of causality.

To understand what approximations are being made, let us start from the general solution for the electric field $E(z,t)$ in a dispersive medium,

$$E(z,t) = \int_{-\infty}^{\infty} A(\omega) \exp\left[i(k(\omega)z - \omega t)\right] d\omega \ , \tag{7}$$

where $k(\omega) = n(\omega)\omega/c$ with $n(\omega)$ given by Eq. (1), and where $A(\omega)$ is a limited-bandwidth Fourier amplitude peaked around the carrier frequency ω_c, which specifies the shape of the incoming wave packet at the entrance face of the cell at $z = 0$. The limited bandwidth of $A(\omega)$ allows us to perform the Taylor expansion

$$k(\omega) = k(\omega_c) + \left[\frac{dk}{d\omega}\right]_{\omega_c}(\omega - \omega_c) + \frac{1}{2}\left[\frac{d^2k}{d\omega^2}\right]_{\omega_c}(\omega - \omega_c)^2 + \ldots , \tag{8}$$

where

$$\left[\frac{dk}{d\omega}\right]_{\omega_c} = \frac{1}{c}\left[n(\omega) + \omega\frac{dn}{d\omega}\right]_{\omega_c} = \frac{1}{v_g} \tag{9}$$

$$\left[\frac{d^2k}{d\omega^2}\right]_{\omega_c} = \frac{1}{c}\left[2\frac{dn}{d\omega} + \omega\frac{d^2n}{d\omega^2}\right]_{\omega_c} . \tag{10}$$

At the exit face of the cell at $z = L$, one finds

$$|E(z{=}L, t)|^2 \approx \left|E\left(z{=}0, t - \frac{L}{v_g}\right)\right|^2 , \tag{11}$$

provided that the dispersion of the wave packet arising from the quadratic term in the Taylor expansion, Eq. (8), is negligible, i.e., provided that

$$\left|\frac{1}{2}\left[\frac{d^2k}{d\omega^2}\right]_{\omega_c}(\Delta\omega_c)^2 L\right| << 1 , \tag{12}$$

where $\Delta\omega_c$ is the bandwidth of $A(\omega)$. This condition, with Eq. (1) and the inequalities $\omega_c << \omega_o$ and $\omega_p << \omega_o$, leads to the restriction (same as for the uninverted medium) on the cell length L,

$$L << \frac{2}{3} \frac{c\,\omega_o^4}{\omega_p^2\,(\Delta\omega_c)^2\,\omega_c} \,, \tag{13}$$

for propagation with little distortion. Since the inequalities $\omega_p << \omega_o$ and $\Delta\omega_c < \omega_c << \omega_o$ hold in realistic situations, the inequality in Eq. (13) is not a severe restriction on L.

If we also stipulate that the gain at the carrier frequency ω_c from Eq. (1) be small, there is a further restriction on L, namely,

$$L << \frac{c\,\omega_o^4}{\omega_p^2\,\gamma\,\omega_c^2} \,. \tag{14}$$

If we choose $\gamma << (\Delta\omega_c)^2/\,\omega_c$, then the little-gain restriction, Eq. (14), is even less severe than the little-distortion restriction, Eq. (13).

In addition, for "superluminality" to be a significant effect, it is necessary that the peak of the "superluminally" propagated wave packet be well resolved (e.g., by Rayleigh's criterion) from that of a wave packet which has propagated at the vacuum speed c, i.e.,

$$\frac{1}{2\,\Delta\omega_c} << L\left(\frac{1}{c} - \frac{1}{v_g}\right) . \tag{15}$$

This condition, with Eq. (1) and the inequalities $\omega_c << \omega_o$ and $\omega_p << \omega_o$, leads to the restriction

$$\frac{c\,\omega_o^2}{\Delta\omega_c\,\omega_p^2} << L \,. \tag{16}$$

If we choose $\Delta\omega_c \approx \omega_c$, we see that the inequalities Eqs. (13) and (16) are consistent with each other, provided that $\omega_c << \omega_o$. If we choose $\Delta\omega_c < \omega_c$, the ratio of upper and lower bounds on L is increased yet more, so that these inequalities are even less restrictive. Let us choose a cell length L which satisfies these restrictions. Then from Eq. (11), we conclude that *any* limited-bandwidth signal satisfying the above inequalitites will be transmitted with little distortion, with near-unity transmission, and with $v_g > c$, i.e., "superluminally." Since there is no information carried by the peak (or peaks) of any limited-bandwidth signal (not necessarily a Gaussian) which is not already present in its forward tail, there is no violation of causality.

To show this in a general way, let us begin from the Kramers-Kronig relations, which assume only *linearity* and *causality* in the response of the medium. The complex linear susceptibility $\chi(\omega)$ obeys the pair of relations[32]

$$\mathrm{Re}\,\chi(\omega) = \frac{2}{\pi}\mathrm{P}\int_0^\infty \frac{\omega'\,\mathrm{Im}\,\chi(\omega')}{\omega'^2 - \omega^2}d\omega' \tag{17a}$$

$$\mathrm{Im}\,\chi(\omega) = -\frac{2\omega}{\pi}\mathrm{P}\int_0^\infty \frac{\mathrm{Re}\,\chi(\omega')}{\omega'^2 - \omega^2}d\omega', \tag{17b}$$

where P denotes the principal value. The first of these relations, Eq. (17a), when applied to the dielectric constant, yields the zero-frequency sum rule

$$\varepsilon(0) = 1 + \frac{2c}{\pi}\int_0^\infty \frac{\kappa(\omega')}{\omega'^2}\,d\omega', \tag{18}$$

where

$$\kappa(\omega) = \frac{4\pi\omega}{c}\,\mathrm{Im}\,\chi(\omega) \tag{19}$$

is the absorption or the gain coefficient of the medium, depending on the sign of Im $\chi(\omega)$. In the case of an inverted two-level atomic medium,

$$\mathrm{Im}\,\chi(\omega) < 0\,,\ \text{or}\ \kappa(\omega) < 0\ . \tag{20}$$

From the zero-frequency sum rule, Eq. (18), it follows that the dielectric constant near zero frequency, $\varepsilon(0) < 1$, can be less than unity, and hence that the index of refraction near zero frequency, $n(0) < 1$, can also be less than unity. Therefore the phase, group, energy, and signal velocities near zero frequency can all exceed the vacuum speed of light, as before. However, since we have now proved this beginning from the Kramers-Kronig relations, it is clear that this result does not depend on the validity of the Lorentz model, nor on that of the two-level model. Also, it is clear from the beginning that causality is not violated.

The Kramers-Kronig argument demonstrates that *in principle* the phase, group, energy, and signal velocities can all be "superluminal" far below the resonance of an inverted two-level atomic medium, without violating causality[33]. However, the infinite-frequency sum rule implies that any real medium must also have spectral regions with absorption as well as gain. Nevertheless, it should be possible in practice to observe "superluminal" effects close to, but outside of, a strong gain line. For example, a sufficiently strong, low-frequency gain line can give the dominant contribution to the zero-frequency sum rule, in spite of stronger absorption lines at higher frequencies. As a second example, for the case of two nearby gain lines of comparable strength, Steinberg and Chiao[34] have shown that between these lines there exists a point with zero group-velocity dispersion, so that dispersionless "superluminal" propagation is possible there. As a third example, within a bandwidth of approximately half the effective plasma frequency on both sides of a strong gain line, but outside the gain bandwidth, Bolda, Chiao, and Garrison[35] have recently shown theoretically that *negative* group velocities should be observable. They propose a specific system in which an experiment can be done. A computer simulation shows that an off-resonance wave packet which enters a gain medium at the entrance face at $z = 0$ generates a transmitted wave packet at the exit face at $z = L$, whose peak leaves the exit face of the cell *before* the peak of the incident wave packet arrives at the entrance face, similar to the negative group-velocity phenomenon observed by Chu and Wong[19] in the context of absorptive anomalous dispersion. There is a backward-propagating wave packet inside the medium generated simultaneously with the transmitted wave packet at the exit face, which propagates backward through the cell and *annihilates* with the arriving incident wave packet at the entrance face. (All *three* wave packets, i.e., incident, backward, and transmitted, have the same shape and amplitude, but the carrier propagates *forward* inside the backward-propagating wave packet.) There is no violation of the conservation of energy, either globally or locally: Since the medium is inverted, it can "loan" energy towards the generation of the transmitted and backward wave packets at the exit face, in just such a way that this energy is "paid back later" by the annihilation of the incident and backward wave packets at the entrance face. After all these interactions are over, the medium is left in its original inverted state, and the final electromagnetic field configuration asymptotically consists of a single transmitted wave packet with the same energy as the incident wave packet propagating to $z = +\infty$.

While these effects cannot be used to transmit any signal faster than c, they are real, physical effects and could in principle be used as an essentially noiseless amplification scheme for the leading edges of smooth electromagnetic pulses. We use the term "noiseless" because they could function at either the classical or the quantum level, amplifying even single-photon wavepackets without significant contributions from spontaneous emission[36]. For this reason, they may be useful in devices containing other *subluminal* elements, to decrease the overall delay time of smooth pulses, not beyond the speed of light, but closer to it than would otherwise be possible.

These predicted "superluminal" effects are different from the subluminal pulse-reshaping phenomena which arise from the propagation of laser pulses *inside* the gain band[25, 37]. Here we are *outside* of the gain band, in a spectral region where the medium is essentially transparent. They are also different from the off-resonance group-velocity-changing effects in transparent spectral regions next to an absorption line[38, 39]. Here instead of the absorbing, subluminal case, we have considered the inverted, "superluminal" case. (Of course, the subluminal case poses no potential conflict with relativity, whereas the same is not true of the "superluminal" case, which makes it the more important one to understand.) Moreover, these effects are different from the previously observed "superluminal" group velocities of wave packets propagating inside an absorption band discussed above[19]. Here the transmitted wave packets are unattenuated; furthermore, the *energy* and *signal velocities* can now also be "superluminal." These effects are also different from the phenomenon of "superluminal" tunneling of wave packets discussed in the preceding section of this paper, for the same reasons. We believe that they may be similar to a recently observed "superluminal" nonlinear pulse-reshaping phenomenon[40], although there is some ambiguity as to the question of causality in this last case (due to the slow buildup of the nonlinearities used and to the absence of a unique transit length in their two-dimensional geometry), and although the authors mistakenly attribute their effect to a violation of the Kramers-Kronig relations.

An open theoretical issue is how to provide a microscopic description of these effects by means of a fully second-quantized treatment, which could predict the behavior of off-resonance *single-photon* wave packets interacting with the inverted two-level atomic medium. We would then be able to answer the question: Do single-photon wave packets propagate through this medium "superluminally"? Also, the role of spontaneous emission in these effects can then be clarified. We are also planning to do experiments to demonstrate these striking phenomena.

We would like to thank E. L. Bolda, J. L. Bowie, I. H. Deutsch, J. C. Garrison, D. M. Greenberger, B. E. Johnson, A. N. Kaufman, G. Kurizki, M. O. Scully, A. E. Siegman, A. Szöke, and C. H. Townes for helpful discussions. This work was supported by ONR under Grant No. N00014-90-J-1259.

References

1. L.A. MacColl, Phys. Rev. **40**, 621 (1932).
2. L. Eisenbud, Ph. D. Thesis, Princeton University, 1948 (unpublished)
3. E.P. Wigner, Phys. Rev. **98**, 145 (1955).
4. L. Brillouin, *Wave Propagation and Group Velocity* (Academic Press, New York, 1960).
5. J. D. Jackson, *Classical Electrodynamics*, (Wiley & Sons, New York, 1975), p. 302.
6. E.H. Hauge and J.A. Støvneng, Rev. Mod. Phys. **61**, 917 (1989).
7. R. Landauer, Nature **341**, 567 (1989).
8. A. Ranfagni, D. Mugnai, P. Fabeni, G.P. Pazzi, G. Naletto and C. Sozzi, Physica **B175**, 283 (1991).
9. A. Enders and G. Nimtz, J. Phys. I France **2**, 1693 (1992).
10. R.Y. Chiao, P.G. Kwiat and A.M. Steinberg, Physica **B175**, 257 (1991).
11. A.M. Steinberg, P.G. Kwiat and R.Y. Chiao, "Measurement of the single-photon tunneling time", summitted to Phys. Rev. Lett. (1993).
12. C.K. Hong, Z.Y. Ou and L. Mandel, Phys. Rev. Lett. **59**, 2044 (1987).
13. A.M. Steinberg, P.G. Kwiat and R.Y. Chiao, Phys. Rev. Lett. **68**, 2421 (1992).
14. E. Yablonovitch and K.M. Leung, Physica **B175**, 81 (1991).
15. M. Büttiker and R. Landauer, Phys. Scr. **32**, (1985).
16. A.M. Steinberg, P.G. Kwiat and R.Y. Chiao, Phys. Rev. **A45**, 6659 (1992).
17. Th. Martin and R. Landauer, Phys. Rev. **A 45**, 2611 (1992).
18. T.K. Gaylord, G.N. Henderson and E.N. Glytsis, J.Opt.Soc.Am. **B10**, 333 (1993).
19. S. Chu and S. Wong, Phys. Rev. Lett. **48**, 738 (1982); Phys. Rev. Lett. **49**, 1293 (1982).
20. Y. Aharonov and L. Vaidman, Phys. Rev. **A41**, 11 (1990).
21. N.W.M. Ritchie, J.G. Story and R.G. Hulet, Phys. Rev. Lett. **66**, 1107 (1991).
22. D. Bohm, Phys.Rev. **85**, 166 (1952).
23. C.R. Leavens, Sol.St.Comm. **76**, 253 (1990).
24. A. M. Steinberg and R. Y. Chiao, "Tunneling delay times in one and two dimensions", summitted to Phys. Rev. A (1993).
25. C. G. B. Garrett and D. E. McCumber, Phys. Rev. Lett. A **1**, 305 (1970).
26. S. C. Bloch, Am. J. Phys. **45**, 538 (1977).

27. R. Y. Chiao, Phys. Rev. A. **48**, No. 1 (July, 1993).

28. J. A. Stratton, *Electromagnetic theory*, (McGraw-Hill, New York, 1941), p. 334.

29. M. Born and E. Wolf, *Principles of Optics*, (Pergamon Press, Oxford, 1975), p. 23.

30. R. W. Boyd, *Nonlinear Optics*, (Academic Press, New York, 1992), p. 209.

31. D. C. Burnham and R. Y. Chiao, Phys. Rev. **188**, 667 (1969).

32. L. D. Landau and E. M. Lifshitz, *Electrodynamics of Continuous Media*, (Pergamon Press, Oxford, 1960), p. 256.

33. Recently, Bolda, Chiao, and Garrison have proved two general theorems based on the Kramers-Kronig relations showing that for *any* dispersive medium, abnormal (i.e., superluminal, infinite, or negative) group velocities must exist at some frequency, and that at the frequency where the attenuation (or gain) is maximum, the group velocity must be abnormal (or normal).

34. A. M. Steinberg and R. Y. Chiao, (in preparation)

35. E. L. Bolda, R. Y. Chiao and J. C. Garrison, (in preparation)

36. Spontaneous emission near zero frequency is suppressed due to the facts that Im $\chi(\omega) \to 0$ as $\omega \to 0$ [see Eq. (17b)], and that the phase space volume for the spontaneously emitted photon rapidly vanishes as $\omega \to 0$, so that it can contribute an arbitrarily small amount of noise to the forward tail of the signal.

37. L. Casperson and A. Yariv, Phys. Rev. Lett. **26**, 293 (1971).

38. D. Grischkowsky, Phys. Rev. A **7**, 2096 (1973); IEEE J. Quant. Elect. **QE-10**, 723 (1974); Appl. Phys. Lett. 25, 566 (1974); J. K. Wigmore and D. Grischkowsky, IEEE J. Quant. Elect. **QE-14**, 310 (1978).

39. A. E. Siegman, *Lasers*, (University Science Books, Mill Valley, California, 1986), Ch. 9.

40. Y. L. da Silva, Y. Silberberg and J. P. Heritage, Opt. Lett. **18**, 580 (1993).

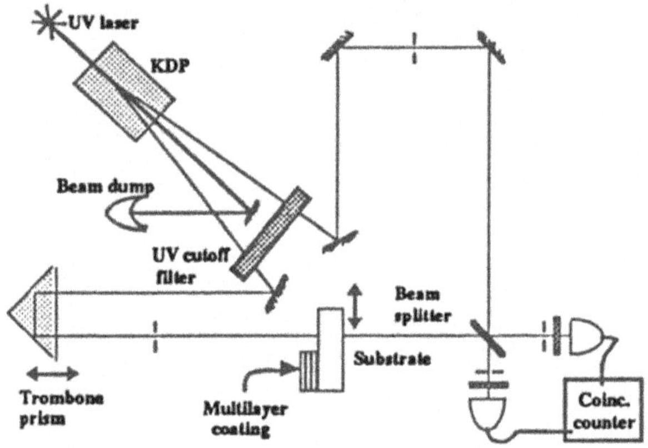

Fig. 1. Experimental setup

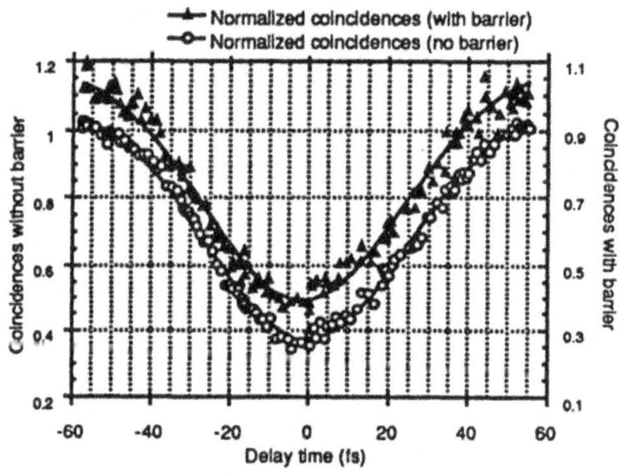

Fig. 2. Coincidence profiles with and without the tunnel barrier essentially map out the single-photon wavepackets.

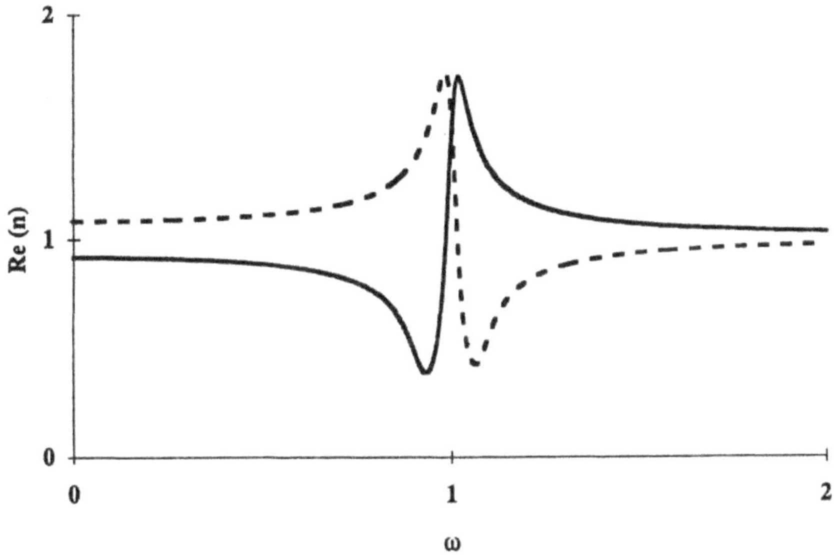

Fig. 3: The real part of the refractive index for an inverted two-level atomic medium (solid line) calculated from Eq. (1) as a function of normalized frequency ω / ω_0 , compared with that for an uninverted medium (dashed line). (For this graph, $\omega_p / \omega_0 = 0.4$, and $\gamma / \omega_0 = 0.048$). Note that as $\omega \rightarrow 0$, the slope of both these curves vanishes. Therefore *all* frequency components sufficiently near zero frequency travel with essentially the same phase velocity, and thus in this spectral region the medium is essentially dispersionless. It follows that in this region, the phase, group, energy, and signal velocities for the *inverted* medium are all equal to each other, and are therefore all "superluminal." Causality is not violated however (see text).

Quantum Cryptography with Quantum Interferometry

Artur K. Ekert
Merton College & Physics Department, Oxford University
Oxford, OX1 4JD, U.K.

G. Massimo Palma
Istituto di Fisica dell' Universita'
via Archirafi 36, I-90123 Palermo, Italy.

Abstract

We provide a brief review of quantum interferometry methods in quantum cryptography. We focus on cryptosystems which are based on non-local nature of quantum entanglement and which can be implemented using two-particle interferometers.

1 Introduction

In this brief review we focus on practical implementation of quantum cryptography [1] using quantum interferometry. The two interferometry based experiments which has been performed so far (July 1993) [2, 3], are quite encouraging and can pave the way not only to various comercially available quantum cryptosystems, but also to some interesting experiments in much wider area of quantum computation.

The structure of this paper is as follows. We start with a short introduction to the key distribution problem in classical cryptograpy and outline how quantum cryptography solved this problem. In Section 4 we describe a secure key distribution scheme exploiting quantum entanglement and in Section 5 we discuss its implementation based on two-particle interferometry. Finaly in Section 6 we analyse an experimental optical realisation of the interferometric key distribution.

2 Key Distribution Problem in Cryptography

Cryptography, despite a colourful history that goes back to 400 BC, only became part of mathematics and information theory during this century, in the late 1940s, mainly due to the seminal papers of Claude Shannon [4]. Today, one can briefly define cryptography as a mathematical system of transforming information so that it is unintelligible and therefore useless to those who are not meant to have access to it. However, as the computational process associated with encryption and decryption is always performed by physical means, one cannot separate the mathematical structure from the underlying laws of physics and quantum cryptography provides a good example of this fact.

Originally the security of a cryptotext depended on the secrecy of the entire encrypting and decrypting procedures; however, today we use ciphers for which the algorithm for encrypting and decrypting could be revealed to anybody without compromising the security of a particular cryptogram. In such ciphers a set of specific parameters, called *key*, is supplied together with the plaintext as an input to the encrypting algorithm, and together with the cryptogram as an input to the decrypting algorithm. The encrypting and decrypting algorithms can be publicly announced; the security of the cryptogram depends now entirely only on the secrecy of the key, and this key, which is very important, consists of any *randomly chosen*, sufficiently long string of bits. Once the key is established, subsequent communication involves sending cryptograms over a public channel which is vulnerable to total passive interception (e.g. public announcement in mass-media). However in order to establish the key, the two users, who share no secret information initially, must at a certain stage of communication use a reliable and a very secure channel.

However difficult from the technological point of view it might be, *in principle* any *classical* private channel can always be passively monitored, without the sender and the receiver being aware that any eavesdropping has taken place. This is a consequence of the more general fact that, classical physics allows us, at least in principle, to perform measurements which provide full knowledge about all possible observables of any physical object without disturbing these observables. All carriers of information are after all physical objects and any information, including a cryptographic key, is encoded by modification of one or few of their physical properties. The laws of classical theory allow us to measure and to know all of them precisely. This is not the case when the nature of the channel is such that quantum theory is needed for its proper description.This fact is the key idea behind quantum cryptography.

3 Quantum Cryptography

Conventional cryptography provides no tools for provable security of the key distribution and any classical encoding is vulnerable to passive interception. Such interception may be difficult from the technological point of view but is perfectly allowed by the laws of classical physics and therefore the two legitimate users can never be sure that there exist only two and no more copies of the key.

In a simplistic way, any totally passive eavesdropping can be viewed as a two-stage process. The first stage amounts to making a copy of the carrier of information and the second to reading from the cloned copy (or copies) a set of values of observables that are used for the encoding of the key. The intercepted original carrier is sent over to the legitimate receiver, who is unable to check whether the carrier has been intercepted or not because the state of the carrier has not been altered during the cloning process. According to quantum theory the first stage of passive eavesdropping cannot in general be completed [5]. Quantum cloning will give a positive result leaving the state of the original intact only if the adversary knows in advance that the carrier of information is in a quantum state picked up from a particular set of orthonormal states. If this is not the case, the adversary will not be able to construct a cloning device; a device of this sort would effectively violate the uncertainty principle enabling measurements of non-commuting observables on different copies of the carrier. This shows that coding based, for example, on non-orthogonal states or quantum correlations which cannot be cloned gives a chance to detect eavesdropping.

Quantum cryptography, which employs quantum mechanics to improve the security of communication, was originally proposed by Stephen Wiesner, [6]. Theoretical models for quantum key distributions based on the uncertainty principle have been analysed by Bennett and Brassard [7] and models based on quantum correlations have been proposed by Ekert [8]. The first practical implementation of a cryptosystem based on the uncertainty principle has been performed by Bennett, Bessette, Brassard, Salvail and Smolin [9]. Here we will describe a practical realisation of a quantum channel which distributes the key and protects it against eavesdropping using the nature of quantum correlations. Let us start, however, with a theoretical idea for an absolutely secure key distribution based on quantum entanglement.

4 Secure Key-Distribution Protocol

4.1 Raw Quantum Transmission

Let us analyse a quantum channel which consists of a source that emits pairs of spin $\frac{1}{2}$ particles, in a singlet state. The particles fly apart along the z-axis, towards the two legitimate users of the channel, say Alice and Bob, who, after the particles have separated, perform measurements on spin components along one of three directions, given by unit vectors $\vec{a}_i$ and $\vec{b}_j$ $(i, j = 1, 2, 3)$ respectively for Alice and Bob. For simplicity both $\vec{a}_i$ and $\vec{b}_j$ vectors lie in the x-y plane, perpendicular to the trajectory of the particles, and are characterized by azimuthal angles: $\phi_1^a = 0, \phi_2^a = \frac{1}{4}\pi, \phi_3^a = \frac{1}{2}\pi$ and $\phi_1^b = \frac{1}{4}\pi, \phi_2^b = \frac{1}{2}\pi, \phi_3^b = \frac{3}{4}\pi$. Superscripts "a" and "b" refer to Alice's and Bob's analysers respectively, and the angle is measured from the vertical x-axis. The users choose the orientation of the analysers randomly and independently for each pair of incoming particles. Each measurement, in $\frac{1}{2}\hbar$ units, can yield two results: $+1$ (spin up) and -1 (spin down) and can potentially reveal one bit of information.

The quantity

$$E(\vec{a}_i, \vec{b}_j) = P_{++}(\vec{a}_i, \vec{b}_j) + P_{--}(\vec{a}_i, \vec{b}_j) - P_{+-}(\vec{a}_i, \vec{b}_j) - P_{-+}(\vec{a}_i, \vec{b}_j) \tag{1}$$

is the correlation coefficient of the measurements performed by Alice along $\vec{a}_i$ and by Bob along $\vec{b}_j$. Here $P_{\pm\pm}(\vec{a}_i, \vec{b}_j)$ denotes the probability that result ± 1 has been obtained along $\vec{a}_i$ and ± 1 along $\vec{b}_j$. According to the quantum rules

$$E(\vec{a}_i, \vec{b}_j) = -\vec{a}_i \cdot \vec{b}_j. \tag{2}$$

For the two pairs of analysers of the same orientation ($\vec{a}_2$, $\vec{b}_1$ and $\vec{a}_3, \vec{b}_2$) quantum mechanics predicts total anticorrelation of the results obtained by Alice and Bob: $E(\vec{a}_2, \vec{b}_1) = E(\vec{a}_3, \vec{b}_2) = -1$.

Let us also, following Clauser, Horne, Shimony, and Holt[10], define a quantity composed of the correlation coefficients for which Alice and Bob used analysers of different orientation

$$S = E(\vec{a}_1, \vec{b}_1) - E(\vec{a}_1, \vec{b}_3) + E(\vec{a}_3, \vec{b}_1) + E(\vec{a}_3, \vec{b}_3). \tag{3}$$

Again, quantum mechanics requires

$$S = -2\sqrt{2}. \tag{4}$$

After the transmission has taken place, Alice and Bob can announce in public the orientations of the analysers they have chosen for each particular measurement and divide the measurements into two separate groups: a first group for which they used different orientation of analysers, and a second group for which they used the same orientation of their analysers. They discard all measurements in which either or both of them failed to register a particle at all. Subsequently Alice and Bob can reveal publicly the results they obtained but within the first group of measurements only. This allows them to establish the value of S, which if the particles were not directly or indirectly "disturbed" should reproduce result of Eq.(4). This assures the legitimate users that the results they obtained within the second second group of measurements are anticorrelated and can be converted into a secret string of bits — the key. This secret key may be then used in a conventional cryptographic communication between Alice and Bob.

4.2 Eavesdropping

The eavesdropper cannot elicit any information from the particles while in transit from the source to the legitimate users, simply because there is no information encoded there! The information "comes into being" only after the legitimate users perform measurements and communicate in public afterwards. The eavesdropper may try to substitute his own prepared data for Alice and Bob to misguide them, but as he does not know which orientation of the analysers will be chosen for a given pair of particles, there is no good strategy to escape from being detected. In this case his intervention will be equivalent to introducing elements of *physical reality* to the measurements of the spin components. This can be easily seen if we put appropriately modified (by the eavesdropper perfect measurement) correlation coefficients into Eq.(3). We obtain

$$S = \int \rho(\vec{n}_a, \vec{n}_b) d\vec{n}_a d\vec{n}_b [\; (\vec{a}_1 \cdot \vec{n}_a)(\vec{b}_1 \cdot \vec{n}_b) - (\vec{a}_1 \cdot \vec{n}_a)(\vec{b}_3 \cdot \vec{n}_b) + (\vec{a}_3 \cdot \vec{n}_a)(\vec{b}_1 \cdot \vec{n}_b) + (\vec{a}_3 \cdot \vec{n}_a)(\vec{b}_3 \cdot \vec{n}_b)], \tag{5}$$

where $\vec{n}_a$ and $\vec{n}_b$ are two unit vectors (for particles "a" and "b" respectively), oriented along the directions of the quantisation axes for which the eavesdropper acquired information about the spin component of a given particle. This information could be acquired either through a direct, "brute" measurement of the spin components or through a more subtle attack on the source, e.g. substituting a source that produces a state of two spin $\frac{1}{2}$ particles correlated with another quantum system on which the actual measurement will be performed by the eavesdropper. The normalised probability measure $\rho(\vec{n}_a, \vec{n}_b)$ describes the eavesdropper strategy (probability of intercepting a spin component along a given direction for a particular measurement). If only one particle (say "a") is exposed to the measurement performed by the eavesdropper along the direction $\vec{n}_a$, one may put $\vec{n}_b = -\vec{n}_a$ as a particular case in Eq.(5).

Simple calculation for a given orientation of $\vec{a}_1, \vec{a}_3$ or $\vec{b}_1, \vec{b}_3$ gives

$$S = \int \rho(\vec{n}_a, \vec{n}_b) d\vec{n}_a d\vec{n}_b [\sqrt{2} \vec{n}_a \cdot \vec{n}_b] \tag{6}$$

which implies

$$-\sqrt{2} \le S \le \sqrt{2} \tag{7}$$

and contradicts Eq.(4). for any strategy described by the measure $\rho(\vec{n}_a, \vec{n}_b)$. This way it has been shown that the generalized Bell's theorem can have a practical application in cryptography, namely, it can test the safety of the key distribution. It is not a mathematical difficulty of a particular computation, but a fundamental physical law that protects the system, and as long as quantum theory is not refuted as a complete theory the system is secure.

Regarding more refined attacks associated with the faked source of three (or more) correlated particles one may think, for example, about delayed measurement on the third particle which is correlated with the two spin $\frac{1}{2}$ particles. By "delayed" I mean "after the orientation of the analysers has been publicly revealed by Alice and Bob". However, as we want the two particles to be in a pure, singlet state, and Alice and Bob test for it through Bell's theorem then we cannot correlate the third particle with the other two without disturbing the purity of the singlet state.

4.3 EPR Scheme

One does not have to use Bell's Theorem to check for eavedropping. The EPR scheme alone can also be aplied with a slightly different protocol.

Alice and Bob, each choose randomly between two directions characterized by azimuthal angles: $\phi_1^a = 0, \phi_2^a = \frac{1}{2}\pi$ and $\phi_1^b = 0, \phi_2^b = \frac{1}{2}\pi$. After a series of EPR pairs have been send and measured, Alice and Bob announce to each other in public which orientation of the analysers they have chosen for each particular measurement but they keep secret the outcomes of the measurements. They then agree to discard all instances in which they happened to measure along different axes, as well as instances in which the measurement failed due to imperfect detectors. The remaining instances, in which both Alice and Bob successfully measured the spin component along the same direction, ought to be perfectly anticorrelated. In order to verify this they publicly compare their measurement results on a sufficiently large random subset of the undiscarded instances; if they find that the data in the tested subset are indeed perfectly anticorrelated then they can infer that the data in the remaining untested subset is probably also perfectly anticorrelated, and therefore can be used as a cryptographic key.

In this scheme, analogously to the previous scheme, eavesdropper disturbance can be quantified using an appropriate linear combination of correlation coefficients. The corresponding expression for S is now given by

$$S = E(\vec{a}_1, \vec{b}_1) + E(\vec{a}_2, \vec{b}_2), \tag{8}$$

and the eavesdropper intervention makes it

$$-1 \leq S = \int \rho(\vec{n}_a, \vec{n}_b) d\vec{n}_a d\vec{n}_b [(\vec{a}_1 \cdot \vec{n}_a)(\vec{b}_1 \cdot \vec{n}_b) + (\vec{a}_2 \cdot \vec{n}_a)(\vec{b}_2 \cdot \vec{n}_b)] \leq 1, \tag{9}$$

as opposed to the value $S = -2$ for the undisturbed siglet state.

In the next section we explain how this EPR scheme can be implemented using quantum intereferometry.

4.4 Error Correction and Privacy Amplification

Before we describe intereferometry-based realisation of the above schemes let us only mention two important mathematical techniques which are essential for practical implementations of quantum cryptosystems namely error correction and privacy amplification[12].

Even when the particles were not disturbed while in transit from the source to Alice and Bob there always be some errors spoiling perfect anticorrelation between Alice's and Bob's strings of data. For example when we use entangled photons rather than spin 1/2 particles practical problem arises from the fact that available detectors sometimes produce a response even when no photon has arrived. Such "dark counts" are caused by thermal excitation, cosmic rays, and radioactive decays in the photomultiplier's glass envelope. Dark counts and other imperfections in the apparatus lead to errors even when there has been no eavesdropping, and make it impractical for Alice and Bob simply to reject their data whenever they find an error in it. If Bob and Alice find a small number of errors, they must devise a way to correct them and proceed. On the other hand, if they find a large number of errors, indicating significant eavesdropping, they must reject their data and start over.

A variety of techniques are available for Alice and Bob to correct a small number of errors through public discussion, such as the use of error-correcting codes, or schemes in which Alice and Bob partition their data into appropriate-sized blocks and compare parities (*i.e.* whether the block has an even or odd number of ones). However, all these techniques potentially leak information to Eve, who may be listening to the public discussion. Therefore, after the quantum transmission and error-correcting discussion, Alice and Bob have to employ another mathematical trick known as "privacy amplification". We are not going to discuss these techiques here; readers interested in details are refered to[12].

5 From Two-level Particles to Interferometry

Now we want to rephrase our previous consideration using a slightly different language. Instead of the familiar Bohm two-spin system we analyse a two-particle interferometer following the approach of Horn and Zeilinger [13]. In Fig.(1) we show a one-to-one correcspondence between spin 1/2 particles and a single particle (photon) injected into either of the two or into a coherent superposition of the two paths. The two paths together are equivalent to one spin 1/2 particle, with each "taken path" being equivalent to one of the two possible states of the spin 1/2 particle (spin up or spin down along a given direction).

We start our discussion with a two-particle interferometer arrangement of Fig.(2) where a central source emits a pair of particles in opposite directions. The arrangement defines four paths a, a', b, and b' in two directions $a - b$ and $a' - b'$. We label the two particles as particle a and particle b. In this notation the quantum state of the pair is

$$|\Psi\rangle = \frac{1}{\sqrt{2}}(|a\rangle \otimes |b\rangle + |a'\rangle \otimes |b'\rangle), \tag{10}$$

where $|a\rangle$ denotes particle a in beam a, $|a'\rangle$ denotes particle a in beam a', etc.

The experimental arrangements from Fig.(2) can serve as an interferometric version of the EPR-based protocol for the key distribution. Particle a travels to Alice, particle b to Bob. Alice and Bob set up their phase shifters for each incoming particle

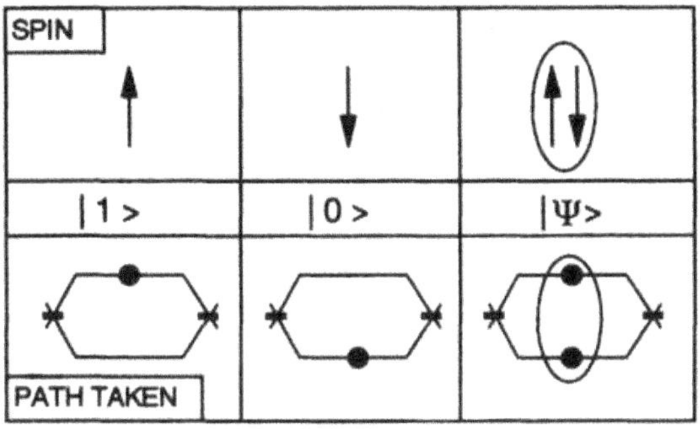

Figure 1: Correspondence between spin 1/2 particles and intereferometry. Spin up is equivalent to "upper path taken", spin down to "lower path taken", and a coherent superposition of spin up and down is equivalent to a coherent superposition of particle taking upper and lower path. If, for example, this particle happens to be a photon, a coherent superposition of two paths is obtained at the output of a beam splitter.

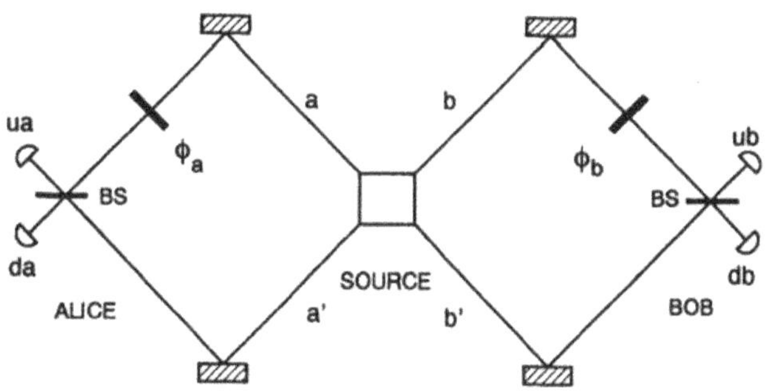

Figure 2: A pair of entangled particles a and b is produced by the source S and travels along directions $a - b$ and $a' - b'$ in the state given by Eq.(?). Particles a and b are deflected and each passes through a phase shifter and then through a beam splitter (BS), ending up in one of the detector paths (u_a or d_a for particle a and u_b or d_b for particle b). There is no single-particle interference here, the count rate at each detector is independent of the phase shifts ϕ_a or ϕ_b, however the two-particle coincident rates ($u_a - u_b$, $u_a - d_b$, etc) depend on the difference $\phi_a - \phi_b$.

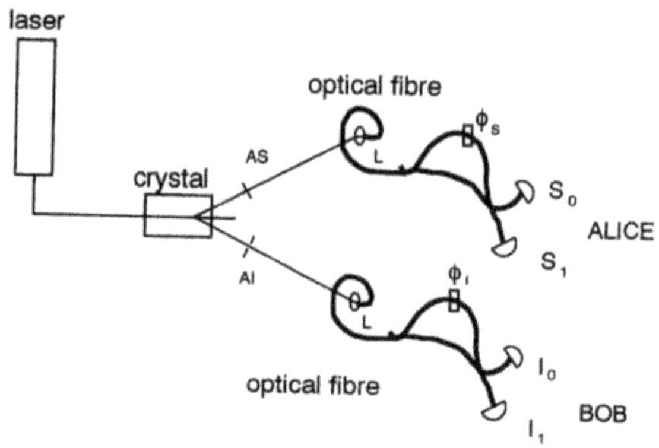

Figure 3: Scheme of the quantum public key distribution apparatus. A short-wavelength laser illuminates a suitably cut nonlinear crystal. Apertures A_S and A_I select photon pair beams which are launched into single mode fibres by lenses L. Identical Mach-Zehnder interferometers are placed in the signal and idler arms of the apparatus and interferometer outputs are viewed by signal S_0, S_1 and idler I_0, I_1 single photon counting detectors. Alice and Bob view these outputs and confirm coincident (within dt) photon detections and relevant phase settings ϕ_S, ϕ_I via a separate public channel.

choosing randomly (and independently from each other) between 0 and $\pi/2$. Assuming symmetric beam-splitters, at the detectors the state given by Eq.(10) will evolve into [14]

$$|\Psi\rangle = \left(\frac{ie^{i(\phi_a+\phi_b)/2}}{\sqrt{2}}\right) \quad [\cos(\phi_a - \phi_b)/2)(|u_a\rangle \otimes |u_b\rangle + |d_a\rangle \otimes |d_b\rangle) \tag{11}$$

$$+ \sin(\phi_a - \phi_b)/2)(|u_a\rangle \otimes |d_b\rangle - |d_a\rangle \otimes |u_b\rangle)], \tag{12}$$

with Alice's and Bob's phase shifters introducing phase shift ϕ_a and ϕ_b respectively. For $\phi_a = \phi_b$ there is a perfect correlation between detector counts, and the counts become completely randomized when $|\phi_a - \phi_b| = \pi/2$.

One can now rewrite the EPR based key distribtion protocol for this intereferometric system simply substituting phase shifts for appropriate angles describing quantization axes for spin componenets. Instead of following this scenario we describe another interferometric set-up for quantum cryptography, the one that has been implemented in practice.

6 Towards Practical Implementations

A scheme of the apparatus is shown in Fig.3. A parametric downconversion source is pumped by a monochromatic short wavelength laser of frequency $2\omega_0$ ($\lambda_0 = 441.6$nm in the experiment). Signal and idler photons are emitted in a broad band cone behind the crystal with pairs satisfying energy and momentum conservation. Photon pairs are selected by placing apertures in the downconverted cone satisfying the phase-matching conditions in the crystal. Signal and idler photons are launched into separate fibre optic cables and propagate to remote Mach-Zehnder interferometers. Each interferometer contains a shorter and a longer path with the difference in transit time over the two paths denoted by ΔT ($\Delta T \simeq 1$ns in the experiment). Signal photon detectors labelled S_1 and S_0 and idler detectors labelled I_1 and I_0 view the four Mach-Zehnder outputs.

This experimental set-up has been analysed in detail by Franson [15], and Rarity and Tapster [16] in connection with Bell's inequalities. Here, we present a simplified but sufficient description of the nonlocal correlation phenomenon that can be observed in this system.

We assume that the output state is a two-photon entangled state of the signal and the idler, which we write as

$$|\Psi\rangle = \int d\omega \quad c(\omega)|1;\omega\rangle \otimes |1;2\omega_0 - \omega\rangle, \tag{13}$$

where we can choose ω to denote the frequency of the signal photon ω_s and consequently, due to energy conservation, the frequency of the idler photon ω_i can be written as $2\omega_0 - \omega$.

Function $|c(\omega)|^2$ is positive in the interval of the order $\Delta\omega$ centered at ω_0. The calculation of the joint probability of registering photons at time $t + dt$ by the signal and idler detectors amounts to summing the probability amplitudes for the photons to have traversed various paths to the detectors and taking the square modulus of the total probability amplitude. If we take into account that $1/\Delta\omega \ll \Delta T$ ($\Delta\lambda \simeq 3$nm in the experiment), then after simple calculations we obtain

$$p(1,1) = p(0,0) = \frac{1}{4}(1 + \cos(\phi_s + \phi_i + \theta)) \tag{14}$$

$$p(0,1) = p(1,0) = \frac{1}{4}(1 - \cos(\phi_s + \phi_i + \theta)), \tag{15}$$

where signal phase shift ϕ_s and idler phase shift ϕ_i can be set up independently in the respective interferometers, and $\theta = (\omega_s + \omega_i)\Delta T = 2\omega_0\Delta T$. In our notation $p(0,1)$ means that the signal photon has been registered by detector S_0 and the idler photon has been registered by detector I_1 within the small time interval $dt \ll \Delta T$. The coefficient of correlation, assuming perfect detection and fixing $\theta = 2k\pi$, $k = 1, 2, 3...$, is given by

$$E(\phi_s, \phi_i) = p(1,1) + p(0,0) - p(0,1) - p(1,0) = \cos(\phi_s + \phi_i). \tag{16}$$

The coefficient depends on the sum of two local parameters ϕ_s and ϕ_i, which results in a non-local character of correlation between the photocounts in the two distant interferometers. We note that for $\phi_s + \phi_i = 0$ there is a perfect correlation between photocounts in the two distant interferometers. Let us now describe how this non-local quantum phenomenon can be used for key distribution.

Two legitimate users of the two distant interferometers, whom we traditionally call Alice and Bob, will set up the local parameters randomly and independently for each incoming photon. Alice, who receives the signal mode, will choose randomly between $\phi_s = 0$ and $\phi_s = \pi/2$, and Bob, who receives the idler mode, will choose randomly between $\phi_i = 0$ and $\phi_i = -\pi/2$. After transmission they reveal publicly the setting of their local parameters, but not which detector registered a photon. They then agree to discard all instances in which $\phi_s + \phi_i \neq 0$, as well as instances in which one or both detectors failed to register a photon due to imperfect quantum efficiency. The remaining instances ought to refer to perfectly correlated photocounts, $\phi_s + \phi_i = 0$. To verify that this is so, Alice and Bob publicly compare the results of the photocounts on a sufficiently large random subset of the undiscarded instances. If they find that the tested subset is indeed perfectly correlated, they can infer that the remaining untested subset is also perfectly correlated, and therefore can form the cryptographic key. In practical realization the test for eavesdropping must be more sophisticated than the subset test described above. A more useful test involves error-correction and hashing techniques[9] [12].

Non-local interference effects of this type have been observed in short-range laboratory experiments with measured correlation coefficients E (Eq. 16) up to 90% [17],[18]. A preliminary experiment has been performed in DRA, Malvern[2], where one beam propagated over 170 meters through a multimode optical fibre before the interferometer, but here the measured correlation coefficient was low, partly due to poor time resolution in the detectors[16]. Such an experiment could easily be improved by using single-mode fibre and in-line fibre interferometers (Fig.3). To ensure high visibility the detector resolution dt must be smaller than the time difference between long and short paths ΔT to discriminate against non-interfering events. In an all fibre apparatus, dispersion effects in the out-of-balance interferometers can be avoided by reduction of bandwidth and operation close to the dispersion minimum (1.3μm in conventional fibres) This choice also has the advantage of low loss but requires further development of suitable photon counting detectors[19]. With careful control of the apparatus, correlation coefficients greater than $E = 0.95$ should routinely be achieved. This implies a high bit error rate of 5×10^{-2}. This can be substantially reduced without compromising security using suitable error correction procedures[12]. The range of a realistic system will depend on losses, which in communication fibres can be as low as 0.17dB/km, coupled with the minimum acceptable data rate. Recent free space communication experiments have demonstrated that propagation losses of some 20dB or more could be tolerated[20].

The analysis of the security of this system is equivalent to the analysis of the security for the EPR-type cryptosystems provided by Ekert[8] and Bennett, Brassard and Mermin [11].

7 Concluding Remarks

We have described here development in one branch of quantum cryptogarphy namely the one that is based on quantum entanglement. However, as we mentioned before, quantum cryptography can also rely on the uncertainty principle. The non-entanglemet cryptosystems can also be implemented using interferometers[21] and some of them are even approaching a prototype status[3]. Using standard low loss, low dispersion communication fibre and polarization insensitive interferometers these system may be capable of key sharing over long (~10-40Km) distances.

Non-interferometric schemes, *e.g.* encoding bits in photon polarization, are also being developed[9, 22]. An extensive bibliography on quantum cryptography has been recently compiled by Brassard[23] and it should be consulted for a more comprehensive picture of quantum cryptography today (July 1993).

8 Acknowledgements

The authors thank Anton Zeilinger for many interesting discussions ranging from quantum interferometry to monastic life in medieval Italy.

References

[1] For a comprehensive, popular introduction to quantum cryptography see C.H. Bennett, G. Brassard, and A.K. Ekert, *Scientific American*, pp.50-57 (October 1992), for more technical introduction see Ref.[9].

[2] A.K. Ekert, J.G. Rarity, P.R. Tapster, and G.M. Palma, *Phys.Rev.Lett* **69**, 1293 (1992).

[3] P.D Townsend, J.G. Rarity, and P.R. Tapster, *Electr.Lett.* **29**, 634 (1993).

[4] C.E. Shannon, *Bell Syst. Tech. J* **28**, 656 (1949).

[5] W.K. Wootters and W.H. Zurek *Nature* **299**, 802 (1982).

[6] S.Wiesner, *Sigact News* **15** (1), 78 (1983).

[7] C.H. Bennett and G. Brassard, *Proceedings of IEEE International Conference on Computers, Systems, and Signal Processing*, Bangalore, India, 175 (1984).

[8] A. K. Ekert, *Phys. Rev. Lett.* **67**, 661 (1991).

[9] C.H. Bennett, F. Bessette, G. Brassard, L. Salvail, and J. Smolin, *Journal of Cryptology*, **5**, 3 (1992).

[10] J.F. Clauser, M.A. Horne, A. Shimony, and R.A. Holt, *Phys.Rev.Lett* **23**, 880 (1969).

[11] C.H. Bennett, G. Brassard, and N.D. Mermin *Phys.Rev.Lett.* **68**, 557 (1992).

[12] C.H. Bennett, G. Brassard and J-M. Robert, *Siam J. Comput*, **17**, 210 (1988).

[13] M.A. Horne, A. Zeilinger, *Symposium on the Foundations of Modern Physics*, ed. P. Lathi and P. Mittelstaedt, p.435 (World Scientific, Singapore, 1985); see also M.A. Horne, A. Shimony, and A. Zeilinger, *Phys.Rev.Lett.* **62**, 2209 (1989).

[14] D.M. Greenberger, M. Horne, and A. Zeilinger, Am.J.Phys. **58**, 1131 (1990).

[15] J. D. Franson, *Phys. Rev. Lett.* **62**, 2205 (1989); **67**, 290 (1991).

[16] J. G. Rarity and P. R. Tapster, *Phys. Rev. A* **45**, 2052 (1992).

[17] J. G. Rarity and P. R. Tapster, *Phys. Rev. Lett.* **64**, 2495 (1990), see also J.G.Rarity, P.R.Tapster, E.Jakeman, T.Larchuk, R.A.Campos, M.C.Teich, and B.E.A.Saleh *Phys. Rev. Lett.* **65**, 1384 (1990).

[18] J.Brendel, E.Mohler and W.Martienssen, *Phys. Rev. Lett.* **66**, 1142 (1991).

[19] B.F.Levine and C.G.Bethea, *Appl. Phys. Lett.* **44**, 553 (1984).

[20] S. F. Seward, P. R. Tapster, J. G. Walker and J. G. Rarity, *Quantum Opt.* **3**, 201 (1991).

[21] C.H. Bennett *Phys.Rev.Lett.* **68**, 3121 (1992).

[22] A. Muller, J. Breguet, and N. Gisin, *Eurphys.Lett.* (to appear).

[23] G. Brassard, Sigact News **24** (1993) (to appear).

Causality and Interference in Electron and Photon Tunneling

Y. Japha and G. Kurizki
Department of Chemical Physics
The Weizmann institute of science. Rehovot, Israel, 76100

Abstract

The fundamental importance of the problem of photonic tunneling is highlighted by its relevance to the long-standing electron tunneling-time controversy. This relevance stems from the analogy between the solution of the Schrödinger equation for an electron tunneling through a one- dimensional barrier and the corresponding solution of the wave equation for a photon that tunnels as an evanescent wave through the interface between different dielectric layers (e.g. glass-air-glass). The proposed definitions of electronic tunneling time contradict each other in general and it is also unclear how to reconcile them with causality or the uncertainty principle. Towards a resolution of this controversy we have performed a comparative study of electron and photon tunneling.

Our calculations yield unambiguous determinations of the tunneling time for a single photon traversing the forbidden air gap between two dielectrics or a photonic periodic structure. We consider pairs of photons produced by parametric down-conversion. One photon of each pair impinges on a tunnel gap, while the other propagates freely prior to detection. By use of the Hong-Ou-Mandel interference technique, the time-of-flight of the tunneling photon can be compared with femtosecond accuracy to the time-of-flight of its twin. We have found a method for a lucid spatio-temporal description of the non-trivial tunneling process. The gap transmission function is expanded in a series of multiple reflections within the gap. Superluminal tunneling time-delays are found. We interpret these delays in terms of the group velocities of the interfering multiply-reflected waves in the gap. Causality is shown to be preserved only by the energy-flow velocities. Group velocity can be superluminal even when the photon crosses the barrier as a propagating wave including the case of normal incidence, due to interference between the multiply reflected waves. The results can be carried over to the problem of electron tunneling times using a one-to-one correspondence we established between spatial photon distribution and electron temporal propagation. An intriguing QED aspect of this work involves the questions of photon localization in space and time and of three-particle quantum correlations.

1 Introduction

The time it takes an electron to tunnel through a barrier has long been a subject of debate[1]. Several definitions were suggested for the traversal time. The main problem facing these definitions is the position-momentum or time-energy uncertainty. Viewing a particle as a flow of probability with a definite value of kinetic energy requires a stationary description, which is inconsistent with the introduction of the dynamical concept of "traversal time". On the other hand, if we describe the particle as localized in time and space, then the process cannot be referred to as tunneling because the momentum or kinetic energy are not well defined.

The most widely used approach is to look at the electron as a wavepacket with a certain spectral and spatio-temporal width. The traversal time can then be defined as the time it takes the peak of this wavepacket to cross from one edge of the barrier to the other. This time is conventionally called the "phase time" [2] although it is associated with group velocity, because it is calculated by using the stationary phase approximation. The peculiar property of the phase time is that it can correspond to superluminal group velocity, i.e., the peak can be transmitted nearly simultaneously. The consistency of this property with relativistic causality has not been discussed thus far. Superluminal pulse propagation has been predicted and observed only in an absorbing medium[3]. but no sufficient explanation was found to the superluminal tunneling.

Another definition for the traversal time is the "dwell time" which is the overall time the electron spends in the barrier regardless of whether it is reflected or transmitted to the other side[4]. The disadvantage of this definition is that it does not distinguish between a transmitted and a reflected particle. It was shown that the time spent by a tunneling particle in the barrier is not a measurable quantity[5]. We can say that within the barrier it is impossible to distinguish between a particle that will eventually be transmitted and one that will be reflected. The definition of the "dwell time" is based on the stationary description of the electron, in which the electron inside the tunneling barrier is described by an evanescent wave. It is difficult to give a dynamical interpretation to an evanescent wave in the way it is done for a traveling wave. One possibility is to describe an evanescent wave as a traveling wave with a complex wave vector, whence an "imaginary traversal time" is used[6]. However, the physical meaning of such a time is obscure. The same is true of the "semiclassical time"[7] obtained by using the imaginary part of the evanescent wave vector to define the group velocity. The difficulty with "semiclassical time" is that it diverges at the critical energy, corresponding to the transition from the allowed to the tunneling region. This divergence seems to contradict the position-momentum uncertainty principle.

Other definitions of the traversal time are based on indirect or auxiliary measurements. For example, the Larmor times are related to the amount of electron spin precession during the barrier traversal[2]. The Büttiker-Landauer time is defined for an electron traversing an oscillating barrier[7]. In the opaque-barrier limit this time coincides with the semiclassical time.

In this paper we suggest a dynamical description which gives a new insight into the temporal aspect of the particle propagation inside the barrier. Part of this insight is obtained from the parallel we draw between the allowed or evanescent propagation of a classical electromagnetic (EM) wave in a dielectric structure and their counterparts for an electron wave function obeying the Schrödinger equation with analogous boundary conditions (Sec 3). This parallel is beneficial for studies of electron temporal tunneling, since from the experimental point of view, EM waves allow much more accurate measurements of time delays than electrons (Sec 2). It should be noted that our analysis for EM waves is based on neglecting absorption and spectral dispersion of the refractive index in the structure , which is analogous to the assumption of a non-absorptive, uniform, rectangular potential barrier for electrons. These assumptions hold to a very good approximation for wavepackets that are spectrally narrow compared to the width of the medium dispersive variation.

We show that interference is the key to understanding the process of tunneling. Causality, which is implicit in EM wave propagation, is reconciled with what appears to be superluminal phenomena in tunneling, by means of this interference. A partial wave expansion is performed, in which every partial wave represents a process of total reflection from the walls of the barrier (Sec 4). We then use the analogy between electrons and photons to first examine the simple case of total reflection (Sec 5) and then tunneling , gaining insight into phase and dwell times (Sec. 7). A straightforward interpretation can be given for the case of a photon traversing a periodic dielectric structure, where it is partially or totally reflected from the interface (Sec 5-7). This interpretation is the basis for a detailed dynamical description of an evanescent wave as a superposition of traveling interfering partial waves which are multiply reflected within a periodic dielectric structure

The second quantization of EM waves introduces an additional level of description into tunneling, which depends on the initial photon-number (boson) distribution, and differs from the description for electrons (fermions). The quantization of evanescent EM waves is more complicated than that of propagating waves, as shown by Mandel et. al. [8].

2 Measurement of tunneling time by two-photon interference

A method for measuring the traversal times of photons through dielectric media has been recently implemented by Chiao et. al. [9] using the Hong-Ou-Mandel interferometer[10]. This method enables one to compare the traversal time of one photon in a medium with the traversal time of a conjugate photon propagating through air. In the Hong-Ou-Mandel interferometer (See Fig. 1) a crystal with a $\chi^{(2)}$ nonlinearity is pumped by a continuous wave (CW) laser. Conjugate pairs of photons (conventionally termed "signal" and "idler") are simultaneously emitted in the process of spontaneous parametric down-conversion. Energy and momentum are conserved during the process. As a result, although the individual photons of a conjugate pair are broadband, their frequencies add up to precisely that of the pump laser. The two photons are recombined by means of two mirrors, so that they impinge on the surface of a 50/50 beam splitter. Coincidence count is recorded when detectors placed at the two output ports of the beam splitter register counts within an electronically set gate window of each other. If there is no delay in the path of one of the photons, the coincidence counter will record a null in the coincidence rate. This null is due to destructive interference between the two Feynman paths leading to coincidence detection, one in which both photons are reflected at the beam-splitter, and one in which both photons are transmitted. If a dielectric or other delay is placed in the arm of the signal photon and the path-length difference between the two arms is scanned by changing the path-length of the idler photon, one obtains a graph of the coincidence rate as a function of the path length. It can then be concluded that the delay of the signal photon is equal to the added path-length of the idler photon needed to obtain a minimum in the coincidence rate. Chiao et. al. have measured, using this method, the time delay for a photon that tunnels across a barrier corresponding to a one-dimensional photonic band gap (a forbidden propagation band) in a periodic dielectric structure. Remarkably, the dip in the coincidence rate appeared earlier than the time it would take the photon to traverse the same path with the velocity of light. In the following we analyze the experiment done by Chiao et. al. and then suggest an interpretation for the results.

The energy-momentum conservation (phase-matching) conditions for the signal and the idler photons are

$$\omega_i + \omega_s = 2\omega_0, \vec{k}_i + \vec{k}_s = 2\vec{k}_0$$

where $2\omega_0, 2\vec{k}_0$ are the frequency and the wave-vector of the pump beam, respectively, $\omega_s(\omega_i), \vec{k}_s(\vec{k}_i)$ are the frequency and the wavevector of the signal (idler) photon. If the photons are z-polarized so that they propagate in a plane perpendicular to the z axis,then the phase-matching conditions determine two directions in space, one for the signal and one for the idler. The wave function of the photon pair exiting the $\chi^{(2)}$ crystal then describes an entangled state of

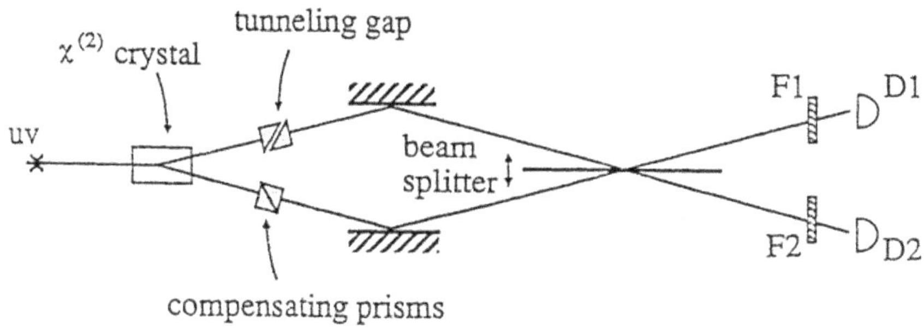

Figure 1: The Hong-Ou-Mandel interferometer as a tool for measuring photonic time delays

the signal and idler photons, which can be written in a simplified way as[11]

$$|\phi\rangle = \int_{-\omega_0}^{\omega_0} d\omega\phi(\omega)a_s^\dagger(\omega_0+\omega)a_i^\dagger(\omega_0-\omega)|0\rangle \tag{1}$$

Here $a_s^\dagger, a_i^\dagger$ are creation operators of a single-mode one-photon state with a given frequency and wavevector of the signal or the idler. They satisfy standard commutation relations with the corresponding annihilation operators $[a_j(\omega), a_k^\dagger(\omega')] = \delta_{jk}\delta(\omega-\omega')$ for $j,k = i,s$. The function $\phi(\omega)$ satisfies a normalization condition $\int_{-\omega_0}^{\omega_0} d\omega|\phi(\omega)|^2 = 2\omega_0$. Each of the photons represents a temporal pulse propagating almost as a plane wave in a certain direction.

The signal photon is now passed through a barrier which consists of a dielectric structure. We denote by d the path-length difference added to the idler photon with respect to the signal photon. The field operator of the signal photon can be written as

$$\Psi_s(x,y,t) = \int_0^{2\omega_0} d\omega a_s(\omega)\varphi_s(\omega,x)e^{i(k_{y,s}(\omega)y-\omega t)} \tag{2}$$

where $a_s(\omega)$ is the ω-mode annihilation operator, and $\varphi_s(\omega,x)$ is a solution of the wave equation on either side of a barrier of length L. It is given by

$$\varphi_s(\omega,x) = \begin{cases} e^{ik_x x} + r\, e^{-ik_x x} & x<0 \\ \sigma e^{ik_x(x-L)} & x>L \end{cases} \tag{3}$$

where r and σ are the reflection and transmission coefficients, satisfying

$$|r|^2 + |\sigma|^2 = 1, \; r\sigma^* + r^*\sigma = 0$$

As reported by Chiao et. al. [9].The coincidence rate for tunneling through a periodic dielectric structure had the shape of a Gaussian shifted forward in time with respect to the Gaussian that would have been obtained if the signal photon had propagated through air. The dip in the coincidence rate was observed at about 2.1 fs, while for a photon crossing the barrier with the speed of light it would have appeared at 3.6fs. This implies that the wavepacket (group) velocity of the signal photon that tunneled through the barrier at was superluminal, about $1.7c$!!

In what follows we shall examine these results by analyzing the source of time delays and frequency shifts. We shall then interpret the astonishing but understandable effect that looks like superluminal propagation.

3 Electron-Photon analogies

In the case of photons, we consider a dielectric medium that varies only in the x direction, choosing a transverse electric (TE) field polarization (perpendicular to the plane of incidence). The classical z-polarized electric field of a monochromatic plane wave can then be written as

$$E_z(x, y, t) = \sqrt{2\pi\hbar\omega}\,\psi(x)e^{i(k_y y - \omega t)} + \text{c.c.}$$

As shown in Sec 2, the classical wave function ψ specifies the photon space-time propagation as revealed by two-photon interference. This propagation can be compared to that of an electron wave function which is a solution of the Schrödinger equation.

We consider three schemes for tunneling:

1. A one-dimensional electron with (average) energy E is incident upon a rectangular potential barrier of height V and width L. We call this scheme an electronic potential barrier (EPB). The transmission function for an electron with energy $E > V$ is given by

$$\sigma = \left(\cos\kappa L - i\frac{k_x^2 + \kappa^2}{2k_x\kappa}\sin\kappa L\right)^{-1} \tag{11}$$

where $k_x = \sqrt{2mE}/\hbar$ and $\kappa = \sqrt{2m(E-V)}/\hbar \equiv \sqrt{k_x^2 - k_0^2}$. This equation applies also to $E < V$ by means of analytical continuation. Any time-dependent solution of the Schrödinger equation in this scheme can be expanded in terms of the stationary solutions with different E.

2. A TE-polarized EM wave is incident at an (average) angle θ upon an air gap of width L placed between two dielectric media with refractive index n. We emphasize (see Introduction) that the refractive index is purely real and frequency-independent, which is only justified for wavepackets of limited bandwidth. The wave tunneling across the gap at θ larger than the total internal reflection angle θ_c is called frustrated total internal reflection (FTIR). As opposed to the previous scheme, this one has two spatial dimensions: x

(normal to the boundaries of the gap) and y (parallel to the gap). Nevertheless, we have noted the possibility of drawing a parallel between the two schemes by taking the y dimension in FTIR to be analogous to the time dimension in EPB. The transmission coefficient in FTIR has the same form as in Eq. 11 with $k_x = \sqrt{n^2\omega^2/c^2 - k_y^2}$ and $\kappa = \sqrt{\omega^2/c^2 - k_y^2} \equiv \sqrt{k_x^2 - k_0^2}$. Every solution for the EM wave equation in this scheme can be expanded in terms of plane-wave solutions with constant k_y. Phase-shift measurements of waves tunneling through FTIR were measured by Carniglia and Mandel[12] and the results indicate the similarity to EPB tunneling (see Sec. 7).

3. A TE-polarized EM wave is normally incident on a one-dimensional periodic dielectric structure with photonic band gaps (forbidden-propagation bands). Such a photonic band structure (PBS) is made of N unit cells each of width $l = L/N$. For simplicity we consider the unit cell to be made of two layers with different refractive indices n_1 and n_2 and widths d_1 and d_2 ($d_1 + d_2 = l$) (Such structures are known as multilayer-coated mirrors). The wave function representing the electric field in this structure can be written in the form

$$\psi_j^{(i)}(x) = a_j^{(i)} e^{ik_i(x-jl)} + b_j^{(i)} e^{-ik_i(x-jl)} \tag{12}$$

where the subscript j denotes the index of the unit cell and the superscript $i = 1, 2$ denotes the layer within each unit cell.

The equation connecting the different layers is[13]

$$\begin{pmatrix} a_{j-1} \\ b_{j-1} \end{pmatrix} = \begin{pmatrix} A & B^* \\ B & A^* \end{pmatrix} \begin{pmatrix} a_j \\ b_j \end{pmatrix} \tag{13}$$

where the coefficients a_j, b_j without superscript refer to the first layer. Here we have

$$A = e^{-ik_1 d_1}(\cos k_2 d_2 - i\frac{k_1^2 + k_2^2}{2k_1 k_2} \sin k_2 d_2)$$

$$B = -i\frac{k_1^2 - k_2^2}{2k_1 k_2} \sin k_2 d_2 \tag{14}$$

and the matrix elements satisfy $|A|^2 - |B|^2 = 1$.

We can find the transmission and reflection coefficients σ_N, r_N if we apply the boundary conditions and multiply the matrix of Eq. 13 N times by itself. We obtain the following form of the transmission coefficient through a PBS with N unit cells for a frequency ω in an allowed band

$$\sigma_N = \{\cos N\kappa l - i[-\text{Im}\alpha/\sin \kappa l] \sin N\kappa l\}^{-1} \tag{15}$$

Here $\alpha(\omega)$ is a linear combination of the coefficients A, B of Eq. 14 and their complex conjugates, which is determined by the boundary conditions, and $\kappa(\omega)$ is the wavevector of the eigenmode of frequency ω inside the PBS. Both α and κ can be explicitly calculated for the specific boundary conditions.

Within a band gap we replace κ by $i\mu$ and obtain

$$\sigma_N = \{\cosh N\mu l - i[-\text{Im}\alpha/\sinh \mu l] \sinh N\mu l\}^{-1} \tag{16}$$

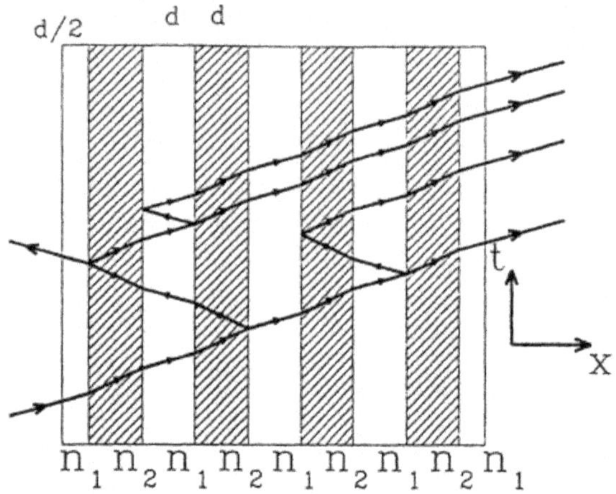

Figure 2: A schematic picture of the propagation of a wave inside a PBS. The picture shows only some of the multiply reflected partial waves

It can be seen that Eq. 15 for PBS is of the same form as Eq. 11 for EPB or FTIR on making the appropriate replacements. However, for the ensuing analysis it is crucial to note the following distinction between the two schemes: although a photon with frequency within a band gap will exhibit total reflection or tunneling in a PBS, we still find traveling waves within each layer of the structure, instead of evanescent waves in FTIR. The decay of the wave as we go deeper into the PBS is manifested by the diminishing of the coefficients a_j, b_j as a function of the layer index j. Consequently, we have a "macroscopic" scale of evanescent waves and a "microscopic" scale of traveling waves inside each layer (See Figure 3).

We conclude that in all three cases (EPB,FTIR,PBS) the wave function in the case of a rectangular barrier at $0 < x < L$ may be represented by:

$$\psi(x) = \begin{cases} e^{ik_x x} + r\, e^{-ik_x x} & x < 0 \\ \chi(x) & 0 < x < L \\ \sigma e^{ik_x(x-L)} & x > L \end{cases} \tag{17}$$

where r and σ are the reflection and transmission coefficients as in Eq. 3 above. In all three cases, the reflection and transmission coefficients can be written in the form

$$\sigma = \begin{cases} (\cos\kappa L - i\coth\eta\sin\kappa L)^{-1} & \text{allowed regime} \\ (\cosh\mu L + i\cot\xi\sinh\mu L)^{-1} & \text{tunneling} \end{cases} \tag{18}$$

$$r = \begin{cases} (\cosh\eta + i\cot\kappa L\sinh\eta)^{-1} & \text{allowed regime} \\ (\cos\xi - i\coth\mu L\sin\xi)^{-1} & \text{tunneling} \end{cases} \tag{19}$$

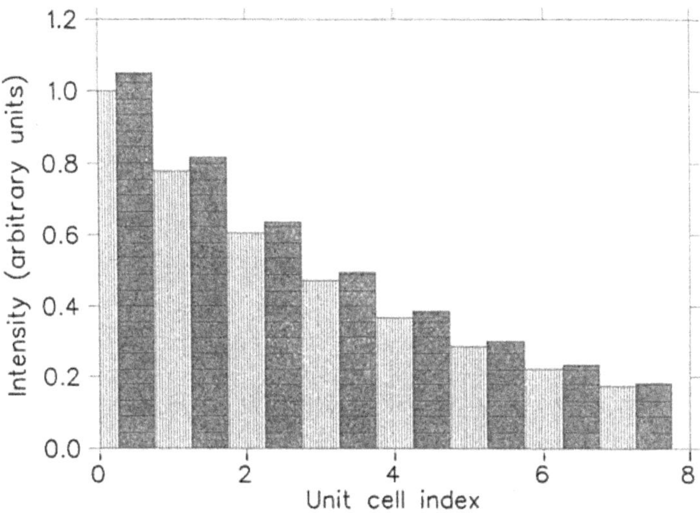

Figure 3: "Microscopic" and "Macroscopic" structure of the waves in a PBS in the case of total reflection. The graph shows the intensity of the forward propagating wave in each layer. The intensity of the backward propagating wave in layer 1 is the same with another phase

where the parameters $k_x, \kappa, \mu, \xi, \eta$ are given in Table 1 for the three cases. In the transition from the allowed regime to tunneling we replace κ by $i\mu$ and η by $-i\xi$. When we replace σ by r we set $\kappa L \to i\eta$ or $\mu L \to -i\xi$. Hence one can write the reflection coefficients in terms of the transmission coefficient

$$r = \begin{cases} e^{-\eta}(1 - e^{i\kappa L}\sigma) & \text{allowed regime} \\ e^{i\xi}(1 - e^{-\mu L}\sigma) & \text{tunneling} \end{cases} \tag{20}$$

and the limit of an infinite barrier (total reflection in the case of tunneling) can be obtained by setting $\sigma = 0$ in Eq. 20.

For EPB and FTIR the wave function $\chi(x)$ inside the barrier is given by:

$$\chi(x) = \begin{cases} ae^{i\kappa x} + be^{-i\kappa x} & \text{allowed regime} \\ ae^{-\mu x} + be^{\mu x} & \text{tunneling} \end{cases} \tag{21}$$

where a and b are determined by the boundary conditions.

In PBS the wave function $\chi(x)$ takes the same form as $\psi(x)$ in Eq. 12.

The correspondence between EPB and FTIR is clearly better than their analogy to PBS. The former correspondence can be first presented for the simpler case of total reflection. The coefficients for total reflection are obtained by taking the limit $L \to \infty$ in Eq. 19. The reflected wave has the same amplitude as the incident wave and it assumes an additional phase ξ, as given by Table 1. This phase shift represents a delay time of the reflected wave with respect to the incident wave in the case of EPB and a lateral shift of the light beam in

	EPB	FTIR	PBS
dimensions	$x - t$	$x - y$	$x - t$
Fourier variable	$E/\hbar$	k_y	ω
tunneling condition	$V > E$	$\theta > \theta_c$	$\lvert\text{Re}\alpha\rvert > 1$
incident wavevector k_x	$\sqrt{2mE}/\hbar$	$\sqrt{\omega^2 \frac{n^2}{c^2} - k_y^2}$	$\frac{\omega}{c}$
group velocity/angle	$\hbar \partial k_x / \partial E = v_g^{-1}$	$\partial k_x / \partial k_y = \tan\theta$	
prop. wave-vector κ in barrier	$\sqrt{2m(E - V)}/\hbar$ $= \sqrt{k_x^2 - k_0^2}$	$\sqrt{\frac{\omega^2}{c^2} - k_y^2}$ $= \sqrt{k_x^2 - k_0^2}$	$\cos^{-1}(\text{Re}\alpha)/l$
tunneling attenuation constant μ	$\sqrt{2m(V - E)}/\hbar$ $= \sqrt{k_0^2 - k_x^2}$	$\sqrt{k_y^2 - \frac{\omega^2}{c^2}}$ $= \sqrt{k_0^2 - k_x^2}$	$\cosh^{-1}(\text{Re}\alpha)/l$
reflection attenuation $e^{-\eta}$	$\frac{k_x - \kappa}{k_x + \kappa}$	$\frac{k_x - \kappa}{k_x + \kappa}$	$\sqrt{\left\lvert\frac{\text{Im}\alpha - \sin(\kappa l)}{\text{Im}\alpha + \sin(\kappa l)}\right\rvert}$
reflection phase-delay $e^{i\xi}$	$\frac{k_x - i\mu}{k_x + i\mu}$	$\frac{k_x - i\mu}{k_x + i\mu}$	$\sqrt{\frac{\text{Im}\alpha + i\sinh\mu l}{\text{Im}\alpha - i\sinh\mu l}}$
probability/current density (y component)	$\rho = \lvert\psi(x)\rvert^2$	$S_y = \frac{c}{4\pi} E_z B_x$ $= \hbar c^2 k_y \lvert\psi(x)\rvert^2$	
current density (x component)	$\frac{\hbar}{m}\text{Im}\{\psi^*(x)\psi'(x)\}$	$S_x = -\frac{c}{4\pi} E_z B_y$ $= \hbar c^2 \text{Im}\{\psi^*(x)\psi'(x)\}$	
group time/shift	$\tau_g = \frac{2}{\mu v_g}$	$\Delta_g = \frac{2}{\mu}\tan\theta$	
dwell time/length	$\tau_D = \frac{k^2}{k_0^2}\tau_g$	$\Delta_D = \frac{k^2}{k_0^2}\Delta_g$	

Table 1: Comparison between the various parameters in three schemes: Electronic potential barrier (EPB), frustrated total internal reflection (FTIR) and photonic band structures (PBS)

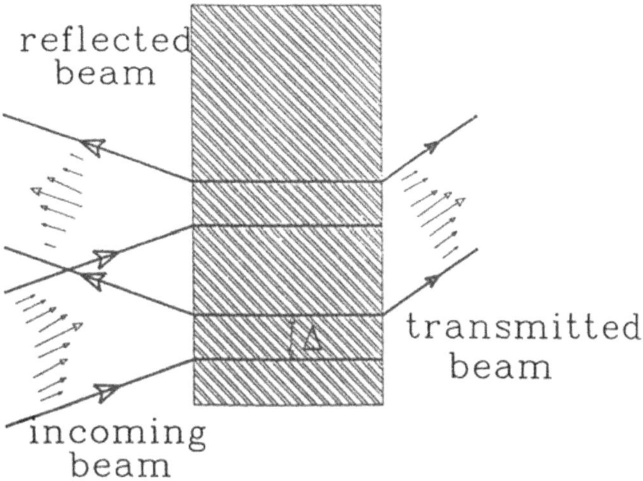

Figure 4: The shifts of the reflected and the transmitted waves for an incoming (Gaussian) light beam impinging on a finite air gap. The Goos-Hänchen shift is the shift for total reflection, i.e., infinite barrier

the y direction in total internal reflection of an EM wave. The latter shift has been known for many years as the Goos-Hänchen shift[14].

There are two principal ways to obtain this shift. One way is to consider an incident Gaussian beam of light which is a superposition of plane waves with a distribution of angles around some angle θ, that is larger than the critical angle for total reflection, as shown in Figure 4[15]. The shift of the reflected beam, which is obtained in the stationary phase approximation, is the derivative of the phase delay ξ with respect to k_y. This yields the "group shift"

$$\Delta_g = -\frac{\partial \xi}{\partial k_y} = \frac{2}{\mu} \tan\theta \tag{22}$$

This is analogous to a delay time of a reflected electron which is obtained in the same method, by taking the derivative of ξ with respect to $\omega = E/\hbar$[16]

$$\tau_g = \frac{\partial \xi}{\partial \omega} = \frac{2}{\mu v_g} = \frac{2m}{\mu k} \tag{23}$$

The other method to obtain the shifts is by using energy conservation considerations[17]. It is based on the assumption that the total energy flux in the x direction on a strip that has the width of the Goos-Hänchen shift is equal to the total flux in the y direction crossing a semi-infinite plane perpendicular to the y direction inside the barrier, as shown in Figure 5. We call this shift the "dwell shift". It represents the average path traversed by a wave in the y

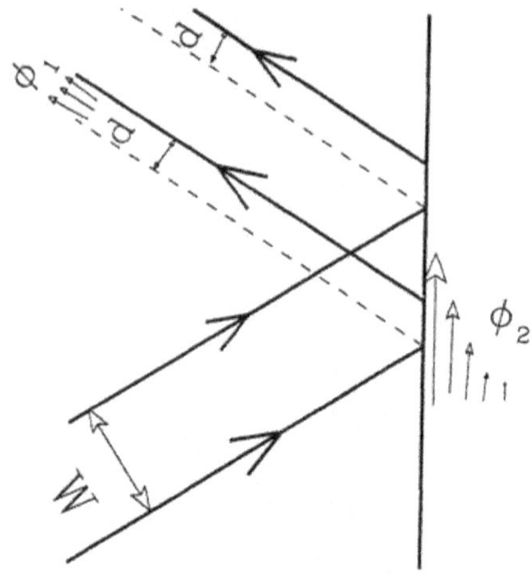

Figure 5: The energy conservation consideration in the calculation of the dwell length in the Goos-Hänchen effect. A plane wave is incident on the surface between two media with refractive indices n_1, n_2. We look at a width W of the beam. The flux ϕ_2 inside the barrier must be equal to the flux ϕ_1 in a strip of width d in the reflected wave. $\Delta_D = d/\cos\theta$ is the dwell shift

direction inside the barrier until it is reflected back. The dwell shift is given by

$$\Delta_D = \frac{1}{S_x^+}\int_0^\infty S_y(x)dx = \frac{k^2}{k_0^2}\frac{2}{\mu}\tan\theta = \frac{k^2}{k_0^2}\Delta_g \tag{24}$$

where S_x^+ is the x component of the Poynting vector of the incoming wave, S_y is the y component of the Poynting vector inside the barrier and $k_0^2 = (\omega^2/c^2)(n^2-1)$ as given in Table 1.

This result is analogous to the dwell time for an electron totally reflected from a potential step[2], which is given by

$$\tau_D = \frac{1}{j^+}\int_0^\infty \rho(x)dx = \frac{2km}{k_0^2\mu} = \frac{k^2}{k_0^2}\tau_g \tag{25}$$

where $k_0^2 = 2mV/\hbar^2$.

Experiments for measuring the Goos-Hänchen shift are described in the literature[14] The striking conclusion from the FTIR-EPB analogy suggested here is that such measurements can be immediately translated into reflection or tunneling times for electrons, which are not directly accessible to experiment.

4 Partial wave expansion

We look for an intuitively clear interpretation of the process of tunneling. The first question we ask is: What is the traversal time for an electron through a

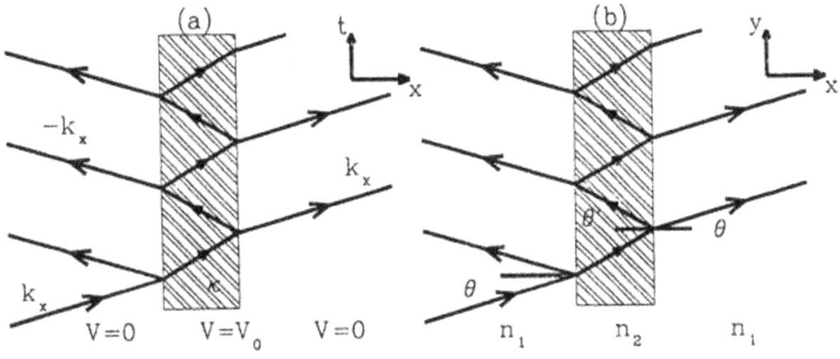

Figure 6: A schematic picture of an electron (a) traversing a potential barrier. and of (b) an EM wave traversing an air gap in the allowed region. The picture shows the transmission and reflection of the electron or the EM wave with multiple reflection inside the barrier

potential barrier if the energy of the electron is larger than the barrier height? The process is shown in Figure 6. The electron has certain probability amplitudes for being transmitted to the other side without internal reflections or being reflected 2,4,6 etc. times inside the barrier before being transmitted. The time it takes the electron to be transmitted (regardless of how many internal reflections it experienced) can be defined as an average over the times $L/v', 3L/v', 5L/v'$ etc., where v' is the propagation velocity inside the barrier, $v' = \sqrt{2(E-V)/m}$, each weighted by the corresponding transmission amplitude. Since these amplitudes interfere, the total average time can turn out to be even smaller than the shortest time L/v', as we will show below in Section 5.

To formulate this picture mathematically we expand the transmission coefficient Eq. 18 in partial waves

$$\sigma = (1 - e^{-2\eta})e^{i\kappa L} \sum_{n=0}^{\infty} (e^{-\eta} e^{i\kappa L})^{2n} \tag{26}$$

Here $e^{-\eta}$ is the amplitude for reflection in the boundary between the two media. The first term in the expansion represents a wave that crosses the first boundary with amplitude $1 + e^{-\eta}$, then propagates a distance L inside the barrier with x-component κ of the wave vector and finally crosses the other boundary with amplitude $1 - e^{-\eta}$. The second wave does the same, but before exiting the barrier it is twice reflected from the boundary walls (with an amplitude $e^{-\eta}$), thus traveling an additional distance of $2L$ inside the barrier. Higher order reflection terms are similarly interpreted. The amplitude for a wave transmitted through the barrier is thus represented as a sum of amplitudes for waves undergoing multiple reflections at the barrier boundary.

The partial wave expansion in the tunneling regime is obtained by analytical continuation of Eq. 26, i.e., by the replacement $\kappa \to i\mu, \eta \to -i\xi$

$$\sigma = (1 - e^{2i\xi})e^{-\mu L} \sum_{n=0}^{\infty} (e^{i\xi} e^{-\mu L})^{2n} \tag{27}$$

However, the interpretation of Eq. 27 is not at all straightforward. It is a sum of strange "paths" where every transmission through a boundary adds an amplitude $1 \pm e^{i\xi}$ and every reflection adds a phase ξ. The propagation inside the barrier is characterized by an "imaginary wave vector" $i\mu$, which means that the wave inside the barrier does not change its phase as a function of the distance but changes its amplitude.

The reflection coefficient r can be also expanded in a similar way. The easiest way is to use Eq. 20 and Eq. 26 to make the expansion

$$r = \begin{cases} e^{-\eta} - (1 - e^{-2\eta})e^{\eta} \sum_{n=1}^{\infty} (e^{i\kappa L} e^{-\eta})^{2n} & \text{allowed regime} \\ e^{i\xi} - (1 - e^{2i\xi})e^{-i\xi} \sum_{n=1}^{\infty} (e^{-\mu L} e^{i\xi})^{2n} & \text{tunneling} \end{cases} \tag{28}$$

The first term in the expansion of r in the case of tunneling is $e^{i\xi}$. This term represents total reflection.

Each term in the transmission and reflection coefficients expansion in the case of tunneling represents consecutive total reflections off the barrier walls. This is why we first have to discuss the problem of total reflection making use of the electron-photon analogies mentioned above. The description of tunneling will then emerge as the sum over many such total reflections. The nature of evanescent waves in total reflection from the dynamical point of view will be discussed in the next section.

5 Wavepacket propagation in total reflection

Total reflection of an EM wave at the boundary between two dielectric media has been thoroughly investigated, both analytically and experimentally, over many years. [14] Various studies have shown that the incoming wave enters into the less dense medium as an evanescent wave and is reflected after a short propagation distance inside the barrier, first in the direction parallel to the boundary and then back, in the direction opposite to that of incidence. Surprisingly enough, the following simple treatment of an evanescent wave in the stationary phase approximation has not been attempted thus far. Here and hereafter we consider the temporal evolution of a totally-reflected electron, which is equivalent to the spatial (y-axis) evolution of its EM wavepacket counterpart, by the analogy in Table 1. We then compare the results to those obtained for time-dependent total reflection of an EM wave in PBS.

For a monochromatic plane wave totally reflected from a barrier, the wave function outside and inside the barrier is given by

$$\varphi(\omega,x)=\begin{cases} e^{ikx}+e^{i\xi}e^{-ikx} & x<0 \\ (1+e^{i\xi})e^{-\mu x} & x>0 \end{cases} \tag{29}$$

Now assume that we send a Gaussian wavepacket concentrated around an energy $E_0 = \hbar\omega_0$ of temporal width Δ. Then the wave function inside the barrier becomes

$$\psi(x,t)=\int d\omega e^{-i\omega t}g(\omega-\omega_0)e^{-\mu x}(1+e^{i\xi}) \tag{30}$$

where $g(\omega)$ is a Gaussian.

We expand the variables μ and ξ to second order in $\omega - \omega_0$ and integrate over the frequency range from $-\infty$ to ∞ (as explained in Sec. 8, this can only be done for a classical EM wave, but it is a good approximation for a photon or electron wavepacket if $\omega_0 \gg \Delta$). This ω integration yields

$$\psi(x,t)\approx e^{-i\omega_0 t}(2\pi\Delta_x^2)^{-1/4}\exp(\frac{\mu_0'^2x^2}{2\Delta_x^2})\exp(i\frac{\mu_0' xt}{\Delta_x^2})e^{-\mu_0 x}\times$$

$$\left\{\exp[-\frac{t^2}{2\Delta_x^2}]+e^{i\xi_0}\exp[-\frac{(t-\xi_0')^2}{2\Delta_x^2}]\right\} \tag{31}$$

where

$$\Delta_x^2=\Delta^2+\mu_0'' x$$

The prime $()'$ denotes a derivative with respect to ω and the subscript $()_0$ implies that the function is taken at ω_0. The derivatives of μ and ξ for EPB and FTIR are given in Table 2. We have assumed here that the temporal width of the wavepacket, Δ, is large enough so that $\Delta \gg \xi_0''$ and have therefore neglected the second order term in the ξ expansion. We must also stress that the approximation leading to Eq. 31 is valid for small enough values of x. Since μ_0'' is always negative in the cases of interest we have a narrowing of the temporal Gaussian as we go deeper into the barrier.

The shape of the envelope of the "forward" evanescent wave inside the barrier is shown in Figure 7 at several times. The shape of the "backward" wave is the same, except for a time delay and a different phase.

In the range of small x we thus have two waves penetrating into the barrier. The first wave starts to travel back in the negative x direction when the peak of the wavepacket arrives at the boundary of the barrier. The second wave starts traveling backwards after a delay time $\tau = \xi_0'$. At the boundary, the first wave is in phase with the incoming wave and the second wave is in phase with the reflected wave. A schematic description of the evanescent wave inside the barrier is shown in Figure 8.

In the case of photonic band structures (PBS) the wave function in the j'th unit cell is given by

$$\varphi_j^{(1)}(\omega,x)=e^{-j\mu l}[a_0e^{i(k_1x-jl)}+b_0e^{-i(k_1x-jl)}] \tag{32}$$

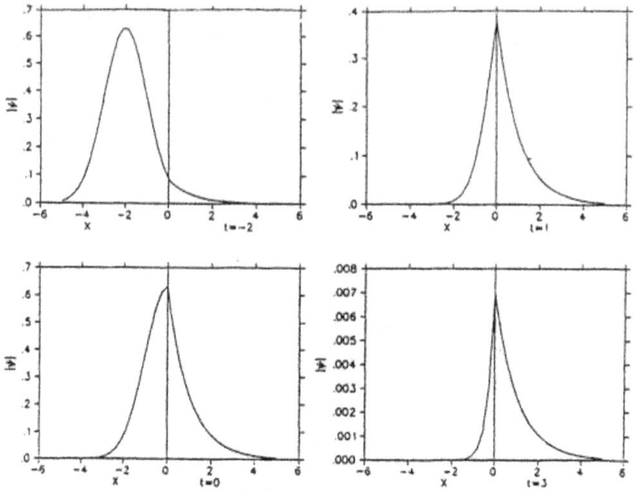

Figure 7: The envelope of the "forward" evanescent wave inside an infinite barrier in total reflection as a function of the distance from the boundary of the barrier. The picture shows the evanescent wave and the incoming wave to the left of the barrier. The reflected wave and the "backward" evanescent wave inside the barrier are the same, but with a time delay equal to τ_g, the delay time for total reflection

In the case where the refractive index of the first layer is similar to the refractive index outside the band structure we have $a_0 = 1$ and $b_0 = e^{i\xi}$. In the more general case where $n_1 \neq 1$ we have

$$a_0 = \frac{1 + \lambda e^{i\xi}}{1 + \lambda}, \quad b_0 = e^{i\xi}\frac{1 + \lambda e^{-i\xi}}{1 + \lambda}$$

where $\lambda = (n_1 - 1)/(n_1 + 1)$ is the reflection coefficient for the boundary between air and the first layer.

For simplicity we assume here a phase difference of ξ between a_0 and b_0.

If the incoming wavepacket is in the shape of a Gaussian as before we have inside the band structure

$$\psi_j^{(1)}(x,t) \approx e^{-i\omega_0 t} e^{-j\mu_0 l}(2\pi\Delta_j^2)^{-1/4} \exp(\frac{\mu_0'^2 j^2 l^2}{2\Delta_j^2})\times$$

$$\left[\exp(i\frac{\mu_0' j l t_-}{\Delta_j^2})\exp(-\frac{t_-^2}{2\Delta_j^2}) + e^{i\xi_0}\exp(i\frac{\mu_0' j l t_+}{\Delta_j^2})\exp[-\frac{(t_+ - \xi_0')^2}{2\Delta_j^2}]\right] \tag{33}$$

where

$$\Delta_j^2 = \Delta^2 + \mu_0'' j l$$

$$t_\pm = t \pm (x - jl) n_1 / c$$

and n_1 is the refractive index of layer 1.

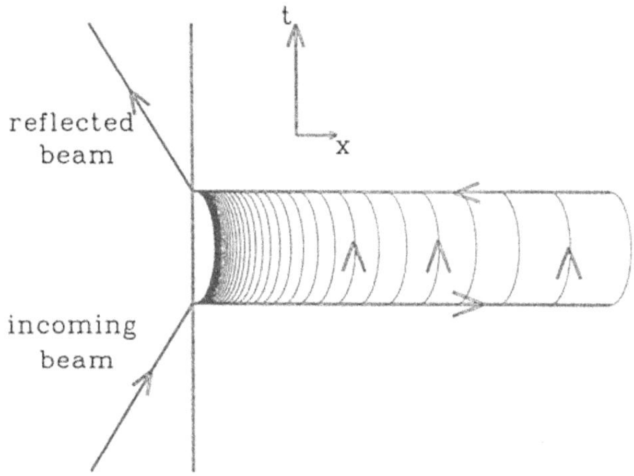

Figure 8: Schematic description of the two "counter-propagating" evanescent waves in total reflection. The double lines describe the center of the time Gaussians inside the barrier. The "forward" evanescent wave transfers energy to the "backward" wave inside the barrier

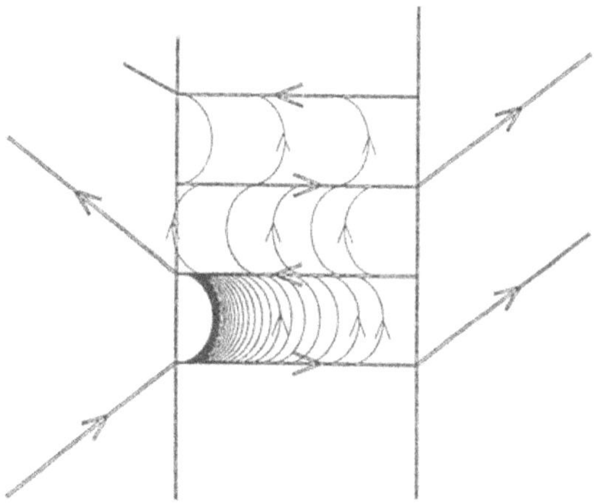

Figure 9: Schematic description of the process of tunneling through a finite barrier from the point of view of group propagation of a wavepacket. The reflected and transmitted waves are constructed from the evanescent waves of the appropriate direction inside the barrier. Every reflection inside the barrier adds a shift (in time or space) to the wave

Equation (33) is similar to Eq. 31 if we substitute jl for x. In the latter case the first "macroscopic-scale" wave consists of "microscopic-scale" waves traveling forward inside each layer, while the second "macroscopic" wave represents "microscopic" waves traveling backwards inside each layer. However, in our interpretation, every such forward or backward traveling "microscopic" wave is in its turn a superposition of many multiply-reflected waves inside the band structure.

6 Wavepacket propagation in tunneling

We now extend the discussion to the case of tunneling. The same method that yields Eq. 31 is used to obtain the shape of the wave function of a particle that tunnels through the barrier. We use the partial wave expansion Eq. 26 and Eq. 17, and take an incoming Gaussian wavepacket. We find that the wavepacket for $x > L$ is

$$\psi(x,t) \approx \exp\{-i\omega_0[t-(x-L)/v_\varphi]\} \sum_{n=0}^{\infty} \frac{1}{(2\pi\Delta_n^2)^{1/4}} \times$$
$$\exp\{i\nu_n[t-(x-L)/v_g]\} e^{2in\xi_0} e^{-(2n+1)\mu_0 L} \exp(\frac{\Delta_n^2 \nu_n^2}{2}) \times$$
$$\left\{ \exp\{-\frac{[t-\delta\tau_n-(x-L)/v_g]^2}{2\Delta_n^2}\} - e^{2i\xi_0} \exp\{-\frac{[t-\delta\tau_{n+1}-(x-L)/v_g]^2}{2\Delta_n^2}\} \right\} \tag{34}$$

where

$$\Delta_n^2 = \Delta^2 + (2n+1)\mu_0'' L - ik_0''(x-L)$$

$\nu_n = (2n+1)\mu_0' L/\Delta_n^2$ is the frequency shift of the n'th Gaussian, $\delta\tau_n = 2n\xi_0'$ is its time delay, $1/v_\varphi = k_0/\omega_0$ is the inverse of the phase velocity and $1/v_g = k_{x0}' \equiv \frac{\partial k}{\partial \omega}\Big|_{\omega_0}$ is the inverse of the group velocity.

The transmitted wave function is a train of Gaussians arriving at the point $x = L$ at times $\delta\tau_n$. The temporal width of each Gaussian is changed after the passage through the barrier. In the cases of interest μ_0'' is negative and this means that the wavepacket becomes narrower after the passage through the barrier. The imaginary part of Δ_n is due to the broadening of the wavepacket because of dispersion, as in free space. The n'th component of the wave function undergoes a reduction in intensity, which is expressed by the exponential term $e^{-(2n+1)\mu_0 L}$, and a reduction ν_n in frequency.

A schematic description of the wave propagation inside and outside the barrier is shown in Figure 9

7 Group and dwell times

In this section we review the various tunneling times/shifts in terms of the tunneling variables μ, ξ in all three analogous schemes described in Sec 3. We

	EPB	FTIR
μ'	$-m/\mu$	k_y/μ
μ''	$-m^2/\mu^3$	$-\omega^2/\mu^3c^2$
ξ'	Eq. 23	Eq. 22

Table 2: Comparison between derivatives of μ and ξ in EPB and FTIR

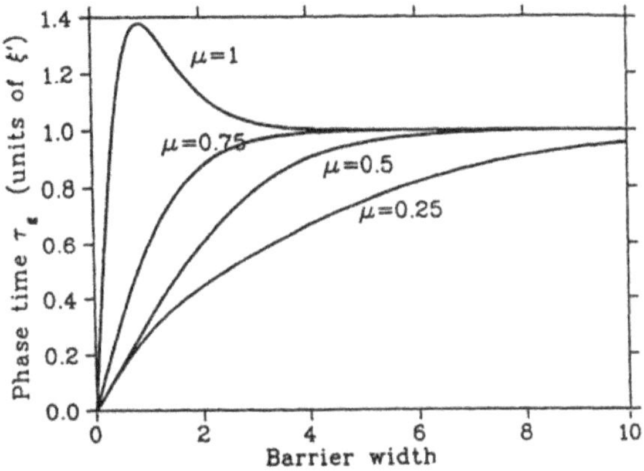

Figure 10: The tunneling phase time τ_g as a function of the barrier width L. For large L the tunneling phase time becomes almost independent of the barrier width

study their connection to the total reflection times/shifts and the partial wave expansion. In this section we use the EPB notation. The discussion applies to FTIR as well, with the appropriate replacements according to Table 1.

In order to calculate the stationary phase time we consider the transmission function of Eq. 18 to be $\sigma = |\sigma|e^{i\phi}$, where $\tan\phi = -\cot\xi\tanh\mu L$ and obtain the phase time as the derivative of ϕ with respect to the frequency ω

$$\tau_g = \frac{\partial\phi}{\partial\omega} = \frac{1}{2}\frac{\xi'\sinh 2\mu L - \mu' L\sin 2\xi}{\sin^2\xi + \sinh^2\mu L} \tag{35}$$

The phase time as a function of the barrier width L is shown in Figure 10. The results are similar to experimental results obtained for the phase shift of the tunneling waves in FTIR[12]

The dwell time for tunneling is obtained by integrating the probability density $\rho(x)$ over the barrier region and dividing by the incoming current j^+ as in Eq. 25. The values of $\rho(x), j^+$ for the electron and its photonic analogue are given in Table 1. The calculation yields

$$\tau_D = (\frac{1}{j^+}\int_0^L \rho(x)dx = \left(\frac{2}{v_g\mu}\right)\left(\frac{1+\cos\xi}{2}\right)\frac{\sinh 2\mu L - 2\mu L\cos\xi}{2(\sin^2\xi + \sinh^2\mu L)} \tag{36}$$

Note that $2/v_g\mu$ is the phase time τ_g for total reflection and $(1+\cos\xi)/2 = k_x^2/k_0^2$ is the ratio between τ_D and τ_g in total reflection as in Eq. 25. The first two terms in Eq. 36 are thus equal to the dwell time τ_D in total reflection.

Now we study the behaviour of the various tunneling times in two limits:

(1) An opaque barrier, namely $\mu L \gg 1$.

(2) Near the critical point, where $\mu, \xi \to 0$.

In the opaque limit of $\mu L \gg 1$ τ_g of Eq. 35 approaches the phase time for total reflection in Eq. 23. As we concluded in Sec 6, the tunneling wavepacket in the case of an opaque barrier consists of two wavepackets peaked around $t = 0$ and $t = 2\xi'$. The interference between the two wavepackets results in a wavepacket that is peaked around $t = \xi'$. In an opaque barrier the dwell time in Eq. 36 approaches the dwell time for total reflection in Eq. 25. We observe the fact that in an opaque barrier both the phase and dwell traversal time become independent of the width l of the barrier. this means that we can expect in principle superluminal velocities for tunneling wavepackets.

Near criticality $\mu \to 0$, $\xi \approx -2\mu/k_x \to 0$, $\xi' \approx -(2/k_x)\mu'$ and $\tau_g \to -4\mu\mu' L/3k$. For dwell time near the critical point we have

$$\tau_D \to \left(\frac{L}{v_g}\right)\frac{1+k_x^2L^2/3}{1+k_x^2L^2/4} \tag{37}$$

where μ' for EPB and FTIR is given in Table 2. In both schemes it is proportional to $1/\mu$. This implies that both τ_g and τ_D converges near the critical point. This conclusion is contrary to the result for total reflection, where the phase time/shift for reflection is divergent near criticality. The partial waves that construct the tunneling wave are also divergent but they construct a wavepacket that is not diverging near the critical point. By contrast, the Büttiker-Landauer semiclassical time (see Introduction)[7] diverges at the critical point, and is therefore unphysical under such conditions. Diverging of the tunneling time near the critical point means that the tunneling particle stays an infinite time in the region of the barrier, and therefore its momentum in this region must be equal to zero. This contradicts the position-momentum uncertainty principle, which implies that a particle with definite momentum cannot be defined in a limited position in space.

We conclude that near the critical point the behaviour of the traversal time of the total wave behaves much different than the traversal times of the partial waves. Therefore the partial wave expansion can be used as a perturbation expansion only for an opaque barrier far from criticality.

8 Quantization and photon localization

In the quantum mechanical case, in contrast to the classical case, the wave function $\psi(t)$ should be replaced by an operator $\Psi(t)$ which contains only positive frequencies

$$\Psi(t) = \int_0^\infty d\omega a(\omega) e^{-i\omega t}$$

where $a(\omega)$ is a photon annihilation operator as in Eq. 1. A state $|1\rangle_\phi$ containing one photon is described by

$$|1\rangle_\phi = \int_0^\infty d\omega \phi(\omega) a^\dagger(\omega)|0\rangle$$

where $|0\rangle$ is the vacuum state and the normalization condition is

$$\int_0^\infty |\phi(\omega)|^2 d\omega = 1$$

A detection of the photon at time t measures the quantity

$${}_\phi\langle 1|\Psi^\dagger(t)\Psi(t)|1\rangle_\phi = \left|\int_0^\infty \phi(\omega)e^{-i\omega t}\right|^2$$

Here the integral is a function of t which contains only positive frequencies. It is a well known mathematical property of Fourier transforms that if $\phi(\omega)$ is defined only for positive frequencies then its Fourier transform $\psi(t)$ behaves for large t like the inverse of some power of t. This means that the photon measured in photodetection cannot fall off like a Gaussian or an exponential at large times. This sets a limit on the possibility of localizing the photon. A photon or any state that is detected by regular photodetection that measures only positive frequencies cannot be localized as a Gaussian or an exponential. This applies to any particle, e.g. an electron.

In the treatment of classical waves used in the previous sections we always used a Fourier transform of the EM wave which contains a distribution of real positive and negative frequencies. If one wants to treat the problem quantum mechanically it must be remembered that single photons or any distribution thereof registered in photodetection, cannot be fully localized as in classical waves.

9 Conclusions

The description of tunneling in a PBS can elucidate the process of tunneling in other structures. A dynamical interpretation of the process of total reflection and evanescent waves can be given based on the analysis of section 5. A decaying wave is usually a result of dissipation inside a medium. Ordinary dissipation means that energy is transferred from the incoming flux to the medium. This differs from the case of an evanescent wave, where all the energy is recovered by the reflected wave. A dynamical interpretation can describe this process of energy transfer as lasting a certain time which is proportional to the thickness of the back-reflection layer inside the medium. In EPB the duration of this energy transfer is revealed by a time delay of the reflected wave, whereas FTIR it is manifested by a lateral shift known as the optical Goos-Hänchen shift. In the case of tunneling a fraction of the incoming wave exits the barrier on the other side with a time delay which equals the time delay of the energy transfer to the reflected wave. More complicated phenomena in tunneling are a consequence of interference between multiple reflections inside the barrier.

References

[1] For a review article see E. H. Hauge and J. A. Støvneng, Rev. Mod. Phys. 61, 917 (1989); For recent papers see H.A.Fertig, Phys. Rev. Lett. 65, 2321 (1990) and Ref. [5].

[2] M. Büttiker, Phys. Rev. 27, 6178 (1983)

[3] C. G. B. Garrett and D. E. McCumber, Phys. Rev. A 1,305 (1970); S. Chu and S. Wong, Phys. Rev. Lett. 48,738 (1982).

[4] The concept first introduced by F. T. Smith, Phys. Rev. 118, 349 (1960). Application to our problem see Ref. [2].

[5] R. S. Dumont and T. L. Marchioro, Phys. Rev. A 47, 85 (1993)

[6] D. Sokolovsky and L. M. Baskin, Phys. Rev. A 36, 4604 (1987) and also Ref. [1] and [5]

[7] M. Büttiker and R. Landauer, Phys. Rev. Lett. 49, 1739 (1982); ibid. Physica Scripta 32, 429 (1985).

[8] C. K. Carniglia and L. Mandel, Phys. Rev. D 3, 280 (1971).

[9] R. Y. Chiao, P. G. Kwiat and A. M. Steinberg, Manuscript for the invited talk by R. Y. Chiao at the International Symposium "Analogies in Optics and Microelectronics,", Eindhoven, the Netherlands, May 3, 1991; ibid. To be published.

[10] C. K. Hong, Z. Y. Ou and L. Mandel, Phys. Rev. Lett. 49, 2044 (1987); A. M. Steinberg, P. G. Kwiat and R. Y. Chiao, Phys. Rev. Lett. 68, 2421 (1992)

[11] C. K. Hong and L. Mandel, Phys. Rev. A 31, 2409 (1985).

[12] C. K. Carniglia and L. Mandel, J. Opt. Soc. Am 61, 1035 (1971).

[13] P. Yeh, A. Yariv and C-S Hong, J. Opt. Soc. Am. 67, 423 (1977).

[14] For an extensive review see H. K. v. Lotsch, Optik 32 (1970); For mea surements see v. H. Wolter, Z. Naturfsg. 5a, 276 (1950); Other reviews in H. Rosu, Internal report, 1993.

[15] A. K. Gathak and K. Thyagarajan, Contemporary optics, Chp. 11, Plenum Press, New-York, 1978;

[16] Cohen-Tannoudji, Quantum Mechanics, vol. 1,

[17] R. H. Renard, J. opt. Soc. Am. 54,1190 (1964)